THE TENTACLES
OF PROGRESS

THE TENTACLES OF PROGRESS

Technology Transfer in the Age of Imperialism, 1850–1940

DANIEL R. HEADRICK

New York   Oxford
OXFORD UNIVERSITY PRESS
1988

Oxford University Press

Oxford   New York   Toronto
Delhi   Bombay   Calcutta   Madras   Karachi
Petaling Jaya   Singapore   Hong Kong   Tokyo
Nairobi   Dar es Salaam   Cape Town
Melbourne   Auckland

and associated companies in
Beirut   Berlin   Ibadan   Nicosia

Library of Congress Cataloging-in-Publication Data

Headrick, Daniel R.
  The tentacles of progress.

  Bibliography: p.
  Includes index.
  1. Imperialism—History.   2. Technology transfer—
History.   3. Great Britain—Colonies—History.
4. Tropics—Economic conditions.   I. Title.
JC359.H39   1988   338.9'26   87-5617
ISBN 0-19-505115-7
ISBN 0-19-505116-5 (pbk.)

9 8 7 6 5 4 3
Printed in the United States of America

Preface

Many people contributed to this book, though they may not have known it at the time. I wish to thank them all.

My thanks go first to the many archivists and librarians without whom this book would not exist, especially those at the following institutions: in Chicago, the libraries of the University of Chicago and Roosevelt University; in France, the Bibliothèque Nationale, the Archives Nationales and its Section Outre-Mer, the Bibliothèque d'Afrique et d'Outre-Mer, the Institut de Recherches Agronomiques Tropicales, and the Muséum National d'Histoire Naturelle; and in Britain the India Office Records and Library, the British Library, and the Royal Botanic Gardens at Kew.

I also owe a great debt to the many people who gave me their support and encouragement, and much valuable information, among them Margaret Anderson, André Angladette, Ralph Austen, Georges Ballard, Bertrand de Fontgalland, Alexander de Grand, Fritz Lehmann, William McCullam, William H. McNeill, Phyllis Martin, David Miller, Joel Mokyr, David Northrup, the late Derek de Solla Price, Joel Putois, Theodore W. Schultz, and Gary Wolfe.

I also wish to thank my editors at Oxford University Press, Nancy Lane, Rosalie West, and Joan Bossert.

And finally I wish to express my gratitude to the National Endowment for the Humanities, whose generous financial support gave me the time to write this book.

This book is dedicated to Cecil and the memory of Edith, and to Sol and Gertrude.

Chicago   D.R.H.
October 1986

Contents

Indianizing the Steel Industry
Conclusion
Notes

THE TENTACLES
OF PROGRESS

1

Imperialism, Technology, and Tropical Economies

Technology and Western-Tropical Relations

A great hope arises every few years that the poorer countries of the world can develop their economies with the help of capital and technology from the industrial nations. Then, after awhile, despair sets in, and the fate of the less developed nations once again seems grim. Such recurring bouts of expectation and disillusionment go back well over a century. As early as 1853, at the start of a great wave of European investments in the tropics, Karl Marx predicted:

> I know that the English millocracy intend to endow India with railways with the exclusive view of extracting at diminished expense the cotton and other raw materials for their manufactures. But when you have once introduced machinery into locomotion of a country, which possesses iron and coals, you are unable to withhold it from its fabrication. You cannot maintain a net of railways over immense country without introducing all those industrial processes necessary to meet the immediate and current wants of railway locomotion, and out of which there must grow the application of machinery to those branches of industry not immediately connected with railways. The railways system will therefore become, in India, truly the forerunner of modern industry. This is the more certain as the Hindus are allowed by British authorities themselves to possess particular aptitude for accommodating themselves to entirely new labour, and acquiring the requisite knowledge of machinery.[1]

Almost a century later, in the year of Indian independence, the economic historian T. S. Ashton concluded his book *The Industrial Revolution, 1760–1830* with the rueful remark:

There are today on the plains of India and China men and women, plague-ridden and hungry, living lives little better, to outward appearance, than those of the cattle that toil with them by day and share their places of sleep by night. Such Asiatic standards, and such unmechanized horrors, are the lot of those who increase their numbers without passing through an industrial revolution.[2]

Despite a century of European rule, India had not become industrialized as Marx had expected, nor had any other non-Western country except Japan. Yet they were not untouched by the upheavals of that century. Theirs was no simple delay in some inevitable process of industrialization, but another, less welcome evolution: the transformation of traditional economies into modern underdeveloped ones.

The period with which this book is concerned—from the mid-nineteenth century to the mid-twentieth—stretches approximately between these two statements. It does not coincide with the usual periodization of history based on political and military events in Europe, which distinguishes sharply between the nineteenth century (i.e., 1815–1914) and the twentieth (1914 to the present). For Europe, 1914 was a significant turning point, marking the end of a century of peace and the beginning of a new Thirty Years' War. For Asia and Africa, however, 1914 was the middle of the colonial era. This era was one of unprecedented change. Though there had been many empires in the past, never before had one civilization overwhelmed all the others and set them on an entirely new course.

The beginning of our era, in the mid-nineteenth century, is approximate. While the "scramble for Africa" has gained great notoriety, Western pressure on Africa and Asia was already apparent earlier in the century; witness the Opium War (1839–42), the explorations of Africa, the war of the Indian Rebellion (1857–58), and the openings of the Suez Canal (1869). What these events have in common is not only the fact of Western intrusion, but the technical innovations which gave Europeans the power to intrude. And at the other end of our era, it was World War II, not World War I, which set off the disintegration of the European colonial empires.

This era, the "new imperialism," coincides with the creation of modern underdeveloped economies in Asia and Africa. While these two processes have often been linked, their relationship remains unclear. A consideration of the technologies involved can shed some light on this question.

The history of technology once consisted only of nuts-and-bolts stories of great inventors and famous engineers. Today, technologies are no longer viewed as "externalities" that arise fortuitously from the minds of geniuses, but as an intrinsic part of the culture and economy of every society. And the task of the social historian of technology is to study the economic and cultural context in which innovations arise and, in turn, their impact upon the societies in which they appear.

Within this contextual tradition, our approach is to reverse the question. Not only does every technology exist in a social context, all events and all social situations occur in a technological context. Given an event or situation, we may ask, What is that technological context, and what part do technological changes play in it? This approach is familiar to those who study the Great Discoveries, the Industrial Revolution, and the winning of the American West. It needs to be applied to the study of the European empires as well.

In the relationships between technological change and European imperialism, we can begin by distinguishing five interactions. The first of these concerns the penetration of Asia and Africa by Europeans and the conquest of colonial empires. What distinguishes the new imperialism from its many predecessors is that it was so swift, thorough, and cheap. In a few years, roughly the half-century before 1914, the major Western powers conquered Africa, Oceania, and large parts of Asia, and they did so at a very small cost in European men and money.

This sudden scramble for territories aroused much interest, not only among contemporaries, but also later among historians and political theorists. Their interest has focused mainly on the motivations of the imperialists and their involvement with Western politics and economics. Yet there is no reason to believe that late nineteenth-century imperialists were any more strongly motivated than their predecessors. The reason for their sudden success was a shift in technology, similar to the development of oceangoing ships some four centuries earlier. Until the nineteenth century relations between Europe, on the one hand, and Asia and Africa, on the other, were determined by their technological balance. At sea, Europeans were almost invincible, as they had been for centuries. Their efforts to penetrate inland, however, were restrained by their numerical inferiority in Asia and their vulnerability to disease in Africa. In 1800, after three centuries of lurking offshore, Europeans could claim only a few footholds in Asia and Africa, mainly harbors and islands.

The new ability of Europeans in the nineteenth century to conquer other continents stemmed from a relatively few inventions. The first was the application of steam and iron to riverboats, starting in the 1820s. By the 1860s iron-hulled steamships appeared regularly in Asian and African waters. Constant improvements in firearms, from muskets to machine guns, gave small European-led units an overwhelming advantage over their African and Asian enemies. Quinine prophylaxis reduced the death rate among Europeans in the tropics, especially in Africa. Steamships, railways, and telegraphs allowed Europeans to control their newly acquired colonies efficiently. With these tools, Europeans brought about the shift in global relations we call the new imperialism. Having dealt with these questions in another book, I will not return to them here.[3]

Whatever the motives of the imperialists—the long-standing debate on this question is still going strong—the territories they added to the Western empires were soon incorporated into the world economy. This incorporation, in fact, predated the scramble and included territories that were never officially annexed; for that reason, it has been called the "imperialism of free trade."[4] Hence the second of our five interactions is the impact of an expanding Western economy on world trade. In the nineteenth century, the industrialization of the Western nations stimulated a growing demand for the products of the tropics. Falling transport costs made it increasingly worthwhile to ship cheap bulky commodities. New industries in the West required new raw materials from the tropics: cotton and indigo for cloth, palm oil to grease machinery, copper and gutta-percha for electric and telegraph lines, tin for canned goods, and rubber for clothing and automobiles. In addition, an affluent and demanding Western clientele consumed increasing amounts of sugar, tea, coffee, cocoa, and other tropical goods. These goods had long been supplied to Western buyers by non-Western producers and traders, but in the course of the nineteenth century, these suppliers failed to keep up with the growing demand. Enough conflicts ensued, from the Opium War to the opening of the Niger, that after the midcentury, the "imperialism of free trade" was supplanted by wars of conquest.[5]

Once in control of an area, Western colonialists were not content to administer their new subjects and tax the existing economy, as previous conquerors had done. Instead, they strove to increase production and lower the costs of tropical products by applying Western industrial and scientific methods. Thus our third interaction is a massive transfer of technology from the West to Africa and Asia. This transfer stimulated a growth in tropical production and in international trade. Meanwhile, greater security, more regular food supplies, and a growing demand for labor in turn stimulated the growth of tropical populations. The transfer did not, however, diversify the tropical economies, nor did it significantly raise per capita incomes. In other words, the tropical economies grew, but did not develop.

The fourth consequence of technological change was cultural as well as economic. Colonized people were not mere objects at the hands of the colonizers. The invasion of their countries by a technologically more advanced culture awakened not only the well-known movements for national independence, but also a desire to obtain more Western products and share in the benefits of Western technology. The Western invasion created new desires among tropical consumers. Railways and telegraphs built by the Europeans for their own benefit were soon flooded with Asian and African customers. In the twentieth century, motor vehicles, televisions, and modern weapons have become irresistible but barely affordable temptations for the peoples of poor countries. Along with the demand for devices has arisen a parallel demand for technological knowledge. Thus Western technology flowed to Asia and Africa, first pushed upon the colonies by Europeans and later pulled by the awakening demands of Asians and Africans. It reversed the age-old pattern of world trade in which the Western peoples craved the goods of the East, but had little but bullion to offer in exchange. Starting in the mid-nineteenth century, Asia and Africa imported ever-increasing quantities of manufactured goods from the West.

Meanwhile, other Western innovations, stimulated by war, politics, and science, have had a fifth, and more ominous, impact on the tropics. Almost all the technological changes which affected the relations between the West and the tropics originated in the West or from the work of Western scientists and engineers; they were developed for the benefit of the West, or of some sections of Western society, with scant regard for their long-range impact on the tropics. The Western talent for technological innovation could easily turn against the tropics. Already in the nineteenth century, Westerners had shown a propensity for finding substitutes for goods in short supply. During the Napoleonic Wars, beet sugar replaced cane sugar on the European continent; by the turn of the century it had all but ruined the economy of the West Indies. In the second half of the nineteenth century, aniline dyes replaced indigo and other natural colorings; iron and steel ships ruined the teak shipbuilding industries of South Asia; and petroleum replaced palm oil in the lubrication of machin-

Table 1.1 World Trade and Industrial Production, 1820–1948

Period	World Trade	World Industrial Production
1820–1840	+2.81%	+2.9%
1840–1860	+4.84	+3.5
1860–1870	+5.53	+2.9
1870–1900	+3.24	+3.7
1900–1913	+3.75	+4.2
1913–1929	+ .72	+2.7
1929–1938	−1.15	+2.0
1938–1948	0	+4.1

ery. In the twentieth century, synthetic rubber cut into the market for natural rubber, chloroquine was substituted for quinine, and synthetic fibers replaced silk. Other tropical products such as gutta-percha, sisal, kapok, jute, guano, and copra saw their markets shrink or vanish. The industrial world's demand for tin was held in check by electrolytic plating, and later by plastics and aluminum. In these and other ways, Western scientists and engineers have prevented the demand for tropical products from growing in proportion to the growth in industrial production or in tropical populations.[6]

The tropics were not the theater of war until 1941. Yet long before that they felt the effects of military and economic warfare in the West through the disruptions of trade and the demands that industrial nations made on them. It is instructive to compare the growth of industry in the first half of the twentieth century with the failure of world trade to keep up. According to Walt W. Rostow, world trade and world industrial production grew at the annual rates shown in Table 1.1.[7] Through wars and depressions alike, industrial production continued to grow, albeit erratically; in contrast, world trade, which had almost trebled from 1880 to 1913, ceased growing and fluctuated around its 1913 level until the late 1940s. In other words, the industrial nations were becoming less and less dependent on world trade, while the tropical countries had little or no industry to fall back on when trade faltered.[8]

The relentless advance of Western technology did not just leave the tropics behind; it widened the gap between them and the industrial nations, as the demographer Nathan Keyfitz explained:

The resentment of the poor whose independence has turned to dust will hardly be diminished by the reflection that after all it was they who wanted independence in the first place, inspired by a nationalism borrowed from the west. Their cries of exploitation, that had some semblance of justification in the nineteenth century, must become ever less convincing as they see that they suffer from the opposite of exploitation—markets have altered so that no colonial exploiter is able to put their labor to a profit. They are worse than exploited—they are irrelevant. The exploited could always strike against those who were making a profit on them; what recourse is there for the unexploitable, locked out of the market by the shift of prices?[9]

The Transfer Process

Modern technology changes constantly, here by leaps, there by small adaptations. Scholars, like the general public, have been fascinated with the creative act of invention and the first phases of innovation. Less exciting perhaps, but just as important, is the transfer process: how an innovation spreads geographically and culturally, and how it is adapted to new environments.

Why do technologies migrate? At first sight this may seem a simple utilitarian process: a technique, having proved its value in one place, is adopted by people in another place who think it may be useful to them as well. In a perfectly free market with perfect access to information, the spread of innovations would result from calculations of expected marginal returns. But in the real world so many factors distort the process that a whole field of scholarship, diffusion research, has arisen to investigate transfers within advanced industrial societies.[10]

The transfer of technology from one society to another, and especially from one civilization to another, is of an altogether higher order of complexity, and no theory has yet emerged to encompass it all. Yet certain basic premises are clear. First of all, the transfer of technology is not one process but two. One of these is the relocation, from one area to another, of equipment and methods, along with the experts to operate them. The other is the diffusion from one society to another of the knowledge, skills, and attitudes related to a particular device or process. Thus we will distinguish the geographic relocation from the cultural diffusion of technology. This will help us understand the process and impact of technology transfers in the past, with its hopes and disappointments.

Technology does not flow of its own accord from "advanced"

to "backward" areas. History is replete with examples of long contacts between societies that did not result in technological transfers; for instance, during centuries of trade contacts across the Sahara, the Islamic religion spread into the Sudan, but the North African plow and wheel did not. Technology transfers, like other cultural changes, involve the decisions of individuals and the reactions of the societies they live in.

Let us first consider the transfer agents: exporters, importers, and migrants. Exporters of technology, such as salesmen, foreign aid officials, and engineers on contract, specialize in the geographic relocation of technology, for the value of the devices they export depends in part on the lack of appropriate substitutes in the importing country, and their expertise depends on the absence of native experts. In contrast, importers of technology—students, purchasing agents, spies—are more likely to seek its cultural diffusion and to spread a particular device or process as a means of developing human capital and complementary technologies in their home countries.

A third kind of transfer agent, the migrant, is simultaneously an exporter and an importer of technology. Only rarely did immigrants bring with them the tools of their trade; their contribution to the process was mainly a cultural one. Within the West, many groups of migrants—sevententh-century Huguenots, nineteenth-century immigrants to America, refugees from Nazi Germany—were associated with the spread of technology. In the history of the Americas, the contribution of immigrants was more important than that of foreign investments, imported machines, or visiting experts. In fact, in countries of European settlement, immigrants and foreign investments played complementary roles.[11] Skilled Westerners rarely settled in the tropics, however, but came for a few years, segregated themselves from the native society, then returned home. Whatever the importance of immigrants for the development of white settler areas, they were not a significant factor in most tropical countries.[12]

Hence we are left with four basic categories of transfer: the geographic relocation of technology by Western experts; its relocation by non-Western importers; its cultural diffusion by Western experts; and its diffusion by non-Western importers.

As technology is transferred from one society to another, it encounters both resistance and support. These influences are exerted at both ends. Societies which possess a technology that others want occasionally try to ban its export for military or political reasons; spies and businessmen see to it that they seldom succeed. Having learned this lesson, industrial societies generally encourage the export of the technology, with appropriate safeguards, rather than try to suppress it. Yet the success of such efforts hinges more on the ability and willingness of the importing society to accept the technology than on the support of the exporting society. Let us therefore concentrate on the importing society, and consider it from the technological and economic standpoint.

One important consideration is the gap between the technological level of the exporting and the importing country. This is especially important to underdeveloped countries trying to catch up with the complex, ever-changing technologies of the West. As the economist Paul Bairoch has pointed out, the machines and processes of the first industrial revolution were simple enough that craftsmen with preindustrial skills could understand, copy, and improve them. Hence these technologies spread easily from Britain to other Western countries, sometimes in the minds of emigrant workers. By the late nineteenth century, however, industrial technology was no longer within the reach of craftsmen, but required a knowledge of engineering and science. Illiteracy now obstructed the transfer of technology: "It may seem at first sight that if intellectual underdevelopment is a handicap, the underdeveloped countries are no worse off than the Western countries were. Unfortunately, that is not the case, **for if man has perhaps remained the same, the tools have totally changed.**"[13] The gap between preindustrial crafts and industrial technology had grown so wide that the traditional skills were useless in the process of modernizing: "The condition which, from a technical point of view, allowed the diffusion of the English industrial revolution, can in no way play the same role today."[14] The underdeveloped countries therefore had to turn to the advanced countries for their capital equipment, thereby losing the benefits—multiplier effects, backward linkages, externalities—of the capital goods industries which played so large a role in the industrialization of the West: "This technical complexity makes the equipping of industry depend almost exclusively on the outside, and thereby one of the essential mechanisms for the diffusion of progress disappears."[15] Modern machinery, however, is not only more complex but also much more expensive than its antecedents; thus it costs far more to equip one workplace with twentieth-century machines than with those of the early nineteenth century. In other words, as time goes on, underdevelopment make industrialization increasingly difficult.[16]

Yet it is astonishing how, given the right incentives, the most complex technologies can be transferred quickly to the most primitive environments; oil wells, mining camps, and satellite tracking stations are but a few examples. All that is required is a transportation system that links the transferred technology to its home environment from its new location. Transportation technologies are by nature the easiest to connect to their point of origin; hence they are among the first to penetrate new environments. The railways of the nineteenth century, like the airlines of today, were extensions of distant financial centers, factories, and consumers. Such enterprises can operate for decades as alien enclaves, linked to distant suppliers and customers with little or no local articulation. Modern transportation permits the geographical relocation of technologies with little cultural diffusion or linkages with the local economy. In the worst cases they may provide a substitute rather than an incentive for local development.

The cultural and political aspects of technology transfer are even more complex than the economic or technical ones, as international businessmen and foreign aid officials know. Some Western technologies, such as automobiles and radios, have aroused almost universal enthusiam and even provoked revolutions of rising expectations. Those that threaten to change the culture, like contraceptives, often encounter resistance or censorship. Most often the introduction of a new technology will not meet with one kind of reaction, but with several at once; in other words, it will provoke conflicts within the importing society.

Transfer agents have frequently sought the help of the authorities in the form of subsidies, special taxes or tariffs, police protection, and other reassurances. At other times a government, learning of a foreign technology, will take the initiative in encouraging or restricting its import. In any case, the transfer of technology soon becomes politicized.

The attitude of a government or ruling elite will depend on its nature and on how it views the potential impact of the new technology. Its reaction will be conditioned by its assessment of risks and benefits in four areas: its domestic power, prestige, and security; its international position; the impact on the population at large; and the personal wealth and comfort that members of the ruling elite expect to enjoy from the new technology.

The reactions of non-Western governments to Western technology have varied from near-total rejection (as in Tokugawa Japan) to enthusiastic adoption (as in Meiji Japan). Governments based on traditional elites have approached the question with a mixture of fascination and wariness, and they have sought those imports which would strengthen their position vis-à-vis domestic rivals or foreign powers without triggering dangerous social or cultural upheavals. Attempts at selective modernization have led more than one regime to its downfall, from nineteenth-century Egypt and the Manchu dynasty of China to the Shah of Iran. Reconciling conflicting goals is especially difficult for conservative regimes, which often fall back on the most limited range of technology transfers: buying modern weapons for their armed forces and allowing foreigners to extract their mineral wealth to pay for those weapons.

Thus many factors favor the geographic relocation of technology over its cultural diffusion. Exporters of Western technology have an interest in selling machines, spare parts, and expertise. For importers, geographic relocation, while costlier, is faster and entails fewer risks than cultural diffusion. And governments need foreign equipment and experts, but shy away from the political dangers involved in cultural or social changes. Thus risk aversion leads to technological dependency.

The cultural diffusion of technology, in contrast, is a much more difficult task. It takes a willingness to accept changes, a strong political cohesiveness, and a common vision of the future. Western societies, facing less culture shock, have readily imported industrial technologies. Among non-Western societies, only Japan had the requisite cultural and social base. Others have had to undergo political revolutions first.

The Setting and the Argument

In the European colonial empires we find many of the elements common to all cases of technology transfer: the distinction between geographic relocation and cultural diffusion, the role of importers and exporters of technology, the cultural and economic matrix of the importing society, and the politicizing of technology. One major difference, however, separated the colonies from independent non-Western countries like Japan, Ethiopia, or China. In the colonies the ruling elites were both a small minority with an insecure mandate and representatives of a technologically more advanced society. Hence their complex and often vacillating attitude toward technology transfer.

Western technology did not reach the tropics in a continuous stream, but in the form of discrete projects: a railway, a plantation, a telegraph network. While none of them triggered an industrial revolution, some were successful in the eyes of their promoters and gave colonialists reasons to boast. Others, like the Cape-to-Cairo and Transsaharan railways and the East African groundnut scheme, were acknowledged failures. In seeking to understand why Western technology had such limited consequences in the tropics in the long run, I have had to select examples from among the myriad cases of technology transfer. The ones I have chosen were the large and successful projects, those of which the Europeans were proudest, rather than the mistakes.

In this narrative, India occupies the place of honor and provides half the examples. This is, in my opinion, a fair reflection of India's weight in the European colonial empires. In area, Britain's Indian Empire was as large as Europe outside of Russia, or one-third the size of Africa. More important than size was India's population, which rose gradually from 255.2 million in 1867–72 to 305.7 million in 1921, then shot up to 389 million in 1941. Africa had less than half as many people as India at any point in our period. The Dutch East Indies had one-tenth as many people in 1850, one-eighth as many in 1900. And of all of the inhabitants of the British Empire in 1910–11, including the United Kingdom, fully three-quarters lived in India.[17]

The size and population of India were reflected in its place in the world economy. In 1913 the export trade of India was equal to that of the Dutch East Indies, Malaya, Nigeria, and South Africa combined, or 63 percent greater than that of sub-Saharan Africa. From 1860 to 1910 India's trade with Britain (its first trading partner) grew threefold, remaining equal to British trade with China, South Africa, and Australia combined.[18]

India's place in the British Empire and in the world economy appears also in the figures on foreign investment given by Herbert Feis and others. Of the £1,789 million in long-term publicly issued British capital invested in the empire up to February 1914, India and Ceylon received £379 million, or 21.3 percent. The only parts of the world that received more British investments were the United States (£755 million), Latin America (£757 million), Canada and Newfoundland (£515 million), and Australia and New Zealand (£416 million), in other words regions with predominantly European populations. In comparison, France invested £160 million in all its colonies, and Germany invested some £145 million in all of Asia and Africa, most of it outside the German colonial empire.[19]

India was also the strategic center of the British Empire and the envy of all other colonial powers. Until the late nineteenth century, Africa was an obstacle between Europe and India rather than an object of European ambitions. Throughout the period of their rule, British statesmen worried about the security of India. It was these concerns that led them to conquer lands on the approaches to India, from South Africa and Egypt on the west to Malaya on the east, and to build naval bases at Singapore, Aden, Cape Town, and Mauritius. The same worries also led Britain into numerous wars and adventures in Afghanistan, Tibet, Burma, China, and Mesopotamia. The possession of India not only determined the direction of British imperialism but also provided the manpower and financing for many of these wars.

Finally, there is the chronology of conquest. Because India was under colonial rule for many decades before Africa, Egypt, Malaya, or Indochina, the influx of Western technologies started there much earlier and reached a higher degree of complexity. Similarly, education and political consciousness reached higher levels sooner in India than elsewhere in the European colonial empires. Hence, from the viewpoint of technology transfers, India was the forerunner and model for other colonies.

Earlier, we noted that Western technology interacted with the tropics in five ways: new means of penetration and conquest, the increasing Western demand for tropical products, the export of European technologies to the tropics, the rising demand of Asians and Africans for Western technologies, and finally, the substitution of synthetics for tropical products. This book focuses on the third and fourth of these interactions.

We will begin by examining the major technologies that Europeans brought to their colonies for the purpose of linking them to the world market (i.e., shipping, railways, telegraphs, and cities) and of increasing their production of trade goods (mines, plantations, irrigation systems). Here our theme will be the geographical relocation of new technologies from Europe to the tropics or, in a few cases, within the tropics.

The cultural diffusion of Western technologies to non-Western peoples became an issue in the early years of the twentieth century. This phenomenon, in turn, divides into two others: European efforts to impart their technological culture to their subject peoples through education and apprenticeships, and Asian and African efforts to acquire this culture on their own. These issues will constitute the last two chapters.

The argument of this book lies in the contrast between the successful relocation of European technologies under colonialism and the delays and failures in spreading the corresponding culture. The cause of this contrast was the unequal relationship between the tropical colonies and their European metropoles. In order to obtain the full benefit of Western technology through its cultural diffusion, Africans and Asians had first to free themselves from colonial rule and then—a more arduous task—learn to understand, and not just desire, the alien machinery.

Notes

1. Karl Marx, "The Future Results of British Rule in India," *New York Daily Tribune,* August 8, 1853, p. 5.

2. T. S. Ashton, *The Industrial Revolution, 1760–1830,* rev. ed. (New York, 1964), p. 111.

3. Daniel R. Headrick, *The Tools of Empire: Technology and European Imperialism in the Nineteenth Century* (New York, 1981).

4. John Gallagher and Ronald Robinson, "The Imperialism of Free Trade," *Economic History Review,* 2nd ser., no. 6 (1953): 1–15.

5. David Kenneth Fieldhouse, *Economics and Empire, 1830–1914* (Ithaca, N.Y., 1973).

6. Currently, the search is on for cheap substitutes for petroleum; when they are found, the impact on the tropics will be devastating.

7. Walt W. Rostow, *The World Economy: History and Prospect* (Austin, Tex., and London, 1978), p. 67.

8. The same trend can be described another way. Since the late 1920s, imports (excluding petroleum) to Western Europe and the United States from primary producing countries have fallen from 3.5 to 2.5 percent of their combined gross national products. See Ragnar Nurkse, *Problems of Capital Formation in Underdeveloped Countries and Patterns of Trade and Development* (New York, 1967), p. 183.

9. Nathan Keyfitz, "National Population and the Technological Watershed," *Journal of Social Issues* 23, no. 1 (January 1967): 76. He makes the same point in his more recent book *Population Change and Social Policy* (Cambridge, Mass., 1982), pp. 56–58.

10. See Everett M. Rogers with F. F. Shoemaker, *Communication of Innovations: A Cross-Cultural Approach* (New York, 1971) and Patrick Kelly and Melvin Kranzberg, *Technological Innovation: A Critical Review of Current Knowledge* (San Francisco, 1978).

11. Warren C. Scoville, "Minority Migrations and the Diffusion of Technology," *Journal of Economic History* 11 (1951): 347–60; Ragnar Nurkse,

"International Investment To-Day in the Light of Nineteenth-Century Experience," *Economic Journal* 64 (December 1954): 744–58.

12. On the migrations of British engineers to the rest of the world, see R. A. Buchanan, "The Diaspora of British Engineering," *Technology and Culture* 27 (July 1986): 501–24.

13. Paul Bairoch, *Révolution industrielle et sous-développement* (Paris, 1963), p. 169; boldface in the original.

14. Ibid., pp. 174–75.

15. Ibid., p. 207. On the critical role of the capital goods industry in Western industrialization, see Nathan Rosenberg, "Factors Affecting the Diffusion of Technology," *Explorations in Economic History* 10, no. 1 (Fall 1972): 3–33 and "Economic Development and the Transfer of Technology: Some Historical Perspectives," *Technology and Culture* 11, no. 4 (October 1970): 550–75.

16. Bairoch, pp. 193–99.

17. Kingsley Davis, *The Population of India and Pakistan* (Princeton, N.J., 1951), p. 27; E. A. Wrigley, *Population and History* (New York and Toronto, 1969), p. 205; A. J. H. Latham, *The International Economy and the Underdeveloped World, 1865–1914* (London and Totowa, N.J., 1978), pp. 104, 112, 117; Lilian C. A. Knowles, *The Economic Development of the British Overseas Empire, 1763–1914*, vol. 1: *The Empire as a Whole* and *The British Tropics* (London, 1924), pp. 515–17.

18. Latham, p. 66. Adam Willis Kirkaldy, *British Shipping: Its History, Organisation and Importance* (London and New York, 1914), appendix 14.

19. Herbert Feis, *Europe: The World's Banker, 1870–1914*, 2nd ed. (New York, 1965), pp. 23, 55–56, 74; Latham, pp. 53–60; Gerald S. Graham, "Imperial Finance, Trade, and Communications, 1895–1914," in *Cambridge History of the British Empire*, vol. 3: *The Empire-Commonwealth, 1870–1919*, ed. E. A. Benians, James Butler, and C. E. Carrington (Cambridge, 1959), p. 458; Samuel Berrick Saul, *Studies in British Overseas Trade, 1870–1914* (Liverpool, 1960), p. 206.

2

Ships and Shipping

The new global economy of the nineteenth century was created by the application of iron, steel, and fossil fuels to transportation. Improvements in transportation, and concomitant advances in communications, affect the world economy in several ways.

By lowering the cost of freight, they make it worthwhile to carry goods further than before, or to carry cheaper goods over a given distance. This widens the market for all products and makes it increasingly profitable for each region to specialize in those products in which it has a comparative advantage. Cheaper transportation also reduces the prices of many goods. For goods for which the demand and supply are elastic, the effect will be to increase the volume of freight. Economies of scale, both in production and in transportation, will in turn lead to lower costs. Thus volume may rise and costs drop still further. This was, by and large, the trend of the world market in the years 1860–1913, especially for the products of the tropics.

Another kind of transportation improvement is greater speed. Even when a faster transportation system costs more over a given distance than a slower one, it leaves capital tied up in transit for a shorter time; this benefit weighs most in the case of high value-density goods, that is, those that cost a lot in relation to their bulk. Thus faster transportation quickens the pace of finance. Finally, technological advances make transportation and communications systems safer, more reliable, and more punctual. This reduces insurance and inventory costs and permits businesses to plan ahead more efficiently. It also reassures the timid, lowering the threshold of entrepreneurship. After the mid-nineteenth century, international trade ceased being the preserve of privateers and merchant-adventurers.

Transportation systems were not only services, they were also industries, and these industries were very concentrated. Given the remarkable freedom of trade and navigation in effect before World War I, a world market based on comparative advantage would have arisen anyway. But the pattern of advantages was skewed by the control which a few countries, Britain in particular, exercised over the transportation and communications industries. This control gave those who exercised it two advantages. It gave them access to the best opportunities in world trade and to a number of ancillary activities such as insurance, banking, brokerage, and warehousing. And it created backward linkages to the manufacturers of ships, railway materiel, telegraph cables, machinery, steel, and other industrial goods. The effects of these linkages, and the various efforts that were made to redress the balance, form an essential part of our story.

The industrialization of transportation and communications began in the first half of the nineteenth century with the invention of railways, the telegraph, and the iron-hulled steamboat. Its impact on the world economy, however, had to await a number of improvements. It is wrong to think of technological change as consisting of a sudden "invention" followed by a lengthy process of "diffusion." Rather, invention, development, and diffusion go on simultaneously. Thus, in the nineteenth century, transportation technologies underwent constant improvement. Steam engines became more efficient, iron replaced wood and steel replaced iron, and the telegraphs became faster, cheaper, and more widespread. By 1914 the first industrial transportation and communications network was in place throughout most of the world. Today's ships, railways, and telegraphs are seldom faster, cheaper, or better run (in many cases they are much worse) than they were before World War I.

Just as the steam, steel, and telegraph system was being perfected and installed all over the world, the radio, the automobile, and the airplane came along to speed things up still further. Under the pressure of technical ingenuity, the transformation of the world has never slackened its pace. Our purpose, however, is not to celebrate Western technical ingenuity, but to investigate its impact on tropical Asia and Africa, and in particular the coincidence between the era of steam and steel and the new imperialism.

The various parts of the transportation and communications network—shipping, railways, telegraphs, canals, port cities, and so on—grew simultaneously and interdependently. For the sake of clarity, however, we will consider them separately, beginning with ships and shipping.

Characteristics of World Shipping

What impressed contemporaries most about the new ships of the nineteenth century was their speed. Attention focused mainly on the North Atlantic, where shipping lines, supported by the patriotic fervor of rival nations, vied for the blue riband. On other routes, prestige gave way to multiple calculations involving the costs of higher speeds, the requirements of mail contracts, and the impact of technological change on the competitive positions of various shipping lines.

On the routes to the east, the acceleration of service was especially noticeable between Europe and India. In the eighteenth century, lumbering East Indiamen commonly took from 5 to 8 months to travel between London and Calcutta; a record of sorts was set in 1789 when the *Stuart* spent 14 months traveling from Amsterdam to Calcutta. By the early nineteenth century some sailing ships were making the journey in as little as 2 months, only to be held up, sometimes for weeks, trying to sail up the Hooghly River to Calcutta. In any event the monsoons ruled out hasty return trips, and letter writers expected to wait 2 years for an answer. It is not hard to understand the enthusiasm with which steamships were greeted in the British community in India and among those merchants in Britain who traded with India. Transit times between India and Britain fell to a month for mail in the 1830s, for passengers and valuable cargoes in the 1840s, and for ordinary freight after 1869. It continued to fall until World War I. A measure of the increased speed is found in the contracts between the British Post Office and the Peninsular and Oriental Steam Navigation Company: the times allotted to the mail ships declined from 6 weeks (England to Calcutta) in the 1840s to 11½ days (Brindisi to Bombay) in 1908–15. By the outbreak of World War I, mail, passengers, and fast freight made the voyage from London to Calcutta in 2 weeks.[1]

Much the same was true of other routes. The voyage between England and Sydney or Melbourne took 125 days in the early nineteenth century. Better sailing ships and charts brought the average time down to 92 days in the 1850s. In 1881 the *Aberdeen* made the voyage in 42 days. And by the beginning of this century the trip took a month. The China trade was similarly speeded up. Since the days of the Opium War (1839–42), the costly and perishable tea of China gave every incentive to speed. Clippers, the swiftest sailing ships, made the trip in 90 to 105 days. In 1868 the *Agamemnon,* a compound steamer, shortened it to 68 days. In 1882 the *Stirling Castle* went from London to Hangkow in just under 29 days; this remained the standard speed thereafter. The trip between the Netherlands and Java, which had taken a year in the seventeenth century, still took 100 to 120 days, on average, in the 1850s. Then it began to fall. By the Overland Route across Egypt, such a trip took 42 days in 1859; in 1900, using the Suez Canal, it took about a month, and in the 1920s only 3 weeks.[2]

The speeding up of shipping was certainly welcome news to travelers and letter writers and deserved all the attention it received. But from an economic point of view, the cost of shipping had a much more profound impact. The first ships to trade across the oceans only made a profit by carrying the most valuable of cargoes: spices, silk, bullion, and the like. By the eighteenth century it was worth transporting sugar, cotton, tea, opium, and manufactured goods across the oceans; but it still cost some £30 a ton to carry barrels of wine from Europe to India in 1795. After a period of great fluctuations in the early nineteenth century, freight rates fell erratically until World War I. The cost of shipping wool from Australia to Europe declined by half between 1873 and 1896; the freight for Indian jute dropped by 75 percent from 1873 to 1905. For different cargoes and periods, the decline averaged anywhere from 60 to 95 percent. By the 1920s freights on long voyages cost only one-twentieth of the value of the average cargo.[3]

In a paper read before the Royal Statistical Society in 1937, L. Isserlis brought together index numbers for both tramp shipping freight rates and British wholesale prices from 1869 to 1936; Table 2.1 shows the general trends and the relationship between these two indices.

What do the figures in Table 2.1 tell us? First, a secular decline in tramp freight rates from 1873 to 1908 is apparent, and it appears again from 1918 to 1934. In between these two periods, World War I both boosted demand for shipping and cut the supply, thus causing a major inflation. In peacetime, however, rates not only declined, but did so more than wholesale prices. In fact, the decline in rates contributed much to the decline in prices and to the rising standard of living of the British people, as contemporaries realized. As the economist Michael Mulhall wrote in 1881: "Perhaps the secret of prosperity has been the development of the carrying trade, by land

Table 2.1   Index Numbers of Tramp Shipping Freight Rates and Whole-sale Prices 1870–1935 (1869 = 100)

Year	Yearly		Seven-Year Moving Averages		
	Freight Index	Price Index	Freight Index	Price Index	Freight/Price Ratio (percent)
1870	103	98	—	—	—
1875	99	98	102	101	101.0
1880	87	90	86	88	97.7
1885	63	73	69	76	90.8
1890	64	73	65	71	91.5
1895	56	63	58	65	89.2
1900	76	77	60	69	87.0
1905	51	73	50	74	67.6
1910	50	80	57	81	70.4
1915	199	110	318	126	252.4
1920	374	256	390	181	215.5
1925	110	139	122	132	92.4
1930	93	99	101	101	100.0
1935	88	86	—	—	—

Source: L. Isserlis, "Tramp Shipping Cargoes, and Freights," *Journal of the Royal Statistical Society* 101, pt. 1 (1938), p. 122, table VIII. Tramp freight rates were more competitive, and usually lower, than freight rates on scheduled liners.

and sea, which has risen 53 per cent, and cheapened all the products of industry by placing the producer and consumer in closer relations than before."[4]

Falling costs, accompanied by improvements in the quality of shipping—safety, speed, reliability, and better handling—brought new commodities into the world trade network. Some were simply low value-density products like jute, ores, coal, wool, and petroleum. Others were low-cost but perishable products which were for the first time worth shipping from distant continents. Refrigeration, introduced in the 1880s, made it possible to ship meat from Australia and South America and butter from New Zealand. Thanks to a faster journey and better cleaning it was even worth shipping wheat from India to Europe.[5]

One final and all-important aspect of shipping was its volume. In the sixty years from 1850 to 1910 the world's merchant fleets grew from 9 to almost 35 million net registered tons, a fourfold increase. Britain's share remained remarkably constant throughout that period, from 40 to 50 percent of the world total.[6]

While some international trade was not seaborne, the increase

Table 2.2 Index Numbers of World Trade (base year: 1913)

Years	Index Number	Years	Index Number	Years	Index Number
1850	10.1	1891–95	48	1921–25	82
1860	13.9	1896–1900	57	1926–29	110
1870	23.8	1901–05	67	1930	113
1876–80	30	1906–10	81	1931–35	93
1881–85	38	1911–13	96	1936–38	106
1886–90	44	1913	100	1938	103

Source: Walt W. Rostow, *The World Economy: History and Prospect* (Austin, Tex. and London, 1978), p. 669.

in the volume of world trade gives a fair approximation of the importance of shipping.

Table 2.2 shows a fairly steady increase in the volume of world trade from 1850 to 1913 (during which time it multiplied tenfold), followed by a period of inflation due to the war and the Depression, so that by 1938 world trade was barely above what it had been twenty-five years earlier. Insofar as that trade was seaborne, the increase in volume to 1913 and its fluctuations thereafter were the result of both the demand for shipping and its supply. On the supply side two factors were at work: technological advances in shipbuilding and related infrastructures, and organizational advances in shipping. Let us now consider them in detail.

Steamships before 1869

Looking back at the late nineteenth century, the historian E. A. Benians wrote in despair: "The period is remarkable for the progress of invention and the application of science to the arts of life. Invention more than policy transformed the relations of states. Things were in the saddle and rode mankind."[7]

Benians misunderstood technological change. "Invention" is as much the work of people as "policy." What was in the saddle riding mankind was not "things" but the people who developed, manufactured, and controlled these "things." Nowhere is this more evident than in the case of ships and shipping.[8]

In the 1830s and 1840s, steamers were regarded as fast and punctual, but expensive, alternatives to sailing ships. They burned

so much coal—from 4 to 5 kilograms per horsepower per hour—that
their services were called for only where speed mattered more than
cost: transporting the mails or carrying wealthy passengers across
the Atlantic. On longer routes, steamships had to carry so much coal
that there was little room for cargo. To sail from Bombay to Aden
in 1830, the *Hugh Lindsay* had to fill its hold and cabins and pile
its decks with coal, barely leaving room for the crew and the mail.
Ten years later the first Cunard liner, the *Britannia,* needed 640 tons
of coal to cross the Atlantic, leaving only 225 tons of cargo capacity.

Nor was it economical to refuel steamships en route, since most
of the world's steamer coal came from Britain, and it had to pay
freight to other parts of the world. Therefore the use of steamers as
freighters spread slowly: first to British waters, then along the coasts
of western and northern Europe. By the 1850s steam-powered freighters were used in the Mediterranean and across the Atlantic, but only
for expensive cargoes like cotton or perishables like fruit.

Four innovations in the 1850s and 1860s lowered costs and
improved the competitive position of steamers vis-à-vis sailing ships:
the screw-propeller, the iron hull, the surface condenser, and the
compound engine. The screw-propeller, introduced in 1838, was
especially suited to ocean steamers because paddle wheels were inefficient and vulnerable in high seas; by the 1850s, almost all new
steamships were propeller-driven. The idea of building ships of iron,
like so many other shipbuilding innovations, was discussed and
tested for decades before it became accepted. The advantages of
iron over wood were well known: an iron ship weighed 30 to 40
percent less and had a capacity 15 percent greater than a wooden
one of the same displacement; furthermore, iron ships could be
made larger, safer, and longer-lasting than wooden ones. Yet there
were disadvantages also. Only in 1839 did George Airy, the astron-
omer royal, find a way to adapt the compass to iron ships. Iron was
more prone to fouling than copper-sheathed wood. But most of all,
iron shipbuilding meant replacing an industry of carpenters with one
of ironworkers. For these reasons, over thirty years elapsed from the
launching of the first iron steamer, the *Aaron Manby,* in 1821 until
the British Post Office and Lloyds of London accepted iron ships
as equal to wood in the mid-1850s.

Before the 1830s, ships' boilers were fed seawater. The resulting salt deposits, however, limited the pressure of the steam, required
periodic scraping, and corroded the boilers and pipes. By recycling
distilled water, the surface condenser which Samuel Hall patented

in 1834 promised to prolong boiler life, reduce maintenance, and permit higher pressures, thereby lowering fuel consumption. Not until the 1860s, however, were condensers sufficiently developed to become generally used in marine engines.

Engineers knew that higher pressure was the key to fuel efficiency. Thanks to the surface condenser and to stronger boilers and pipes, marine steam pressures rose from 400–500 grams per square centimeter in the 1830s to 1,400–2,100 grams in the 1850s. These higher pressures permitted yet another improvement, the compound engine. After leaving the cylinder, high-pressure steam still contained considerable energy. The idea of using this remaining pressure in a second cylinder had been tried several times but was not applied to marine engines because of the problem of salt deposits. In 1854 the Scottish engineer John Elder put a compound engine with a surface condenser into the coastal steamer *Brandan*. He then built a series of steamers for the Pacific Steam Navigation Company, which operated where coal was especially expensive. Nonetheless, compound engines remained exotic and experimental in the eyes of most shipbuilders until 1865. That year Alfred Holt, founder of the Ocean Steam Ship Company, or Blue Funnel Line, sent a compound-engined steamer nonstop from London to Mauritius, almost 14,000 kilometers. This was a major turning point in marine technology, for it made steam competitive with sail in the cargo trades of Africa and Asia.

For cargo ships, the trend appears clearly in the coal consumption figures. The earliest steamers burned 4 kilograms of coal per hour for every indicated horsepower. By the 1850s coal consumption had fallen to 2 kilograms and by the mid-1860s, to 1.6 kilograms. Ships of the 1850s that burned 30 to 40 tons of coal a day to carry 1,400 tons of cargo were replaced in the 1860s by ships that needed only 14 tons of coal to transport 2,000 tons of cargo.

The Suez Canal

The progress of shipbuilding alone would eventually have brought steam freighters into the Indian Ocean and Far Eastern trades, just as it had brought them into the North Atlantic and Mediterranean. But this gradual trend was broken by an abrupt discontinuity: the opening of the Suez Canal on November 17, 1869.

The story of the Suez Canal is full of ironies. It was proposed by Frenchmen from Napoleon Bonaparte to Ferdinand de Lesseps, enthusiastically backed by the Khedives of Egypt as a means of shaking off Ottoman tutelage, and built by French money and Egyptian labor. And all the while the British government opposed it, finally acquiescing to a fait accompli with the greatest reluctance. Yet it did little for France, and led to the defeat of Egypt. But it encouraged shipping, and especially the maritime power of Great Britain, beyond the most sanguine predictions.

The Suez Canal has stimulated a vast literature, and we need not tarry on the motives and maneuvering that led up to it, nor on its construction and financing.[9] What is relevant here is its consequences. The effect of the canal was to shorten the distance between Europe and the East, as Table 2.3 indicates. The impact of the canal was felt most strongly on the trade between India and Europe and less so, but still significantly, between China and Europe. It was not able, however, to divert the European-Australian trade, because the small advantage it conferred did not compensate for the high tolls on canal traffic.[10]

The canal did not immediately capture all the shipping that the logic of geography might dictate, for several reasons. At first it did not handle ships efficiently. Being 6 meters deep and 22 wide at the bottom, it was too small for the largest steamers of the time. Ships could only pass each other at certain points, and all traffic stopped at night. As a result, the 130-kilometer-long passage took 60 hours.

To handle new and larger ships more efficiently, the canal underwent a series of improvements almost from the beginning. From 1887 on, ships were equipped with electric headlights to allow night travel. This cut the transit time to 16 hours, and later to less than 14. The British Post Office finally allowed the mails to stay aboard ship instead of going by train across Egypt to be put on another ship. The

Table 2.3 Shipping Distances between London and the East

	Via the Cape	Via the Canal	Percent Saved
London-Bombay	19,755 km	11,619 km	41
London-Calcutta	22,039	14,970	32
London-Singapore	21,742	15,486	29
London-Hong Kong	24,409	18,148	26
London-Sydney	23,502	22,493	4

canal itself was deepened and widened several times; by 1924 it was 45 meters wide at the bottom, 73 at the top, and 10.5 deep.[11]

As important as the canal itself was the kind of ships that could use it. Dangerously fickle winds in the Red Sea and prohibitive towing costs through the canal kept sailing ships away; of the 5,236 ships that used it between December 1, 1869 and April 1, 1875, only 4.5 percent were sailing ships.[12] Among steamers, the most modern, those equipped with compound engines, benefited the most from the long-distance trades to the East. The result was a boom in steamship building. Even then, it was several years before the world's shipping fleets and the habits of shippers adjusted to the new geography of the sea.

To drum up support for his project, Ferdinand de Lesseps had predicted that the Suez Canal would handle 3 million tons of traffic a year. Thirteen years later, as the canal neared completion, his promises had swelled to the unbelievable figure of 6 million tons; eventually, traffic through the canal went much higher than even de Lesseps had dared dream.

At the time the canal was being dug, many contemporaries expected it would reverse a 400-year-old trend and give back to the Mediterranean nations the maritime position they had enjoyed before the Great Discoveries. Nothing of the sort happened; to everyone's surprise, the canal only reinforced Britain's domination of world shipping. As Table 2.4 shows, British ships accounted for four-fifths of the total tonnage passing through the canal in the 1880s, and Britain's share, though slowly declining from that peak, did not fall below one-half until the 1930s. The reason, as we shall see, was a complex mixture of economics, technology, and politics.[13]

The Suez Canal cut distance, time, and costs; it contributed to the growth of world trade; and it encouraged the further development of marine technology and of shipping organization. Its influence on world trade was second only to the invention of the steamship itself. Let us consider these various changes, beginning with the technical ones.

Shipbuilding after 1869

No comparable period in history saw such rapid advances in shipbuilding, or such a complete transformation of the world's shipping fleet, as the years between the opening of the Suez Canal and World

Table 2.4 Suez Canal Tonnage (1869–1940)

Year	Suez Tonnage	British Tonnage	British as % of Total Tonnage
1870	436,609	289,235	66.25
1875	2,009,984	2,181,387	74.18
1880	3,057,422	3,446,431	79.33
1885	6,335,753	4,864,049	76.77
1890	6,890,094	5,331,095	77.37
1895	8,448,383	6,062,587	71.76
1900	9,738,152	5,605,421	57.56
1905	13,134,105	8,356,940	63.63
1910	16,581,898	10,423,610	62.86
1915	15,266,155	11,656,038	76.35
1920	17,574,657	10,838,842	61.67
1925	26,761,935	16,016,439	59.85
1930	31,668,759	17,600,483	55.58
1935	32,810,968	15,754,818	47.96
1940	13,535,712	7,449,913	55.04

Source: D. A. Farnie, *East and West of Suez 1854–1956* (Oxford, 1969), pp. 751–52.

War I. At the beginning of that era most of the world's ships were small wooden sailing vessels; by the end, almost all were metal-hulled steamers. And most of these new ships were British; between 1890 and 1914 Britain built two-thirds of the world's ships.

While the high drama of technological change first took place on the North Atlantic, the tropical routes soon followed, especially the main axis of European-tropical trade, the British route to India. Here the "standard" level of technology (if not always the state of the art) could be found in the ships of the Peninsular and Oriental Steam Navigation Company (P&O). A list of their best ships at different points in time, shown in Table 2.5, may therefore serve as a guide to changes in merchant marine technology.

Technological advances in shipbuilding continued in the two directions laid out before 1869: hulls and engines. One such advance was the use of steel. Shipbuilders had experimented with steel as early as 1862 when the blockade-runner *Banshee* crossed the Atlantic. But the only steel then available, made by the Bessemer process, was too uneven to seriously challenge wrought iron. Not until the Siemens-Martin process was perfected in the 1870s was a suitable steel available, and only in the early 1880s did it become cheap enough for merchant ships. After that, steel quickly replaced iron

Table 2.5 Fastest P&O Liners, 1842–1912

Year	Name	Mate-rial	Length (meters)	Displ. (tons)	Power (hp)	Engine type	Speed (knots)	Pas-sengers
1842	*Hindostan*	wood	71.65	1800	520	simple	10	200
1852	*Himalaya*	iron	103.70	3500	2000	simple	13	400
1870	*Khedive*	iron	110.00	4000	3000	compound	14	500
1881	*Rome*	iron	136.50	5000	4600	compound	14.5	610
1887	*Victoria*	steel	142.50	6522	7500	triple	15	690
1894	*Caledonia*	steel	152.00	7558	11000	triple	16.5	810
1903	*Moldavia*	steel	158.75	9500	14000	quadruple	17	916
1912	*Maloja*	steel	173.50	12500	16000	quadruple	18	560

Note: Like most liners, these ships carried both passengers and freight; only on the North Atlantic were there pure passenger liners.

Source: Georges Michon, *Les grandes compagnies anglaises de navigation* (Paris, 1913), p. 37.

in the shipbuilding industry, for the transition threatened no established crafts or hoary traditions as had the transition from wood to iron. And the advantages of steel were readily acknowledged: a 15 percent saving in weight, greater resilience, and better resistance to corrosion. By 1885 half of all new ships were made of steel, and by 1900, practically all.[14]

Steel also made its way into the engines and boilers. Once again, the goal was to reduce fuel consumption by means of higher pressures. Using mild steel boiler plates, engineers were able to raise pressures to 6 kilograms per square centimeter in the late 1870s, then to 9 kilograms in the early 1880s, to 12 in the 1890s, and finally to 14 by the turn of the century. Higher pressures in turn allowed them to wring more energy out of the steam by adding yet a third cylinder to the engine.

The first ship equipped experimentally with a triple-expansion engine was the *Prepontis,* built by R. Napier and Company of Glasgow in 1874. The ship that really introduced triple expansion into the merchant marine, however, was Napier's *Aberdeen,* built in 1881. With a steam pressure of 9 to 10 kilograms, it burned only 600 to 700 grams of coal per horsepower per hour, one-third to one-quarter as much as the compound engines of the 1850s. In such an engine, the energy released by the burning of a sheet of paper was sufficient to move one ton over one kilometer. The result was a shipbuilding boom in the 1890s as various shipping lines replaced their now-obsolete iron-hulled compound-engined steamers with the newer steel ships with triple-expansion engines.

While the *Aberdeen* was exceptionally economical, having been designed for the Australian trade where distances were great and coal was expensive, the trend toward greater fuel efficiency was noticeable throughout the merchant fleet. Average fuel consumption per horsepower per hour, which had been 2,300 grams of coal in 1855, dropped to 1,600 in 1865, to 800 in 1881, and to 700 in 1891. Not only were ships becoming more fuel-efficient, but the price of coal was also declining from an average of 12 shillings, 6 pence per ton in 1867–77 to 9 shillings per ton in 1878–87. The falling cost of energy was one of the major components in the decline of freight rates.

After the turn of the century, a fourth cylinder was added to some marine engines. The extra cost and complexity of quadruple-expansion engines was only justified in fast liners, however. By the first decade of the twentieth century the reciprocating steam engine, which had so transformed the world in the nineteenth, had reached the point of diminishing returns, and was challenged by two newcomers, the steam turbine and the diesel engine, to which we shall return later.[15]

Improvements in engines and hulls were accompanied by an increase in the size of ships. In the 1850s a 200-ton freighter was considered very large, while in 1900 many freighters measured over 7,500 tons. The average gross tonnage of ships passing through the Suez Canal rose from 1,348 in 1870 to 2,877 in 1890 and to 5,086 in 1910. By 1914 there were many ships of over 20,000 tons.[16] The size of freighters was important for two reasons. One was the naval architects' rule of thumb according to which the energy required to move a ship at a given speed varied as the two-thirds power of displacement; or, put another way, a ship twice as large as another consumes only 1.6 times as much fuel. The size of the crew was also less than proportional to the size of the ship; thus, as ships grew larger, the number of crew members per 100 tons of displacement shrank from 1.6 in 1855 to 1 in 1891.[17]

In the first three decades of this century, shipbuilding once again changed, partly of its own momentum, and partly under the pressure of war and reconstruction. In the race to build ever faster warships and transatlantic liners, engineers and metallurgists created boilers able to withstand pressures of over 80 kilograms per square centimeter. Even quadruple-expansion engines could not use such pressures efficiently. The steam turbine, introduced in 1894 by Charles Parsons, overcame the limitations of reciprocating engines and propelled his *Turbinia* at a then unheard of 34 knots. Because the energy needed to move a ship varies with the square of the speed, this technical achievement was far too costly for the merchant marine, except on luxury liners. In 1910 Parsons solved the dilemma with a gearbox which allowed a fast-rotating turbine to turn a propeller slowly. From then on, even slow freighters could benefit from the fuel efficiency of turbines.

Just as the geared turbine was threatening the supremacy of the reciprocating steam engine the diesel engine appeared. For most shipping firms—the P&O, British India, Glen, and Nederland lines, for example—the switch to diesels took place in the mid-1920s.[18] Since diesels require oil, while steam turbines could be built to burn either oil or coal, the competition between the two types of engines was clouded by the relative prices of coal and oil, which fluctuated in the uncertain economy of the war and its aftermath. In Britain, which had abundant coal and a long tradition of building marine steam engines, both diesels and steam turbines gained at the expense of reciprocating engines.[19] In other merchant marines, oil replaced coal. Even in steam-powered ships, oil offered many advantages over coal: it could be stowed in parts of the ship not suitable for cargo; it had a greater caloric density, which meant ships required less refueling; it demanded far less labor; and supplies were more evenly distributed throughout the world. As a result, many new steamers were built, and old ones converted, to burn oil. The gross tonnage of oil-burning ships rose from 1.75 million in 1914 (86 percent of which were steamers) to 36.25 million in 1939 (58 percent steamers).[20]

Ships' hulls also evolved, though not as dramatically as they had in the nineteenth century. While the general structure changed little, two trends already felt in the nineteenth century continued to influence ship design. Special-purpose ships were built to carry products not easily transported in ordinary freighters: refrigerator ships for meat, butter, and fruit; and tankers for the growing petroleum industry. And larger ships replaced smaller ones; thus the average size of ships using the Suez Canal increased over 60 percent between 1910 and 1938, from 5,086 to 7,747 tons.[21]

Tropical Harbors

To accommodate the increasing size and number of ships and the growing volume of trade, the ports of the tropical seas had to undergo a parallel transformation. Traditional ones like Alexandria, Bombay, Cape Town, and Calcutta grew unrecognizably in the course of the nineteenth century. In addition, the rulers of colonial empires built a whole series of new ports—Port Said, Karachi, Dakar, Singapore, Hong Kong, to name a few—to serve the expanding traffic. The new ports often surpassed their older rivals. By the 1890s Singapore served over fifty regular shipping lines; Hong Kong cleared more shipping than Liverpool, and almost as much as London.[22]

As shipping changed, ports became more differentiated. Some still played the traditional role of outlets for the products of a rich hinterland; Alexandria, Calcutta, and Hong Kong were in this category. Others, with less of a hinterland but located on an important sea-lane, became ports of call like Aden or Port Said, or entrepôts like Singapore or Colombo.

In the early years of steam the need for frequent refueling led to several acts of territorial expansion; Socotra was occupied in 1835 and Aden in 1838 to become coaling stations between India and Egypt. Harbors with little or no hinterland, like Gibraltar, Papeete (Tahiti), Las Palmas (Canaries) and Saint Vincent (Cape Verde Islands) became major bunkering stations for passing steamers. Port Said, at the entrance to the Suez Canal (and roughly halfway between London and Bombay), was the premier coaling station, selling 1.5 million tons of coal in 1900; Montevideo and Las Palmas followed with 1 million apiece. While bunkering was their main business in the steamship era, ports of call provided many other services to passing ships. Brindisi and Taranto in Italy were drop-off and pick-up points for the mails to the East. Ships stopped at ports with cable offices to receive orders and commercial information. Unimportant out-of-the-way islands like Saint Helena and Norfolk were visited by postal steamers from their metropoles, more to show the flag than for economic reasons. Thus was the symbiosis between shipping and empire kept alive in the nineteenth century.[23]

In the new age of steamships, what dictated the importance of a port was trade and not, as in previous centuries, the existence of a natural harbor. Iron, steel, steam power, and concrete made it possible for engineers to build large artificial harbors on practically any coast, although the cost was high. All the world's first-class harbors, where the largest oceangoing ships could safely dock, had to have certain features: lights, buoys, and breakwaters; a minimum depth at dockside of 9.75 meters; dry docks and repair shops; cranes and warehouses; and supplies of food, water, and naval stores.

In the days of sail, Calcutta was the preeminent port of India. Situated on the Hooghly River, the city received both seagoing ships and country boats (later also river steamers) from the rich and densely populated Ganges valley. Such was the geography of the sea before the days of steam that the shortest way from London to Delhi went via Cape Town and Calcutta, and that Europe was closer to Calcutta than to Bombay. Initially steam power was a great boon to Calcutta, for sailing ships often took three weeks to make their way from Diamond Harbour near the mouth of the Hooghly to Calcutta 50 miles upstream, whereas steamers did it in a day or two. It is not surprising that the first steam vessels to appear in Calcutta in 1822–23 were the tug *Diana* and the dredge *Pluto,* nor that the British merchants of Calcutta eagerly encouraged steam navigation by lobbying and offering prizes.

In the long run, however, the evolution of steam shipping diminished Calcutta's relative importance within India. The Red Sea route and the Suez Canal favored Bombay more, as did the growing railway network which linked various parts of the country. By the 1870s, passengers, mail, and the costlier cargoes went from Calcutta to Europe via Bombay. The change in British India's center of gravity became official in 1911, when New Delhi replaced Calcutta as the residence of the viceroys.

Though Calcutta's relative place in India declined, its importance to shipping grew in absolute terms. From the 1830s on there had been numerous proposals to build commercial docks there. Between 1869 and 1881 the port added eight jetties for oceangoing ships. The Kidderpore Docks, completed in 1892 at a cost of £2 million, consisted of a 3.9-hectare tidal basin and a 14-hectare dock area able to take ships drawing over 9 meters. Jetties were also built at Garden Reach, just below the city, for ships up to 200 meters long, and at Budge-Budge, 19 kilometers downstream, for oil tankers. Between the sea and the port, the river had to be constantly dredged to a depth of 9.15 meters. Despite all these works, the harbor facilities proved insufficient in World War I and thereafter. By 1926, almost 1,300 ships a year were docking at Calcutta. King George's

Dock, the largest in Asia and capable of handling 12 ships at once, was added in 1927–28. It was to be the high point of Calcutta's history as a port.[24]

Madras was an example of an artificial harbor. Though the oldest British settlement in India, it was the last major Indian port to be developed and until 1890 had no safe harbor at all. Ships had to anchor 400 meters offshore, and cargo was loaded and unloaded by surfboats, 1.5 tons at a time, when the sea was calm. Wrote one indignant Briton:

> The very skill of the boatmen is one of the difficulties of the port, for their numbers being limited, they are able to set regulations at defiance, and to charge pretty much what they like—twice, four, and six times the legal hire being a common rate, and ten times being by no means uncommon; well knowing that the course which would elsewhere be followed of importing additional men from other localities would be inoperative at Madras, for in consequence of caste prejudices, and the disinclination of natives to adopt customs, or to follow trades which have not been followed by their fathers before them, the ordinary laws of political economy cannot be applied in their integrity in this country, and it does not follow that because there is a greater demand than supply of Masulah boatmen, and a very handsome profit to be reaped by those who might qualify for the occupation, that outsiders will qualify for it, and come forward to break the monopoly.

Faced with this obstruction of "the ordinary laws of political economy," the authorities decided to build a proper harbor. In the 1860s a 335-meter-long pier was extended out to sea, but it only served in good weather. In 1876, after much deliberation, work began on a set of three piers totalling almost 3 kilometers in length, to enclose a harbor. In November 1881 the work was almost finished when a cyclone demolished the outer portions. Further work, costing over a million pounds, finally transformed Madras by 1890 into a first-class port able to take the largest ships in an 81-hectare sheltered harbor.[25]

Many smaller ports could not afford the heavy costs involved in trying to cope with the ever-increasing size of steamers. The increasing differentiation between first-class port cities and sleepy coastal towns was especially noticeable in Africa. According to one geographer, Africa had 88 ports in 1940, of which only 16 were able to handle the largest ships, and only 44 had a depth of 6.1 meters or more and other modern facilities needed by average-size ocean vessels. Of these 44, 5 were in South Africa, 6 in Algeria, 4 in

Tunisia, and 3 in Egypt; tropical Africa lagged far behind. Good natural but undeveloped harbors could be found along desert coasts, in Portuguese colonies which Portugal was too poor to invest in, or in Tanganyika, which Britain slighted in favor of Kenya. Many West African coastal cities—Saint-Louis, Accra, Abidjan, Lome, Cotonou, and others—had no harbor at all before 1940. The few ships that stopped at these ports had to anchor far offshore, while cargoes and passengers were conveyed in open surfboats or in shallow lighters. Not only these towns, but their hinterlands also, were the backwaters of the colonial world.[26]

The New Organization of Shipping

Since the industrial revolution, technological changes have tended toward increasing complexity, both mechanical and organizational. Once a new machine or process is invented, further improvements or adaptations usually involve more parts and more connections to other devices. A more complex machine or process, in turn, costs more, both initially and in use. Complex machines and processes also need specialized inputs—technicians, raw materials, energy, spare parts—and are thus more dependent on outside organizations. Finally, technical innovations are subject to the pressures of obsolescence, for the economic lifespan of machines and processes is often shorter than their physical life.

The evolution of merchant ships followed this pattern of increasing complexity, cost, dependence, and obsolescence. Making the technological changes justify their costs both demanded and caused economic and organizational changes. As ships became more expensive to buy, new ways of financing them were devised: government contracts, limited liability companies, loans by large investment banks. Once a ship was launched, it had to be put to work as efficiently as possible. Gone were the days of the merchant-captains who sailed the seas for years on end in search of freight and customers, following traditions, rumors, and hunches, until they had accumulated enough valuable cargo or treasure to return home. On a steam freighter the captain and crew only navigated; they were not entrepreneurs but employees, and their ships were the floating appendages of organizations half a world away.

To justify their cost, steamers required speed, tight scheduling, and well-organized procedures for loading, unloading, refueling, and maintenance. These in turn required both appropriate infrastructures and large, aggressive, efficient business organizations. Sophisticated ships went hand in hand with sophisticated shipping. Both flourished when their economic rationale—economies of scale—operated most fully. In other words, they simultaneously required and stimulated an increase in the volume of shipping.

Planning ahead required a means of communication faster than the ship itself. While shipping companies date back to the 1840s, their growth was spurred by the spread of submarine telegraph cables after 1860. Telegraphic communications had two major impacts on the shipping industry. The first was to give shippers sufficiently fresh information about markets and prices to enable them to place orders and arrange shipping on the basis of real conditions instead of hunches. This was especially important for bulk products with slender profit margins and fluctuating prices.

Second, the telegraph allowed shipping lines to communicate with their ships in harbor, thereby optimizing the use of their investments. The kind of flexibility which this implies was of course not equally available to all ships. Shipping lines bound by published schedules could not vary their plans very quickly. Tramps, on the other hand, were free to change their routes, cargoes, and destinations according to the opportunities of the moment, as communicated to them by their home offices. Ships often stopped in ports along their way to pick up telegrams with orders and market information. Much cargo was not dispatched to a particular destination but "for orders," with the intention that the freighter would stop at some port of call to learn of its ultimate destination. This practice ended with the use of the radio, which allowed home offices to control their freighters at sea.[27] Because of their flexibility, tramps had lower rates than scheduled liners, and their rates thus formed the wholesale price of world shipping which we saw earlier.

Sailing ships had traditionally belonged to trading firms, groups of merchants who both shipped freight for others and operated on their own account. This mixed and ad hoc business organization could not bear the high cost of steamers, however. In its place a new organization, the shipping line, offered speed and punctuality and thus allowed customers to plan ahead. The first clients to be attracted by the speed and punctuality of steamships were the postal systems, and for these benefits governments were willing to pay heavily. Mail subsidies allowed steam shipping to emerge in the 1840s, when steamers were otherwise uneconomical. Among the first lines founded in the 1830s and 1840s, several grew rich carrying the mails and proclaimed it in their names: the Royal Mail Steam Packet Company, which served South America and the Caribbean; the British and North American Royal Mail Steam Packet Company, also known as the Cunard Line; the French Messageries Maritimes; and the Royal Dutch Mail Line.

The subsidies were costly: in the 1870s, the Peninsular and Oriental Line received £450,000 a year to run weekly steamers to Bombay and fortnightly ones to China and Japan. By the turn of the century the British mail lines were getting over a million pounds a year, over a third of which came from the colonial budgets; India, for instance, was required to cover half the net loss of the P&O mail contracts. In exchange for their favored status and secure profits, mail carriers were subject to stiff government regulations. Their ships had to meet Admiralty specifications, so that in the event of war they could quickly be converted into armed transports. More importantly, they had to be fast; in the words of the postal contract: "The speed of the British ships shall equal the highest speed of the foreign mail ships on the same route." Business and government collaborated in support of empire communications and shipbuilding technology.[28]

The mail contracts were a cushion for a few favored shipping lines. For the rest, however, shipping was a cutthroat business subject to all sorts of fluctuations: technological obsolescence, ruinous competition, swings in the supply and demand of the basic cargoes, and new markets and sources of supply, to name a few. Meanwhile, the tonnage a shipping line had at its disposal was inelastic in the short run, for ships could hardly be used for anything but shipping. So heavy were the investments in steamships that fixed costs usually represented 75 percent of the total cost of each voyage. The result was that freight rates, hence earnings, gyrated wildly while tending downward in the long run. The shipping companies did not take these matters lightly but fought back with cartel arrangements called conferences or rings. The purpose of a conference was twofold: to keep freight rates from falling and to prevent outsiders from muscling in. This was accomplished by offering customers a deferred rebate. This rebate, usually 10 percent, was promised to customers some 6 to 12 months after the date of shipment if they remained loyal and did not in the meantime use any shipping line that was not a member of the conference. If they did ship with an outsider, they would not only lose whatever rebates they had earned, but would later be denied shipping space on conference lines when they might really need it. All but the most reckless shippers bowed under such oligopolistic pressures.

The first conference was formed in 1875 by the lines that served Calcutta, whose revenues had been hurt by a combination of the steamship building boom of the early 1870s, the Suez Canal, and a general trade depression. Two years later, they introduced the deferred rebate system. Lines that served the China coast, where freight rates had fallen from £8 per ton in 1860 to £1 15 shillings in 1878, quickly followed this example. Led by the shipping agent John Swire, the P&O, Ocean Steam Ship Company, Glen Line, Castle Line, and Messageries Maritimes founded the first China Conference in 1879. Its effects were soon felt as its members' freight rates rose, on average, by 16.75 percent in its first year, while the volume of cargo also rose.[29]

The conference system soon spread to other routes. The West Africa–UK Conference began in 1894 with an agreement between the Elder Dempster Line and the Woermann Linie of Hamburg. The Straits Conference of 1897 resulted from an agreement between the shipping agents of Singapore and Penang and the lines that stopped there on their way from China or Japan to Europe.

The conference system was challenged in the British courts in 1885–90, but it was found legal. In 1909 a Royal Commission on Shipping Rings also found them acceptable. Nonetheless, they were vulnerable to the competition of tramp freighters. Tramps, more adaptable to changing trade conditions thanks to the telegraph, were able to undercut the scheduled liners' rates by more than the deferred rebates. In 1900, British tramps accounted for a third of the world's maritime cargo capacity. As a result, conferences were never as successful on the homeward (i.e., Europe-bound) trade in tropical bulk products as they were in the outward trades in manufactured goods. Only in West Africa were the tramps kept at bay, because the conference lines owned the surfboats and lighters which ships needed to load and unload their cargo. Colonial governments, which depended on shipping for all their supplies, were thus forced to pay conference rates. The trade and economic development of West Africa was hampered by such an arrangement.[30]

The Major Shipping Companies

The maritime expert Georges Michon published a list, given in Table 2.6, of the major shipping companies as of 1911, which gives a fair idea of the distribution of shipping business both by nationality and by main region of activity. Of the lines mentioned in Table 2.6, the British had 57 percent of the ships and 56 percent of the tonnage, the Germans had 25 and 28 percent respectively, the French 12 and 11 percent, and the Japanese 6 and 5 percent.

In the East, the first steamship line was the P&O. It was founded in 1834 as the Peninsular Steam Navigation Company, linking Britain to Spain and Portugal. Three years later, it signed a contract with the British government to carry the mails to Gibraltar, then to Malta and Alexandria, for £34,200 a year; at that point it changed its name to Peninsular and Oriental. In 1842 it obtained a contract of £115,000 a year to carry the mails from Suez to Ceylon, Madras, and Calcutta. After that it extended its service throughout the eastern seas: to Penang, Singapore, Hong Kong, and Shanghai after 1845, and in the 1850s to Bombay, Sydney, and Yokohama. In 1867 the P&O received £450,000 a year, almost half of all British mail sub-

Table 2.6   Major Shipping Companies in 1911

Name	Nationality	Total Gross Tonnage	Number of Ships
Hamburg-Amerika	German	1,210,000	180
Norddeutscher Lloyd	German	723,000	128
Peninsular & Oriental[a]	British	538,000	71
Ellermans Lines	British	522,900	124
British India[a]	British	506,000	116
White Star	British	501,000	33
Alfred Holt[a]	British	457,000	70
Cie. Gen. Transatlantique	French	421,000	80
Furness	British	387,000	123
Messageries Maritimes[a]	French	335,000	63
Nippon Yusen Kaisha[a]	Japanese	327,000	76
Cunard	British	323,200	26
Union Castle[b]	British	320,000	44
Elder Dempster[b]	British	286,000	99

[a] Lines that specialized in the South and East Asian trades.
[b] Lines that specialized in the African trades.

Source: Michon, pp. 16–17.

sidies. These subsidies allowed it to purchase the most modern and luxurious ships available: in 1851 the *Himalaya,* then the largest and fastest ship afloat, and later the compound-engined *Mooltan* in 1860 and the triple-expansion liners *Britannia* and *Victoria* in 1887. As its network spread, it increased the size of its fleet from 45 ships and 58,185 tons in 1854 to 71 ships and 538,875 tons in 1912. These vessels were the lifeline of the British military and civil servants in India and the East, for whom the epitome of comfort was a first-class cabin facing away from the sun: port outward and starboard home-ward, or "posh." In addition to the mails, the servants of empire, and a few wealthy tourists, the P&O liners also carried precious car-goes, such as frozen English beef.[31]

While the P&O and a few other mail carriers were quasi-official branches of their respective governments, many other lines got their start with a small mail contract, and later, when more efficient ships became available, were able to grow without further subsidies. The classic example is the British India Steam Navigation Company. This line was founded in 1856 by William Mackinnon, who had come from Scotland to Calcutta as a grocery clerk some ten years before. As its original name—the Calcutta and Burmah Steam Navigation Company—implied, it began with a small mail contract between Cal-cutta and Rangoon. It quickly prospered in the rice and timber trades. By 1862, having extended its routes to Ceylon and the Persian Gulf on one side, and to Malaya and Singapore on the other, it changed its name to British India. The company still obtained mail contracts, for instance the Aden-Zanzibar-Natal contract in 1862, which triggered Mackinnon's later obsession with East African af-fairs. But it never became dependent on them. Instead it specialized in bulky freight and in the transport of migrants between India, Burma, Ceylon, Malaya, and China. It kept ahead of its rivals by using the most modern equipment available. While the P&O resisted the Suez Canal, the British India was the first line to send a ship through the canal after the inauguration ceremonies. In the 1870s it had regular sailings throughout the Indian Ocean and beyond, to Britain, China, the Dutch East Indies, and Australia. By the early 1890s it owned 110 ships, with 39 percent more tonnage than the P&O and routes twice as long.

In the early years of this century, both the P&O and the British India had become huge fleets. In 1914 they merged, creating a com-bined fleet of 186 ships displacing 1,136,000 tons, with another 25 ships under construction. It was the largest fleet in existence, though they kept their separate names. By 1939 they had a total of almost 2 million tons of gross registered shipping, with a presence throughout the Mediterranean, the Indian Ocean, and the Far East, and a quasi monopoly on the foreign trade of India.[32]

Following the example of the P&O and the British India, many other companies entered the growing trade between Europe and the ports of Asia and Africa. British firms led the field by introducing the most technically advanced ships. Thus the brothers Alfred and Philip Holt started the Ocean Steam Ship Company, or Blue Funnel Line, in 1865 by sending compound-engined steamers between China and Britain at record speeds. Having wrested the China trade from the clippers, they went on to gain a large share of the trades of Malaya and the Netherlands East Indies as well. Several other British shipping companies appeared in the Far East after 1869—the Glen, Castle, Ben, and China Mutual lines—but most of them were eventually acquired either by the Holts or by the British India–P&O group. After World War I, only these two remained as the giants in that trade.

In the West African trade, the pioneer line was Macgregor Laird's African Steam Ship Company, founded in 1852 with a mail contract to Lagos and the Bight of Benin. It was later joined by the British and African Steam Navigation Company; their merger in 1890 formed the Elder Dempster Lines, which dominated the trade of British West Africa. On the South African route the first line, the General Screw Steam Shipping Company, failed and was replaced by the Union Steam Ship Company in 1857, followed in 1872 by the Castle Line. Their merger in 1900 formed the Union Castle Line.[33]

There were two kinds of French companies. One was the officially sponsored and subsidized Compagnie des Services Maritimes des Messageries Impériales, renamed Messageries Maritimes in 1871. It got a foothold in the passenger business to the East by offering a Marseille-Alexandria service connecting with the London-Paris-Marseille railway, which saved British travelers several days over the P&O's London-Alexandria ships. In the 1870s, after the opening of the Suez Canal, it ran the fastest passenger liners to India, Indochina, China, and Australia. Since France had far less trade with the East than Britain did, the Messageries survived thanks to special sources of revenue: a hefty mail subsidy (6 million francs or £250,000 in the 1860s), its use as a troop carrier for France's eastern colonies, and the favor of many British travelers who preferred the French style to the frumpy P&O. As a result, the Messageries grew quickly until it owned over half the French merchant tonnage in 1874.

The second kind of French shipping lines were small companies, often branches of trading firms like Maurel et Prom of Bordeaux and Senegal; the Compagnie Cyprien Fabre, which had trading posts in Guinea; the Compagnie Fabre-Fraissinet of Marseille; and later the Chargeurs Réunis of Dakar. These lines suffered from the underdevelopment of French Africa which, except for Senegal, offered little trade until after World War I.[34]

German shipping grew fast from the 1880s until World War I, fueled by the fast-growing maritime commerce of Germany. Two giant firms, the Hamburg-Amerika Linie and the Norddeutscher Lloyd, concentrated on the lucrative international routes to the Americas and the Far East, leaving the unprofitable trade with Germany's poor backwater colonies to smaller companies. Of the colonial lines the most important was the Woermann Linie, founded in 1886 by a Hamburg trading firm which had been sending ships along the West African coast since the 1840s. By the turn of the century it offered regular monthly sailings to Togo, Cameroon, and Southwest Africa, and to various non-German ports along the way. In the last decade before World War I, two new companies entered the East African routes to Tanganyika: the Hamburg-Bremen-Afrika Linie and the Deutsch Ost-Afrika Linie. The growing German interest in world trade is revealed in the names of several new docks opened in the harbor of Hamburg in the 1890s: the Asiaquai, the Indiahafen, the Australiaquai, and lastly the Afrikaquai.[35]

The Causes of British Supremacy

At several points in our story we have noted the overwhelming predominance of British shipbuilding and British shipping. What reasons, other than "perfidy," explain this persistent strength? Partly it was the weaknesses of Britain's rivals. In the first half of the nineteenth century, Americans built better and faster ships more cheaply than the British, but they lost their advantage at the time of the Civil War. The substitution of iron for wood and steam for sails gave the advantage back to Britain, which had more developed iron and mechanical industries. Other rivals, France and Germany in particular, were too involved in the Continental wars which Britain's insularity spared it.

But that is a partial explanation, for technological change renewed merchant fleets faster than wars destroyed them. It was the ability to make merchant ships pay a profit which explains Britain's advantage. Several economic forces favored Britain. The economies involved in large-scale production of similar ships and marine engines kept costs down. The London money market was both more affluent and more attuned to the needs of shipbuilders and shipping companies than any other. The organization of British shipping was more flexible than that of its rivals. Many nations had state-subsidized mail carriers, but only Britain had the huge number of tramps—60 percent of her merchant fleet in 1911—by which British shipping companies ferreted out the unpredictable opportunities of the world market. And Britain's control over most of the world's submarine cables gave British shippers an edge over their rivals in the form of cheaper and faster telegraphic information. In a sense Britain had two merchant fleets. One, composed of conference liners, skimmed the cream off the market by keeping freight rates high for customers who valued predictable and punctual service; the other, the tramps, took the low end of the market by offering cheap rates to customers who valued money more than time.

But reinforcing these man-made advantages was an enormous natural advantage. In an age of coal-burning ships, Britain had the most, the best, and the cheapest steamer coal. Coal was important not only as fuel but also as an export commodity. Normally industrial economies import heavy raw materials and export much lighter manufactured goods. Thanks to coal, which in 1911 constituted nine-tenths of Britain's exports by weight, British shipping was much more evenly balanced; in 1912 British ships passing south through the Suez Canal were loaded, on average, to 72 percent of capacity, while those passing north were loaded to 98 percent. Thus the costs of each voyage could be spread over both the outward and homeward sections. As the maritime historians Kirkaldy and McLeod explained, "We have been able to carry on the trading of the world at comparatively low freights, because our ships have entered and left our ports fully loaded."[36]

The French and Italian merchant marines did not benefit nearly as much from the Suez Canal as they had hoped, because ships heading east from Marseille or Genoa either had to go out in ballast or, if they wished to earn freight on the outward leg of their journey, they first had to stop in Britain to pick up a cargo of coal, salt, or railway iron. Either way, their costs were higher than those of British ships.

Britain was also the first to exploit coalfields in other parts of the world. Coal from Bengal was being used in steamers in the 1830s, from Borneo in the 1840s, and from Natal in the 1860s. Though not as good as Welsh coal, they gave Britain a near-monopoly of the world's steamer coal supplies.

After World War I, the advantages which Britain gained from coal were lost when shipbuilders turned to oil. To be sure, the British Empire had oil reserves in Burma, the Persian Gulf, and Trinidad, and close ties with the Netherlands' oil reserves in Borneo. But in the interwar years other nations—the United States, Mexico, the Soviet Union—had better access to oil than Britain. And oil, unlike coal, did not constitute a valuable outbound cargo for Britain, but another heavy import. The weakening of Britain's maritime supremacy was partly a result of the passing of the age of coal.[37]

Britain's supremacy was also the consequence of owning India. In the period 1814–50, four commodities had dominated India's exports: indigo, raw silk, opium, and cotton. After 1850 raw cotton retained a steady share (about 20 percent) of the export trade, but the three other costlier products were replaced by tea and especially by bulky products such as jute, rice, wheat, oil seeds, and manganese ore. In exchange, India imported European (mainly British) iron, steel, railway materiel, manufactured goods, and coal. In 1892–93 almost 70 percent of India's foreign trade passed through the Suez Canal, almost all of it in British ships.

Not only was India's trade important in quantity and composition, it also balanced the British trade in a way which allowed Britain to remain, until 1914, the center of world shipping, finance, and insurance. India ran a balance of trade surplus with continental Europe which almost balanced out Britain's trade deficit with Europe: India supplied continental Europe with foodstuffs and raw materials, Europe supplied Britain with manufactures, and Britain supplied India with other manufactures and with services, in particular shipping. The system worked smoothly until World War I, but this division of labor between India and Britain had two long-range effects: it allowed Britain to perpetuate industries that were becoming less competitive with those of its industrial rivals, and it made it more difficult to produce in India those goods which Britain could produce and ship more cheaply. As time went on, the British maritime supremacy encouraged both the obsolescence of British industries and the underdevelopment of India.[38]

Notes

1. The mails traveled from London to Brindisi and from Bombay to Calcutta by train, which added a day at each end. Georges Michon, *Les grandes compagnies anglaises de navigation* (Paris, 1913), pp. 41–42; Allister Macmillan, *Seaports of India and Ceylon; Historical and Descriptive, Commercial and Industrial, Facts, Figures, & Resources* (London, 1928), pp. 35–37.

2. Gerald S. Graham, "The Ascendancy of the Sailing Ship, 1850–85," *Economic History Review* 9 (1956): 82; Halford Lancaster Hoskins, *British Routes to India* (London, 1928), p. 418; E. A. Benians, "Finance, Trade, and Communications, 1870–1895," in *The Cambridge History of the British Empire*, vol. 3: *The Empire-Commonwealth, 1870–1919*, ed. E. A. Benians, James Butler and C. E. Carrington (Cambridge, 1959), p. 202; Auguste Toussaint, *History of the Indian Ocean*, trans. June Guicharnaud (Chicago, 1966), p. 212; Francis Edwin Hyde, *Far Eastern Trade, 1860–1914* (London, 1973), pp. 23–25; John de la Valette, "Steam Navigation to the East Indies. Early Efforts—The Establishment of the Netherlands Line—Recent Developments," *Asiatic Review* 30 (1934): 342–50; Michon, pp. 41–42.

3. Hoskins, p. 419; Douglas North, "Ocean Freight Rates and Economic Development, 1750–1913," *Journal of Economic History* 18 (December 1958): 541–45; Charles Knick Harley, "Transportation, the World Wheat Trade, and the Kuznets Cycle, 1850–1913," in *Explorations in Economic History* 17 (1980): 223, table 2; Harold James Dyos and Derek Howard Aldcroft, *British Transport* (Leicester, 1969), pp. 244–45; A. Fraser-Macdonald, *Our Ocean Railways; or, the Rise, Progress, and Development of Ocean Steam Navigation* (London, 1893), p. 102; Francis Edwin Hyde, "British Shipping Companies and East and South-East Asia, 1860–1939," in *The Economic Development of South-East Asia: Studies in Economic History and Political Economy*, ed. C. D. Cowan (London, 1964), p. 44.

4. Michael G. Mulhall, *Balance-Sheet of the World for Ten Years, 1870–1880* (London, 1881), p. 2.

5. Samuel Berrick Saul, *Studies in British Overseas Trade, 1870–1914* (Liverpool, 1960), p. 194, n. 4.

6. Adam W. Kirkaldy, *British Shipping: Its History, Organisation and Importance* (London and New York, 1914), appendix 17; the author estimated one steamer ton as equivalent in carrying capacity to four sailing ship tons. Michon, using net tons, arrives at slightly lower ratios of British to world shipping (p. 5).

7. Benians, p. 181.

8. There is a considerable literature on steamships and marine engines before 1869; James P. Baxter III, *The Introduction of the Ironclad Warship* (Cambridge, Mass., 1933), p. 33; Bernard Brodie, *Sea Power in the Machine Age: Major Naval Inventions and Their Consequences on International Politics,*

1814–1940 (London, 1943), pp. 149–54; Ambroise Victor Charles Colin, *La navigation commerciale au XIXe siècle* (Paris, 1901), pp. 33–61; Charles Dollfus, "Les origines de la construction métallique des navires" in *Les origines de la navigation à vapeur,* ed. Michel Mollat (Paris, 1970), pp. 63–67; Dyos and Aldcroft, pp. 238–43; Charles Ernest Fayle, *A Short History of the World's Shipping Industry* (London, 1933), pp. 231–41; Fraser-Macdonald, pp. 213–31; Duncan Haws, *Ships and the Sea: A Chronological Review* (New York, 1975), pp. 115–57; Kirkaldy, pp. 90–140; Carl E. McDowell and Helen M. Gibbs, *Ocean Transportation* (New York, 1954), pp. 27–28; Thomas Main, *The Progress of Marine Engineering from the Time of Watt until the Present Day* (New York, 1893), pp. 56–69; L. T. C. Rolt, *Victorian Engineering* (Harmondsworth, 1974), pp. 85–97; René Augustin Verneaux, *L'industrie des transports maritimes au XIXe siècle et au commencement du XXe siècle,* 2 vols. (Paris, 1903), 2: 4–39.

9. On the Suez Canal in general, see Douglas A. Farnie, *East and West of Suez: The Suez Canal in History, 1854–1956* (Oxford, 1969), and John Marlowe, *World Ditch: The Making of the Suez Canal* (New York, 1964). On the background to its construction, see Hoskins, pp. 292–345. On the construction itself, see J. Clerk, "Suez Canal," *Fortnightly Review* 5, n.s. (1869): 80–100, 207–25; Louis Figuier, *Isthmes et Canaux* (Paris, 1884), pp. 99–103, 129–64, 172–217; and Lord Kinross, *Between Two Seas: The Creation of the Suez Canal* (London, 1968), pp. 149–53, 215–25.

10. Figures from W. Woodruff, *Impact of Western Man: A Study of Europe's Role in the World Economy, 1750–1960* (New York, 1967), p. 260, table 6/6a. The Panama Canal, which opened in 1914, was also important for world trade; however, it was not on the routes between Europe and any European colony, and therefore did not affect the trades we are considering here. See Adam Willis Kirkaldy and Sir Charles Campbell McLeod, *The Trade, Commerce and Shipping of the Empire* (London and New York, 1924), p. 89.

11. Hoskins, pp. 453–71; Kirkaldy and McLeod, pp. 80–84; Arthur John Sargent, *Seaways of the Empire: Notes on the Geography of Transport* (London, 1918), pp. 50–52; André Siegfried, *Suez, Panama et les routes maritimes mondiales* (Paris, 1940), p. 77.

12. Max E. Fletcher, "The Suez Canal and World Shipping, 1869–1914," *Journal of Economic History* 18 (1958): 558.

13. Fletcher, p. 565; Siegfried, p. 92; Michon, p. 14; Sargent, p. 55.

14. On the introduction of steel into shipbuilding, see A. M. Robb, "Shipbuilding," in *A History of Technology,* ed. Charles Singer, E. J. Holmyard, A. R. Hall, and Trevor I. Williams, vol. 5: *The Late Nineteenth Century, c. 1850–c. 1900* (London, 1958), pp. 352, 373; Colin, pp. 57–58; and Brodie, p. 164.

15. On late nineteenth-century engines see Toussaint, p. 212; Dyos and Aldcroft, p. 242; Fraser-Macdonald, pp. 210–20; Kirkaldy, pp. 130–38; and Main, p. 70. On coal consumption see Charles Knick Harley, "The Shift from Sailing Ships to Steamships, 1850–1890: A Study in Technological Change and Its Diffusion," in *Essays on a Mature Economy: Britain after 1840,* ed. Donald McCloskey (London, 1971), p. 232; and Colin, p. 51.

16. Siegfried, pp. 91–92; Hoskins, pp. 447, 472. Gross tonnage is a measure of volume corresponding to the number of 100-cubic-foot (2.83-cubic-

meter) spaces enclosed; net tonnage equals gross tonnage minus space for crew, engines, etc.; deadweight tonnage measures the number of tons of cargo and fuel a ship can safely hold in the best conditions. Approximately 100 net tons = 160 gross tons = 240 deadweight tons. See L. Isserlis, "Tramp Shipping Cargoes, and Freights," *Journal of the Royal Statistical Society* 101, pt. 1 (1938): 62.

17. Harley, "Sailing Ships to Steamships," pp. 233–34.

18. Hyde, "British Shipping," pp. 40–44; de la Valette, p. 350; Kirkaldy and McLeod, pp. 48–50.

19. Kirkaldy and McLeod, p. 149.

20. William Ashworth, *A Short History of the International Economy since 1850,* 3rd ed. (London, 1975), p. 70.

21. Siegfried, pp. 91–92.

22. Benians, p. 201.

23. W. E. Minchinton, "British Ports of Call in the Nineteenth Century," *Mariner's Mirror* 62 (May 1976): 145–57; Hoskins, pp. 123, 188–211; Fraser-Macdonald, pp. 103–21. See also the map in Kirkaldy.

24. George Walter Macgeorge, *Ways and Works in India: Being an Account of the Public Works in that Country from the Earliest Times up to the Present Day* (Westminster, 1894), pp. 518–19; Macmillan, pp. 35–40.

25. Macgeorge, pp. 514–18. See also Macmillan, pp. 290–91.

26. G. F. Deasy, "The Harbors of Africa," *Economic Geography* 18 (October 1942): 325–42. See Gerald S. Graham, "Imperial Finance, Trade, and Communications 1895–1914," in *The Cambridge History of the British Empire,* vol. 3: *The Empire-Commonwealth, 1870–1919,* ed. E. A. Benians, James Butler, and C. E. Carrington (Cambridge, 1959), p. 476; and Verneaux, 1:314–15.

27. Minchinton, pp. 146–47.

28. Howard Robinson, *Carrying British Mails Overseas* (London, 1964), pp. 234–38, 267.

29. Hyde, "British Shipping," pp. 32–33.

30. Hyde, *Far Eastern Trade,* pp. 25–40; Charlotte Leubuscher, *The West African Shipping Trade, 1909–1959* (Leyden, 1963), pp. 13–19; A. S. Hurd, *The Triumph of the Tramp Ship* (London, 1922), pp. 162–71; David Divine, *Those Splendid Ships: The Story of the Peninsular and Oriental Line* (London, 1960), pp. 156–63.

31. Charles R. V. Gibbs, *British Passenger Liners of the Five Oceans: A Record of the British Passenger Lines and their Liners, from 1838 to the Present Day* (London, 1963), pp. 44–101; Boyd Cable, *A Hundred Years of the P. & O.–Peninsular and Oriental Steam Navigation Company, 1837–1937* (London, 1937). See also Michon, pp. 24–73, and Fraser-Macdonald, pp. 92–107.

32. John S. Galbraith, *Mackinnon and East Africa, 1878–1895; a Study in the 'New Imperialism'* (Cambridge, 1972), pp. 30–42; George Blake, *B.I. Centenary, 1856–1956* (London, 1956). See also Divine, pp. 168–75, and Gibbs, pp. 118–25.

33. Francis Edwin Hyde, *Blue Funnel: A History of Alfred Holt and Company of Liverpool from 1865 to 1914* (Liverpool, 1957); *Far Eastern Trade;* and "British Shipping"; P. N. Davies, *The Trade Makers: Elder Demp-*

ster in *West Africa, 1852–1972* (London, 1973), and "The African Steam Ship Company," in *Liverpool and Merseyside: Essays in the Economic and Social History of the Port and Its Hinterland,* ed. John R. Harris (Liverpool, 1969), pp. 212–38; Marischal Murray, *Ships and South Africa; a Maritime Chronicle of the Cape, with Particular Reference to Mail and Passenger Liners from the Early Days of Steam Down to the Present* (London, 1933); H. Moyse-Bartlett, *A History of the Merchant Navy* (London, 1937), pp. 230–35.

34. Roger Carour, *Sur les routes de la mer avec les Messageries Maritimes* (Paris, 1968); Roger Pasquier, "Le commerce de la Côte Occidentale d'Afrique de 1850 à 1870," in *Les origines de la navigation à vapeur,* ed. Michel Mollat (Paris, 1970), pp. 122–24; Emile Baillet, "Le rôle de la marine de commerce dans l'implantation de la France en A.O.F.," *Revue Maritime* 135 (July 1957): 832–40; Colin, pp. 168–69; Michon, pp. 29–32.

35. Karl Brackmann, *Fünfzig Jahre deutscher Afrikaschiffahrt* (Berlin, 1935); Emil Fitger, *Die wirtschaftliche und techniche Entwicklung der Seeschiffahrt von der Mitte des 19. Jahrhunderts bis auf die Gegenwart* (Leipzig, 1902); Aimé Dussol, *Les grandes compagnies de navigation et les chantiers de constructions maritimes en Allemagne* (Paris, 1908).

36. Fletcher, pp. 565–69; Kirkaldy and McLeod, pp. 51–52.

37. Brodie, pp. 113–17; Kirkaldy and McLeod, pp. 81–83, 159–69; Minchinton, p. 151; Sargent, pp. 21, 59; Michon, pp. 6–19.

38. Fletcher, pp. 571–72; Latham, pp. 68–70; K. N. Chaudhuri, "Foreign Trade and Balance of Payments (1757–1947)," in *The Cambridge Economic History of India,* vol. 2: *c. 1757–c. 1970,* ed. Dharma Kumar (Cambridge, 1983), pp. 804–77.

3

The Railways of India

In 1893 the maritime historian Fraser-Macdonald entitled a book on shipping *Our Ocean Railways*.[1] To Victorians, the analogy between steamships and railways was obvious. Both were powerful machines of iron and steam, swift and punctual means of travel, the most visible of the innovations transforming the world.

The railway era lasted about a century. The first three decades, from 1830 to 1860, were a time of experimentation and rapid growth in Britain, western Europe, and the eastern United States. From 1860 to 1914, the web of steel spread throughout the world, and so did the political, financial, and engineering techniques that had evolved along with it in its early years.

The spread of railways from their North Atlantic birthplace to the nonindustrialized parts of the world was the result of both demand-pull and supply-push. By and large, the demand for railways was strongest in those countries with a large and growing European population, such as eastern and southern Europe, North America, Argentina, and Australia. In other areas of the world the push came from Europeans and Americans who saw in railroads an unprecedented instrument of progress and profit or, as Cecil Rhodes put it, "philanthropy plus 5 percent." If non-Western peoples and rulers resisted the idea, as often happened, the railway promoters knew how to apply guile or force, just as they knew how to overcome natural obstacles with bridges and tunnels. Among non-Western peoples, only the Japanese showed real enthusiasm for railways.

Until 1914, railways were expected to open up new regions to settlement and develop commerce. To many observers, they were *the* key to modernization, progress, and economic development. After

1918 the pioneering aura that railways had once possessed passed on to the automobile and the airplane. With the exception of the USSR, the industrial countries built only a few new lines to fill in the gaps, while they closed uneconomic ones. In underdeveloped parts of the world, new rails were laid only where the demand for them was unequivocal, for instance between a new mine and the nearest harbor. With few exceptions, the world's railways map of 1940 closely resembled that of 1914.

From a geographic point of view, we can distinguish several different railway patterns. The most highly industrialized regions, northwestern Europe and the eastern United States, had a dense web of railways, with very few places more than twenty kilometers away from a rail line. Less industrialized or populated regions—western North America, Russia, India, South Africa, and parts of Australia and Latin America—had railway networks linking the major towns.

These more or less dense networks contrast sharply with the railways of areas that developed late, like Africa; there, lines ran inland from harbors to mining or agricultural areas without connecting to one another, and many major cities, in fact whole colonies, had no rail service at all. In a few colonial areas such as Indochina, French North Africa, and Malaya, a hybrid pattern developed: short lines running inland, connected by one trunk line parallel to the coast.

Transcontinental lines once aroused glittering fantasies. Ever since the linking of the Central Pacific and Union Pacific in 1869 opened up the American West and forged one nation of continental dimensions, enthusiasts proposed dozens of other lines in the hopes of repeating the same political and economic feat. Some succeeded, notably the Canadian Pacific in 1885 and the Transsiberian in 1903. Others remained in the planning phase a few years too long and were upstaged by aircraft and automobiles in the 1920s. The French Transsaharan railway projects, designed to link Algeria with the Niger, called forth more ink than steel.[2] A similar spirit animated Cecil Rhodes and his followers who dreamed of a Cape-to-Cairo railway; unlike the Transsaharan scheme, however, the Cape-to-Cairo aroused little official interest, for Britain was a maritime nation.[3]

It is tempting to give the different railway patterns shorthand labels, such as "advanced" for the rail networks and "colonial" or "underdeveloped" for short lines running inland from a harbor. To do so, however, would be doubly misleading, for colonialism took many forms, and the relationships between railways and the rest of society are exceedingly complex. India, which emerged from colonial rule with a "developed" rail network and an "underdeveloped" economy, is a case in point. It is so important that we shall devote this chapter to it. Before turning to it, however, let us consider in a general way the linkages between railways and society.

Railways are economic beings that create and consume scarce resources. Yet their impact goes far beyond the measurable resources they consume and produce, into the realm of external economies.

The output of railways is transportation, and their investments are usually justified by the improvements they bring: lower costs, higher speeds, and greater reliability. These factors are especially important in Africa and India, areas of poor traditional transportation. Estimates of the cost savings vary from region to region. Edward Hawkins, historian of Uganda, contrasts the cost of porterage from the Kenya coast to Uganda, namely £240 per ton or four shillings per ton-kilometer, with the rail freights of twelve cents per ton-kilometer, thirty times less.[4] In the Western Sudan, railway transport from Badoumbe to Bamako cost between 100 and 350 francs, four to twelve times less than head porterage.[5] In India, bullocks did the work that humans did in Africa, yet the savings brought by railways were similar; the ratio of rail freights to bullock-cart freights was 1 to 8 according to John Hurd, while Vinod Dubey gives a ratio of 1 to 20.[6]

Railways also meant speed. The trip from Mombasa to Uganda took 2 to 4 days by train, instead of a year on foot. Trains traveled at 40 to 100 kilometers per hour, ten times the speed of stage coaches, twenty times that of head-porters, thirty times that of oxen. They were far more dependable and kept running when, as frequently happened in India, the rains turned roads into impassible mud or drought decimated the draft animals.

The productivity of railways came from their enormous efficiencies of scale, however, and this was their weakness. One railway expert calculated that a train carrying 50 tons of freight at 20 kilometers per hour—in other words a tiny one—did the work of 13,333 head-porters.[7] Another estimated that an average train carried as much as 15,000 to 20,000 porters.[8] Seldom was such a demand in place before the rails were laid. Lilian Knowles tells of one such instance: "During the war the French Government bought the whole of certain crops in West Africa. They had to organize the transport of 4,200 tons of cereal furnished by eight districts, involving the employment of 125,000 carriers, who gave altogether 2,500,000 days' work."[9] Most of the time, however, railway promoters pinned their hopes on the prospects of future traffic, in other words on economic growth. This is why railways were financially so risky, and why after 1920 they faced such severe competition from less productive but more flexible road vehicles.[10]

Just as railways provided far more than a new form of an old service, so, on the input side, they consumed goods and services of a sort, or in quantities, which had not been available before. Therefore the impact of railways was felt, by a ripple effect, throughout the society. Their enormous fixed costs and long gestation period required more ready capital than had ever been assembled before. The development of banks and capital markets were a necessary complement to the railways, as was the willingness of savers to invest their funds in speculative enterprises or to lend them to governments. Furthermore, railways needed land in unprecedented amounts and in specific locations. In every country, even liberal Britain, they needed the government's power of eminent domain, and they frequently obtained land free, or at low cost. To build and operate railways demanded engineers and workers with a variety of new skills; this led to apprenticeship programs and technical education on a scale that resembled the raising of armies in wartime. Railways also needed equipment and fuel, both of which in turn demanded an industrial base. Thus at every step we run into the indirect effects of railways on parts of society that are neither their suppliers nor their customers. Banking, education, government, commerce, travel, industry: almost every aspect of society was transformed by the touch of railways.[11]

However, the countries in which railways were built were not isolated. Only the land on which they were built, and the transportation they provided, were specific to a particular location; all the other inputs were mobile. Loans could be raised in Paris or London to build railways in Argentina or Siberia; Chinese navvies were brought to California; locomotives, rails, crossties, and coal were shipped from one end of the earth to another.

This was done because most countries in which railways were built lacked, at the time, the wherewithal to build them. But it was also a matter of politics, for railways, being large enterprises, involved governments at every step. International transfers of skills, capital, and equipment were accompanied, often preceded, by political links. Independent nations like Russia, Argentina, and Spain obtained railways with foreign help through political connections. It is no surprise that in the colonies, whose political ties to a foreign country were tightest of all, the metropoles retained control over railway building and kept many of the benefits for themselves. It is because so many of the factors of production and external benefits of railways are not location-specific that colonial railways are excellent examples of the pitfalls of technology transfer.

Among colonial railway systems, that of India is unique. By its geographic pattern and by every other measure, it ranks among the largest and most advanced networks. Of the many ways to measure a rail network, the simplest is length of track. We can group the world's rail networks before 1947 into three categories. The first consists of the United States alone, with a network which reached a peak of 691,811 kilometers in 1930, more than the next six networks put together. Next came four nations with networks of 60,000 kilometers or more by the early twentieth century: Germany, Canada, India, and Russia/USSR. The third category includes nations with networks of 45,000 kilometers or less, from Britain and France on down.

Though a late starter, India's share of the world's track length quickly rose from 1.3 percent in 1860 to 4.1 percent in 1880, 5.3 percent in 1900, 5.4 percent in 1920, and 5.6 percent in 1940. The growth of the seven longest rail networks is shown in Table 3.1 and Figure 1.

Density is a more complex measure of the penetration of railroads into a society. By the 1920s British India had more tracks and less area, and thus a higher rail density, than South America, Africa, or the rest of Asia. It even had a higher density than the USSR, Australia, or Canada. Of all the world's regions of comparable area, only the United States and Europe had denser networks, as Table 3.2 indicates.[12]

As befitted its size, the Indian rail network was the costliest construction project undertaken by any colonial power in any colony. Of the £1,531 million placed by British investors in foreign railway securities before World War I, £140.8 million went to India and Ceylon; to that we must add the large share of securities which the governments of India and Ceylon floated for state railway construction. According to B. R. Tomlinson, of the £271 million of British capital exported to India before 1911, three-quarters (around £200 million) were invested in railways.[13]

The output of the Indian rail network, that is to say how many passengers and how much freight it carried how far, also places it among the world's largest, though not as clearly as its length. When

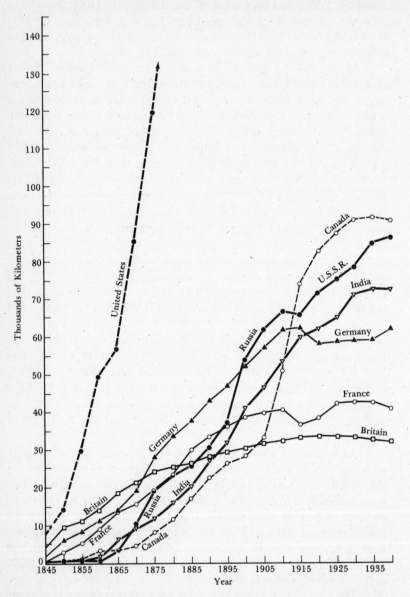

Figure 1 Lengths of the Seven Longest Rail Networks, 1845–1940 (*Sources:* See Table 3.1)

Table 3.1 Comparative Railways Lengths, 1845–1940 (in kilometers)

Year	USA	Russia	Canada	India	Germany	France	Britain
1845	7,456	144	26	—	2,143	875	3,931
1850	14,517	501	106	—	5,856	2,915	9,797
1855	29,569	1,049a	1,411	325	7,826	5,035	11,744
1860	49,286	1,626	3,323	1,542	11,089	9,167	14,603
1865	56,462	3,842	3,605	5,655	13,900	13,227	18,439
1870	85,167	10,731	4,212	8,637	18,876	15,544	21,000
1875	119,243	19,029	7,768	11,751	27,970	19,357	23,365
1880	186,111	22,865	11,635	15,764	33,838	23,089	25,060
1885	258,302	26,024	17,337	20,662	37,571	29,839	26,720
1890	334,979	30,596	22,537	27,227	42,869	33,280	27,827
1895	375,416	37,058	25,895	32,007	46,500	36,240	28,986
1900	416,461	53,234	28,684	40,396	51,678	38,109	30,079
1905	493,728	61,085	33,153	46,084	56,739	39,607	31,456
1910	566,099	66,581	50,579	52,767	61,209	40,484	32,184
1915	629,466	65,100	73,759	59,585b	62,091	36,400	32,650c
1920	654,309	71,600	82,354	61,957	57,545	38,200	32,707
1925	672,613	74,500	86,902	64,707	57,716	42,100	32,849
1930	691,811	77,900	91,062	70,565	58,176	42,400	32,632
1935	674,664	84,400	92,005	72,126	58,841	42,600	32,450
1940	653,356	86,400d	90,979	72,144	61,940d	40,600	32,094

Notes: (a) 1853; (b) 1915–16 and every five years thereafter; (c) approximately; (d) 1939.

Sources: For India: Morris D. Morris and Clyde B. Dudley, "Selected Railway Statistics for the Indian Subcontinent (India, Pakistan and Bangladesh), 1853—1946–47" in *Artha Vijnana* 17, no. 3 (September 1975): 193–96. For Canada: M. C. Urquhart and K. A. H. Buckley, eds., *Historical Statistics of Canada* (Cambridge and Toronto, 1965), pp. 528–32. For the United States: U.S. Bureau of the Census, *Historical Statistics of the United States: Colonial Times to 1970* (Washington, D.C., 1975), pp. 728 and 731. For Europe: B. R. Mitchell, *European Historical Statistics, 1750–1970* (New York, 1975), pp. 581–88.

Table 3.2 Comparative Railroad Densities in 1935 (in kilometers of track per 1000 square kilometers)

Germany	157	United States	72.1	USSR	3.8
Britain	133	India	16.1	Africa	2.4
France	77.6	Canada	9.2		

eager entrepreneurs first promoted railways in India, they envisoned them mainly for freight. In the end, most of the railways' business was transporting passengers, 95 percent of whom bought the cheapest fourth-class tickets. In numbers of passengers, the Indian railways surpassed the Russians until the 1930s and almost equalled the French, though they lagged behind the German and British systems (see Table 3.3). Similarly, the Indian railroads produced more passenger-kilometers than the French, or about two-thirds as many as the German network from 1905 on (Table 3.4). In terms of freight, the Indian railroads were further behind: half to two-thirds the French tonnage, a much smaller fraction of the German and British tonnages (Table 3.5). In ton-kilometers of freight carried, the Indian and French systems show very similar figures after 1905, though far behind Germany and Russia (Table 3.6).

The conclusion is inescapable: judging from all these aggregate figures of size and output, India had one of the world's top rail networks, at least from 1890 on. Yet among nations with large rail networks, India remains a special case for two reasons: it was the only colony among sovereign states; and it was the only one of them that failed to industrialize during the railway boom. Our task is to inquire into the relationship between these two facts.

Table 3.3 Number of Passengers, 1871–1939 (in millions)

Years	India	France	Germany	Britain	Russia/USSR
1871–74	22.0	111.2		416.6	24.0[a]
1875–79	35.1	142.1		526.6	28.5[b]
1880–84	58.6	191.8		636.2	37.3[c]
1885–89	92.1	223.6	358.5[d]	715.2	39.4
1890–94	123.0	287.8	453.3[e]	839.6	49.8
1895–99	147.0	374.9	699.0	995.5	76.6
1900–04	188.0	415.8	922.2	1,151.3	116.6
1905–09	276.6	454.6	1,290.4	1,219.1	148.8
1910–14	392.2	475.8	1,681.5[f]	1,315.1[f]	214.4
1915–19	443.1	345.2			310.8
1920–24	545.6	684.0	2,441.3[g]	1,295.0	116.0
1925–29	609.0	762.6	2,032.0	1,042.9	277.0
1930–34	515.4	710.0	1,516.0	809.0	630.8
1935–39	536.4	558.8	1,783.0[h]	866.6	1,098.8

Notes: (a) 1871–73; (b) 1875–76; (c) 1872–74; (d) 1888–89; (e) 1890–92; (f) 1910–13; (g) 1922–24; (h) 1936–38.

Sources: Morris and Dudley, pp. 206–10; Mitchell, pp. 601–12.

Table 3.4 Number of Passenger-kilometers, 1882–1939 (in millions)

Years	India	France	Germany	Russia/USSR
1882–84	4,879	6,895	7,400	
1885–89	6,475	7,469	8,700	
1890–94	8,281	9,162	12,120	
1895–99	9,425	11,482	16,400	10,200[a]
1900–04	12,275	13,360	21,740	14,600
1905–09	17,591	15,220	29,720	20,200
1910–14	23,497	17,080	38,875[b]	
1915–19	27,781	12,200		
1920–24	30,713	26,560		12,600
1925–29	33,353	27,860	47,520	24,200
1930–34	28,482	26,360	35,940	68,800
1935–39	29,473	23,100	48,200[c]	82,900

Notes: (a) 1897–99; (b) 1910–13; (c) 1935–38.

Sources: Morris and Dudley, pp. 206–10; Mitchell, pp. 601–12.

Table 3.5 Net Tons of Goods Carried, 1871–1939 (in thousands)

Years	India	France	Germany	Britain	Russia/USSR
1871–74	4,231	51,593		181,851	21,418[a]
1875–79	7,755	62,890		207,003	
1880–84	14,460	84,718		252,664	45,818[b]
1885–89	20,601	79,052		272,506	55,150
1890–94	26,996	96,180	218,965[c]	308,474	76,181
1895–99	34,525	109,600	301,589	372,368	106,027[d]
1900–04	45,351	83,200	378,816	435,869	166,000
1905–09	58,011	99,600	492,261	493,496	200,000
1910–14	73,972	117,800	613,064	533,418[e]	163,000[f]
1915–19	84,644	75,300	364,350		91,000
1920–24	92,768	154,000	358,700	318,292	49,000
1925–29	111,723	205,600	499,540	302,588	136,000
1930–34	101,529	175,800	368,760	272,984	270,000
1935–39	120,299	143,600	516,800[g]	284,767	491,000

Notes: (a) 1872–76; (b) 1882–84; (c) 1888–92; (d) 1895–98; (e) 1910–13; (f) 1910 and 1913–14; (g) 1935–38.

Sources: Morris and Dudley, pp. 214–18; Mitchell, pp. 589–600.

Table 3.6   Net Ton-Kilometers of Goods Carried, 1882–1939 (in millions)

Years	India	France	Germany	Russia/USSR
1882–84	4,443	10,797	16,300	
1885–89	5,561	10,096	19,000	
1890–94	6,901	12,184	23,800	
1895–99	8,491	13,700	30,600	30,600[a]
1900–04	11,964	16,000	38,000	42,200
1905–09	15,566	19,100	49,400	52,100
1910–14	22,939	22,300	63,100[b]	64,800[c]
1915–19	32,133	19,300		52,300
1920–24	30,515	30,200	53,111[d]	20,700
1925–29	33,302	39,000	70,000	60,500
1930–34	30,079	34,100	52,800	166,200
1935–39	35,687	28,800	76,300[e]	339,800

Notes: (a) 1897–99; (b) 1910–13; (c) 1910 and 1913–14; (d) 1922–24; (e) 1935–38.

Sources: Morris and Dudley, pp. 214–18; Mitchell, pp. 589–600.

Origins of the Indian Railways

Major public utilities in the nineteenth century required the joint efforts of the private and public sectors. As a result, they have fostered
an intense debate on the question of whether their primary motivation was political or economic. So it was with the Indian railways,
only more than other projects, because there were two governments
involved, the British and the Indian. Yet between the economic and
the political sides of the debate, a third aspect of the Indian railways
has been somewhat neglected. Railroads were, to Victorians (and to
many a nostalgic railfan since), exciting and ingenious mechanisms
that deserved to be built for their own sake, even if that meant finding
commercial or political justifications to obtain the necessary backing.

From their very inception, the Indian railways were the special
creation of British engineers. Among them were John Chapman,
W. P. Andrews, and especially Rowland Macdonald Stephenson, an
entrepreneur and a visionary who dreamed of building a railroad
from England to China, via India. Beginning in 1841, he set about
convincing the East India Company to allow, and to subsidize, railway construction in India. For several years, that "august and dilatory
body" resisted all such pressures, on the grounds that railways were
not the business of government.[14] As Leland Jenks explained,

The negative character of Company rule did not prevail merely because it was oriental, however. . . . It was part and parcel of the governmental pessimism which guided British statesmanship in varying degrees from 1825 to 1874. There being no economic process which government would not mar by its intervention, policy consisted in doing as little as decency would permit.[15]

The Honourable Company was more than merely conservative; it was ponderous, lethargic, and determined to avoid the nineteenth century. Stephenson and other railway promoters had to gather allies influential enough to force the issue. In 1844 Stephenson sailed to India, partly to survey the route from Calcutta to Delhi, partly to drum up support. It was not difficult to persuade the British "influentials" of Calcutta, who had earlier shown their eagerness for riverboats and steamships. He also got the support of a prominent Bengali merchant, Dwarkanath Tagore, who owned a colliery at Burdwan, from where the coal was transported by country boat to Calcutta. His friend, the barrister William Theobald, wrote to Stephenson that Tagore "is very desirous to have a Railway to the Collieries, and would raise one-third of the capital for this portion of the line, if undertaken immediately."[16]

In an article in the Calcutta *Englishman* in 1844, Stephenson described his scheme for major railroad lines linking Calcutta, Delhi, Bombay, Madras, and Calicut. His rhetoric, like that of all railway promoters, appealed to the political and commercial interests he wished to attract:

The first consideration is as a military measure for the better security with less outlay, of the entire territory, the second is a commercial point of view, in which the chief object is to provide the means of conveyance from the interior to the nearest shipping ports of the rich and varied productions of the country, and to transmit back manufactured goods of Great Britain, salt, etc., in exchange.[17]

While Stephenson expressed the standard combination of military and commercial motives, another railway promoter, John Chapman, gave a third motive, one that was to appear with increasing frequency in writings on the Indian railroads by British authors: the uplifting of the Indian people through technology transfer. In a *Letter to the Shareholders of the Great Indian Peninsula Railway,* Chapman expressed "the double hope of earning an honourable competency and of aiding in imparting to our fellow subjects in

India, a participation in the advantages of the greatest invention of modern times."[18]

More importantly, the railroad promoters received the support of the cotton manufacturers of Lancashire and Glasgow, and their members of Parliament. The cotton industry had been hurt by the American "cotton famine" of 1846 and looked yearningly toward India. However, transportation from the cotton districts of the Deccan to the harbors was exceedingly difficult, as bullocks could only travel some 16 kilometers a day and the cotton bales got ruined by rain and dust. As Chapman pointed out, the Lancashire merchants saw a railway from Bombay to the cotton districts "as nothing more than an extension of their own line from Manchester to Liverpool."[19]

The manufacturers in turn had powerful allies, among them Sir Charles Wood, president of the Board of Control, the parliamentary body which supervised the East India Company. He wrote Governor-General Dalhousie:

> If we could draw a larger supply of cotton from India it would be a great national object. . . . It is not a comfortable thing to be so dependent on the United States. . . . If we had the Bombay railway carried into the cotton country it would be the great work which Government is capable of performing with a view to this end.[20]

Added to the cotton interests were other influential lobbies: the London East India houses, the P&O line, the City bankers, the *Times,* the *Economist,* railway journals, and the Midlands cutlery and hardware manufacturers. All of them put pressure on the Court of Directors of the East India Company to reach an agreement with the railway promoters.[21]

In 1849 the Court of Directors gave in. The two lobbying organizations founded in 1845, Stephenson's East India Railway Company (EIR) and Chapman's Great Indian Peninsula Railway Company (GIP), now became limited liability companies. The contracts which the East India Company signed with them included several clauses with long-range implications. While the capital to build the railways was to come from the shareholders, the East India Company guaranteed them a dividend of 5 percent. If profits fell below that figure, the Indian government would make up the difference; if they were higher than 5 percent, the railway and the government would split the excess. Furthermore, during their first ninety-eight years, the railway companies could sell their property to the state for full compensation; otherwise in the ninety-ninth year the state acquired them free. The state also gave the railways free land and other services, in exchange for which it obtained the right to supervise construction and control their rates, fares, and operations.

While the various factions were debating and negotiating in Britain, India waited. The British inhabitants of India were eager but powerless, for India was an autocracy. The one person there who could influence events was the governor-general. It so happened that at this very time, from 1848 to 1856, India was ruled by the most technocratic of all its many governors, the Marquis of Dalhousie. Dalhousie was a brilliant and ambitious politician, a disciple of Jeremy Bentham, and the youngest member of the House of Lords. He had acquired his reputation as head of the Railway Department of the Board of Trade in 1845–46, during the British railway boom. His predecessor in India, Lord Hardinge, had written: "Hitherto, our Rule has been distinguished by building large Prisons; and the contrast with the Mughal Emperors, in this respect of public works, is not to our advantage." Dalhousie was determined to change all that. As soon as he arrived, he set to work to modernize the country as fast as he could. Soon he was writing: "Very large railways projects for all India are in hand, and have been referred to me—a very onerous reference. The electric telegraph for all India is on its way out. Uniform postage for all India is sanctioned and will shortly be put in force. Let no body say we are doing nothing.[22] In addition to railways, telegraphs, and uniform postage, Dalhousie also began canal and road projects and created the Department of Public Works to supervise them; set out to prohibit infanticide and widow-burning and encourage female education; and annexed Oudh (Bihar). In other words, he tried to revolutionize India from above.[23]

Before work on the railways could begin, a number of technical decisions had to be reached. In 1843 George T. Clerk, chief engineer of the Bombay Presidency, had surveyed the difficult route up the Western Ghats or cliffs that separate Bombay from the Deccan. His proposals were confirmed by James Berkley, appointed chief resident engineer of the GIP in 1849. Meanwhile, Rowland Stephenson surveyed the longer but easier route from Calcutta to Delhi via Mirzapur in 1845. His work was confirmed by F. W. Simms, engineer for the EIR.[24]

The engineers had not only the terrain to deal with, but also certain bizarre proposals put forth by well-wishers. Colonel Kennedy, consulting engineer of the government of India for railways, did not think trains could climb gradients of more than 1 in 2,000 and therefore suggested the tracks be laid along coastlines and riverbanks. Colonel Grant of the Bombay Engineers suggested hanging the tracks from chains two and a half meters above the ground, out of the reach of animals.[25]

These madcap ideas were rejected, and the Indian railways were by and large built on the European pattern. Only in one aspect did they differ: they were considerably larger. In the early years there were two competing track gauges in Britain: the standard 1.435-meter gauge, and Isambard Kingdom Brunel's 2.134-meter gauge. In 1845 Simms had recommended to Lord Dalhousie that the Indian gauge be set halfway between the two British gauges, at 1.676 meters. A wider gauge, he believed, would lower the center of gravity of locomotives and thereby reduce "oscillation" and "the danger of trains being blown away in a heavy wind." In 1846 Parliament passed an "Act for Regulating the Gauge of Railways" which set the British gauge at 1.435 meters; this later became the standard gauge in Europe and America as well. Nevertheless, the Court of Directors approved Simms's wider gauge. The next year they decided that all bridges, tunnels, and cuts should be made large enough for double tracks. A wide gauge and double width in turn required gentler curves and larger superstructures. Thus India was given one of the world's largest, hence most expensive, types of railways.[26]

Work on the first two lines began in 1850–51. In April 1853 the first locomotive in India, the "Lord Falkland," pulled a train from Bombay to Thana, a distance of 32 kilometers. A year later the line reached Kalyan, 60 kilometers away at the foot of the Ghats. That year the EIR inaugurated its first line from Howrah, opposite Calcutta, to Hooghly, 37 kilometers away, and in 1855 it stretched 195 kilometers to the Raniganj coal fields.[27]

The attitude of Indians toward the new railways surprised even their promoters. Rowland Stephenson believed that

> the people of India are poor, and in many parts thinly scattered over extensive tracts of country; but on the other hand India abounds in valuable products, of a nature which are in a great measure deprived of a profitable market by want of a cheap and expeditious means of transport. It may therefore be assumed that remuneration for railroads in India must, for the present, be drawn chiefly from the conveyance of merchandise, and not from passengers.[28]

He was wrong. The day after the inauguration of the Bombay-Thana line, the entire train was rented by the merchant Sir Jamsetjee Jee-jeebhoy for a trip with his family. He was soon followed by other Indians, 1,200 a day, 450,000 in the first year. In July 1854 Dalhousie wrote to a friend:

> On the 15th the Railway started most successfully. It has already solved one important problem. Many doubted whether the natives would go on a railway, partly, from timidity, partly from prejudice. The Bombay Railway cleared up the doubt as to the Western population, but still people doubted as to the Bengalees. However, the railway has been crowded for these three days by Calcutta Baboos. It is engaged thousands deep, and they are in the greatest excitement about it, many going even on the tender rather than not go.[29]

A year later, the newspaper *Friend of India* reported:

> The fondness for travelling by the rail has become almost a national passion among the inferior orders; and it is producing a social change in the habits of general society far more deep and extensive than any which has been created by the political revolutions of the last twenty centuries.[30]

Thus began, in the very first days, that aspect of railroad life which no visitor to India has failed to notice: the people's great fondness for travel, and the overcrowding of third-class compartments.

The first two rail lines in India were experimental ventures, authorized by the East India Company with some trepidation. On April 20, 1853, a few days after the inauguration of the Bombay-Thana line, Dalhousie wrote a Minute on Railways to the directors of the East India Company. In it he expressed his hopes that railways would transform the commerce, the politics, and even the society of India:

> A system of internal communication . . . would admit of full intelligence of every event being transmitted to the Government under all circumstances, at a speed exceeding five-fold its present rate; and would enable the Government to bring the main bulk of its military strength to bear upon any given point, in as many days as it would now require months, and to an extent which is at present physically impossible.

> The commercial and social advantages which India would derive from their establishment are, I truly believe, beyond all present calculation. Great tracts are teeming with produce which they can-

not dispose of. . . . England is calling aloud for the cotton which India does already produce in some degree. . . . Every increase of facilities for trade has been attended . . . with an increased demand for articles of European produce in the most distant markets of India. . . . It needs but little reflection on such facts to lead us to the conclusion that the establishment of a system of railways in India, judiciously selected and formed, would surely and rapidly give rise within this Empire to the same encouragement of enterprise, the same multiplication of produce, the same discovery of latent resource, to the same increase of national wealth, and to some similar progress in social improvement, that have marked the introduction of improved and extended communication in various kingdoms of the Western world.[31]

In his Minute, Dalhousie advocated the construction of a network of trunk lines between the presidencies. He endorsed the idea that railroads should be built by private companies, but with a government guarantee and under the supervision of government engineers. It was essentially Stephenson's plan.

Such was Dalhousie's prestige, not only as governor-general of India, but also as a recognized authority on railroad questions, that his Minute formed the basis for the Indian railways for the next seventy years. Shortly after receiving it, the Court of Directors signed contracts with several more companies, all with the same guarantee terms. By 1859 there were six railway companies in India besides the EIR and the GIP: the Madras Guaranteed; the Bombay, Baroda and Central India, or BB&CI; the Scinde Railway, the Eastern Bengal, the Great Southern of India, and the Calcutta and South-Eastern.[32]

On February 28, 1856, just before he left India, Dalhousie wrote a Final Minute to the Court of Directors, outlining the accomplishments of his administration:

> While it is gratifying to me to be thus able to state that the moral and social questions which are engaging attention in Europe have not been neglected in India during the last eight years, it is doubly gratifying to record, that these years have also witnessed the first introduction into the Indian Empire of three great engines of social improvement, which the sagacity and science of recent times had previously given to the Western nations—I mean Railways, Uniform Postage, and the Electric Telegraph.[33]

A year later, the Rebellion (or Sepoy Mutiny, as the British called it) broke out in northern India. It was blamed on the onrush of alien innovations. But Dalhousie was gone by then.

Building the Trunk Lines (1853–71)

In the early years the progress of railroad building was much hindered by the strained relations between the companies' resident engineers and the government's consulting engineers. The latter, usually officers in the Indian army, were assigned to supervise construction and keep down costs. This created considerable friction and slowed down the work so much that only 480 kilometers of track were laid in the first four years.

The Rebellion of 1857 demonstrated the military advantages of railways, as Dalhousie had predicted. No longer did officials think of them as a commercial innovation imposed by London lobbyists on a reluctant East India Company. With the demise of the Company in 1858, railway construction became an imperial priority as well as a shrewd investment.

After the Revolt, a parliamentary committee rebuked the Indian government for delaying construction. To free the contractors from the niggling interference of the government's consulting engineers, a system of postaudits was introduced which greatly speeded up the work, at the same time making it more expensive. To cover great distances quickly, the contractors built at several points at once, hauling rails and sleepers by boat or oxcart instead of using the cheaper "telescopic" method of building out from existing railheads. Though unskilled Indian workers were paid very low wages, their productivity was also low because they used no picks or wheelbarrows, only baskets, so that labor costs were high. British engineers and skilled workers had to be paid at twice the European rate, plus travel expenses. Materiel was all imported, even, for a time, sleepers of Baltic fir. Two-fifths of the capital invested in Indian railways was thus in fact spent in Britain. The government's consulting engineers could only approve the results, not the construction process itself. The combination of guaranteed dividends and the postaudit system of inspection removed all inhibitions on fast and expensive construction.[34]

In the year 1858–59 alone, more track was laid than in all the years before 1858. By 1859, eight more companies had contracts for some 8,000 kilometers of railroads, the main lines of the present network. The first of these trunk lines, from Calcutta to Delhi, was completed in 1866, followed in 1870 by the Bombay-Allahabad line, which connected with the Calcutta-Delhi. The Bombay-Madras line was finished a year later, linking the four major cities of India to one another. By 1871, over 8,000 kilometers of line were open to traffic. (See Figure 2.)

The 1860s were boom years for British engineers and contractors in India. One of them was John Brunton. The son of an engineer, he had begun his career on the London and Birmingham Railway, working under George Stephenson and Isambard K. Brunel. In 1858 he accepted the post of chief resident engineer on the Scinde Railway and moved to India with his family. A brother served under him as district engineer, and a son as assistant engineer; another brother became chief engineer of the Punjab Railway between Multan, Lahore, and Amritsar.[35]

Brunton spent the years 1858 to 1864 building the Scinde Railway, a 173-kilometer-long line from Karachi north to Koltri on the Indus. He then traveled to Britain to recruit twenty more engineers to survey the route of the Indus River Valley Railway between Koltri and Multan. He returned to Britain for good in 1865. Luckily for us, he wrote the story of his life, something engineers rarely do. It was published in 1939 under the title *John Brunton's Book; Being the memories of John Brunton, Engineer, from a manuscript in his own hand written for his grandchildren and now first printed*.[36]

Brunton attributed different characteristics to every ethnic group he met. He thought the Welsh were "such a queer uncivilised people." Indians were either faithful and devoted or untrustworthy, superstitious, and emotional. Their princes were cruel and tyrannical. As the people of Sind were "naturally indolent and devoid of muscular power," he imported workers from Gujerat, "a much superior race." He tells of the reaction of Indians to the first locomotives:

> The natives of Scinde had never *seen* a Locomotive Engine, they had heard of them as dragging great loads on the lines by some hidden power they could not understand, therefore they feared them, supposing they moved by some diabolical agency, they called them Shaitan (or Satan). During the Mutiny the Mutineers got possession of one of the East Indian Line Stations where stood several Engines. They did not dare to approach them but stood a good way off and threw stones at them![37]

Like other Europeans of the time, Brunton was astonished at how quickly Indians took to the railroads:

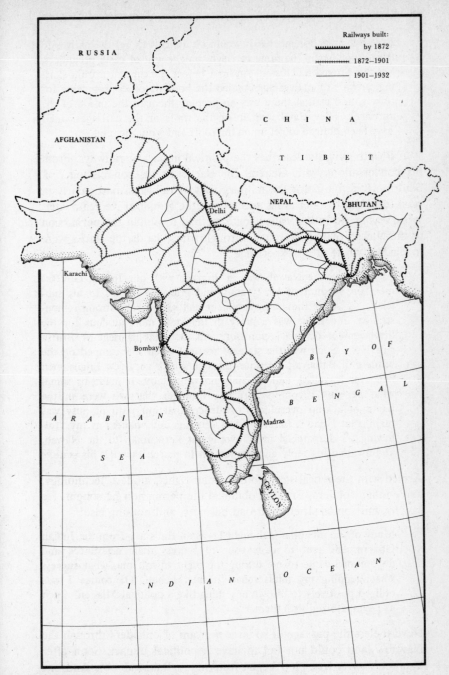

Figure 2 Major Indian Railway Lines, 1872–1932

It was at first thought that it would be difficult to get the natives to travel together in the same carriages on account of caste prejudices, but this proved a delusion. An hour before the time of a train starting, crowds of natives surrounded the booking office clamoring for tickets, and at first there was no keeping them to the inside of the carriages. They clambered up on the roofs of the carriages and I have been obliged to get up on the roofs and whip them off.[38]

Brunton vividly describes the political impact of railways. From his conversations with General Sir Bartle Frere, commissioner in Sind in 1858, he learnt the military uses of railways, particularly in such turbulent frontier areas as Sind and the Punjab. Five years later, in surveying the Indus Valley route, he had to obtain the permission of the Nawab of Bahawalpoor, through whose lands the tracks were to run. He wrote of his experience:

> At this time the Nawab was in disgrace with our Indian Government on account of his tyranny and grievous cruelty to his subjects. . . . Our Indian Government had sent a very strong remonstrance, and there was a party in the State most anxious for the displacement of their oppressor, and the establishment of British rule. . . . Such was the state of matters when I encamped on the borders of Bahawalpoor state—through which very few Englishmen had ever travelled, and certainly no white lady. It made us somewhat nervous I confess. Trusting however that we were in the hands of a kind overuling Providence, that our path of duty was plain, that I had a faithful lot of servants and soldiers as my "following," I despatched my Government Perwannah to the Nawab, and awaited his reply and permission to march through his state.[39]

Armed with the moral strength of human rights, modern technology, and enough soldiers, Brunton obtained the permission he sought. He saw his mission as strictly railroad business, and nothing else:

> Many of the inhabitants thought I was an Emissary from the Indian Government sent to endeavour to redress their grievances—and deputations came to me during the night to ask me, what success was attending my interference in their behalf. Of course I was obliged positively to disown any thing like a political Mission, much as I sympathized with them.[40]

Nevertheless the passage of so large a group of outsiders through the Nawab's land could not fail to have a political impact. Soon after Brunton's party left, plots began festering in Bahawalpoor. And one day a servant of the Nawab poisoned him.

When this happened our Indian Government stepped in, appointed an English officer as Regent—took the young Prince under its care, gave him a first class education, and he now occupies the Musnud of Bahawalpoor. I am told that he by no means follows in his father's footsteps.[41]

Brunton thought of himself as an engineer, not an empire-builder. Yet in his work in India, the flag followed the rails.

The landscape of India presented few problems to the railroad contractors. To be sure, in the drier parts such as Sind and the Punjab, the roadbed had to be protected from high winds by several inches of ballast, and in the Deccan, the monsoon rains could flood the work sites and turn the dusty soil overnight into glutinous mud.[42] On the whole, however, most lines were straightforward and simple to build. Yet two exceptional natural obstacles required unprecedented feats of engineering: the Western Ghats and the rivers.

The Ghats are a precipice at the edge of the Deccan Plateau, where the land abruptly drops 600 meters to the narrow plain of Concan along the Arabian Sea. To lay a double line of tracks up this jagged rocky cliff presented one of the toughest challenges civil engineers had ever faced before the 1860s. George Clerk originally proposed two places where a train could climb the cliff: the Thall Ghat in the direction of Allahabad and Calcutta, and the Bhore Ghat toward Madras. James Berkley, chief engineer of the GIP, began surveying these two routes in 1850.

Work on the inclines took seven years: from 1856 to 1863 for the Bhore Ghat and from 1858 to 1865 for the Thull Ghat. An average of 25,000 people worked on the sites at any one time. The workers were Indian men, women, and children, while the foremen were Englishmen. John Brunton recalled a conversation between Sir Bartle Frere, who had meanwhile become governor of the Bombay Presidency, and a foreman on the Bhore Ghat:

> "Well, my good man, you appear to be the manager here."
> "Yes, Sir" was the reply.
> "And how are you getting on?"
> "Oh, Sir, we are getting on well."
> "How many natives have you under your orders?"
> "Well Sir about 500 on 'em altogether."
> "Do you speak their language?"

"No Sir I dont."

"Well then how do you manage to let these natives understand what they are to do?"

"Oh Sir I'll tell you, I tell these chaps three times in good plain English, and if they dont understand that, I takes the lukri (the stick) and we get on very well."[43]

So steep and broken was the terrain of the Ghats that the two stretches of track required 38 tunnels, 81 bridges, and 14 viaducts. Some gradients were as steep as 37 to 1. Only by using reversing stations on the uphill portion of each incline were the engineers able to keep the gradients from exceeding the climbing ability of steam locomotives. More than rocks and dangerous landslides hampered the work. Water either came down in flash floods, or it was so scarce it had to be hauled up from the valley in oxcarts. Even by Indian standards, sanitation was despicable, and in 1859–60 an epidemic of cholera killed off 30 percent of the work force. Altogether the two inclines cost the then staggering sum of £2 million, almost one-tenth of the cost of all railroad construction in India up to 1870.[44]

The rivers of India were no less formidable an obstacle than the Ghats. They were not only much larger than rivers in Europe, but the monsoons made their flow extremely variable and their channels unstable. In the dry season the Ganges was 600 meters wide and up to 12 meters deep at Benares. When the rains came it could rise 17 meters, flood an area 10 kilometers wide, and scour its bed to a depth of 40 meters. Other rivers were even more violent. In 1841, for example, the Indus rose 33 meters above its normal level. Bridges therefore had to be built to span not just rivers but entire flood plains and to withstand the rushing fury of the monsoon floods.

At a time when Americans were building improbable and dangerous railroad trestles out of timber, the British civil engineers in India insisted on costlier but more durable bridges of wrought-iron trusses resting on masonry piers. Because of their cost, few bridges of this sort were built during the Old Guarantee period. The largest was the one over the Sone River near Delhi which was begun in 1856, interrupted by the Rebellion of 1857, and finished in 1862. It was 1,442 meters long, consisting of twenty-eight wrought-iron spans resting on brick piers sunk to a depth of 10 meters below low water. It carried a double railroad track on its upper level, and a road below, and cost an astounding £330,000.[45]

The Era of State Construction (1870–79)

The first trunk lines were very costly investments. It is estimated that between 1854 and 1869 railway construction demanded £75 million, half of all British investments in India.[46] The promoters anticipated that trunk lines would cost about £5,600 per kilometer of single track and £9,400 for double track. In fact the GIP trunk lines cost £12,500, and the EIR £13,750 per kilometer, while the average for all guaranteed lines was almost £10,625. In comparison, the Australian railways cost £7,500 per kilometer, and the Canadian railways cost £5,300, on average.[47]

There are several reasons for the high cost of the early Indian railways. Among them were the engineering works required to overcome the wide rivers and steep mountains, the heavy-duty double-width bridges and earthworks, the expensive rails and heavy ballasting needed to support the broad gauge, and other examples of technical overbuilding. Because of their high initial cost, the railways paid only small returns on their investment: an average of 3 percent up to 1870.[48] Given their guarantee of 5 percent dividends, the difference had to be made up by the government. To officials wanting to balance the budget and to contemporary critics of the railways, this was an intolerable burden imposed by British capitalists on the Indian people, part of the infamous "drain" of India's wealth. Thus William Massie, finance minister of India from 1863 to 1868, declared: "All the money came from the English capitalist, and so long as he was guaranteed 5 per cent on the revenues of India, it was immaterial to him whether the funds that he lent were thrown into the Hooghly or converted into bricks and mortar. . . . It seems to me that they are the most extravagant works that were ever undertaken."[49] Sir John Lawrence, viceroy from 1864 to 1869, was thinking of railways when he exclaimed: "I know what private enterprise means! It means robbing the Government!"[50] In a minute dated January 1869, he attacked the guarantee system by which "the whole profits go to the Companies, and the whole loss to the Government," because the companies' record of administration was "as bad and extravagant as anything which the worst opponents of Government agency could suggest as likely to result from that system."[51]

For this state of affairs, Lawrence blamed not only the guarantee system of company construction, but also the broad gauge then in use: "Wholly to reject railways for a country which is not able to support the lines of the most costly description is quite unreasonable, and if . . . the expense of the ordinary gauge seems prohibitory, while lines of the narrow gauge would be financially practicable, I should consider it a most mistaken view to reject the narrow gauge line."[52] He therefore proposed that henceforth the state should build all new railroads and do so on the meter gauge. To finance this program along with the many other public works undertaken at the time, he obtained permission to issue bonds in London. His proposal was approved by his successor, Lord Mayo, and by the secretary of state for India, the Duke of Argyll. Thus began the era of state construction, which lasted a decade.

In 1870 the Indian government began building new lines at a rapid pace. It built 3,500 kilometers in ten years, at a cost of £4,000 to £4,500 per kilometer, less than half the cost of the guaranteed lines.[53] Of course the new lines, being narrower, had a lower carrying capacity than the standard 1.676-meter gauge. Despite their lower initial cost, the state lines lost money because the companies had already taken up the most profitable routes between the major cities, while the government's lines into frontier areas or into regions of frequent famines had a social or strategic, rather than a commercial, purpose.

The proper gauge for strategic lines was the subject of disagreement between the civil and military authorities. The Indus Valley Railway from Kotri to Multan and other North West Frontier lines were built to meter gauge in 1871–72. Lord Napier, commander-in-chief of the Indian Army, protested this decision because it required too many reloadings of troops, horses, and materiel. As a result, the frontier lines were rebuilt to the standard gauge, just in time for the Afghan campaign of 1879. Henceforth, the meter gauge was confined to nonstrategic areas in the northeast and the south.

If the original decision to adopt the wider gauge was a costly mistake, Lord Lawrence's meter gauge only compounded it. It saddled India with two systems, each with its own kind of rails, locomotives, rolling stock, and workshops, and no way to shift equipment from one system to another. Even transfers between meter-gauge lines was impossible because they were separated by stretches of standard-gauge track. Freight going between two lines had to be reloaded from train to train, and at every transfer point it was subject to delays, mishandling, and the attention of thieves.[54]

In addition to company and state lines, a third type of railway appeared in the 1870s: the private railway of His Exalted Highness, the Nizam of Hyderabad. Already in the early 1860s the GIP trunk line between Bombay and Madras was built across the lands of the Nizam, 200 kilometers from the city of Hyderabad. The idea of a connecting line, though obvious, did not appeal to everyone equally. The Nizam himself feared that a railway would "upset all orthodox notions" and "make the popular mind gyrate or swing backwards and forwards with a movement like that of children at a fair." Though he "dreaded the British Government and disliked its civilisation," he bowed to British pressure because "he felt that it was the only strong tower where he could in extremity take refuge."[55]

But the Nizam's chief minister, Salar Jang, was determined to modernize Hyderabad, and for that he needed a railway. The problem was capital. Neither the Nizam nor the local moneylenders could come up with the necessary sum of over a million pounds. The Indian government hesitated, preferring perhaps to leave the Princely States in their traditional state of torpid repose. From 1869 to 1875 the government of India pitted its powers of procrastination against the duplicity of Salar Jang. After many complex intrigues involving financial consultant-adventurers, London bankers, and the India Office itself, agents of Salar Jang succeeded in floating a loan on the London money market in 1875, to the dismay of the Indian government. The connecting link between Hyderabad and the GIP's trunk line was the beginning of the world's largest private railway. It was managed for the Nizam by the GIP and, after 1878, by the Railway Branch of the Indian government.

India's first experiment in state railway construction was not only unprofitable, it was unpalatable to powerful interests in London. In 1874 a new secretary of state for India, Lord Salisbury, renegotiated the contracts with the old guarantee companies on even more favorable terms, without consulting the Indian government. He extended the guarantee period from twenty-five to fifty years and cancelled their past debts. He also decreed that the Indian government could only borrow money to build "productive" (potentially profitable) lines, but had to build "protective" (strategic and anti-famine) lines out of its current budget. The Indian Treasury, however, was sorely strained by two series of events. One was the Afghan War of 1878–79, part of Britain's ceaseless attempts to protect India from the Russian menace, but paid for by India not Britain. The other was a series of famines lasting from 1874 to 1879, which killed an estimated 4 million people. Famines were exacerbated by poor transportation. After a bad harvest, Punjabi wheat could still be shipped by rail to Karachi and by steamer to Europe, but it could not be distributed in the hungrier parts of India because the bullocks that pulled the carts had starved too. The Famine Commission recommended in 1880 "that 5,000 miles of line were urgently needed, and that the country could not be held to be safe from such calamities in the future until the Indian railway system could show an aggregate of 20,000 miles."[56]

The decline in revenues to the treasury, however, made it impossible for the government to undertake on its own even a minimal program of famine railroads. Not that it would have done so anyhow, for the railway companies had been campaigning for a decade against state construction, in the name of private enterprise. In 1879 the secretary of state for India ruled that henceforth the Indian government could only build strategic lines to the northern frontier. Elsewhere, railroad building was returned to the private domain.[57]

The New Guarantee Period (1880–1914)

The New Guarantee period from 1880 to 1914 was an era of vigorous construction. (See Figure 2.) During those years the rail network spread throughout the subcontinent, growing from 15,764 kilometers in 1880 to 59,585 kilometers in 1915–16. Though Lord Mayo had declared in 1869 that "the question of the gauge of railways has been settled and must not be reopened," his successor Lord Northbrook reopened it anyway.[58] Henceforth half the new construction was standard gauge, the other half, meter gauge or narrower.

Nowhere did the British Empire exhibit its glory more ostentatiously than in its railway stations. And among railway stations the very epitome of imperial extravagance was Victoria Terminus in Bombay, opened on Jubilee Day in 1887. It was built of Italian marble in a blend of Gothic, Indo-Saracenic, and Venetian styles, with a dome copied from Westminster Abbey. Other stations of that period imitated everything from Roman baths and Alpine chalets to Mogul tombs. In an empire that worshipped the eclectic, nothing was certain except monumental size and a baroque idea of beauty which our own, simpler civilization has not yet learned to appreciate. Fortunately the great stations of India soon began serving a practical function as well, as shelters for thousands of India's homeless and uprooted.

As before, construction was in the British tradition of engineering, that is to say of the finest quality, and therefore expensive. In 1901 a manager of the Eastern Bengal State Railway, Col. W. V. Constable, toured the United States by train. He commented that only a few American railroads were "equal to the first class Indian roads"; most American railroads, especially in the West, would have to spend "hundreds, perhaps thousands, of millions of dollars . . . to bring them up to European or Indian standards."[59]

One reason the American railroads seemed worse than their British or Indian counterparts was the difference in the rails themselves. British rails were carefully made, with a precise sectional profile designed to fit exactly on cast-iron chairs, which were in turn bolted to the crossties. American rails were not uniform and were secured with spikes hammered into the ties. Where British rail inspection was rigorous, Americans were lax. American rail mills produced an enormous output of rails at a lower cost than British rails. America needed cheap transportation in a hurry, whereas Britain could afford to sacrifice quantity for quality. India got British quality at British prices; it had no choice in the matter.[60]

Technically, the main achievement of the period was bridging the rivers of India. Engineers now used steel to erect longer spans and dug deeper piers to support them. The Dufferin, Landsdowne, Jubilee, and Upper Sone bridges were especially noteworthy. The Dufferin Bridge was built across the Ganges at Benares by the Oude and Rohilkund Guaranteed Railway in 1881–87. Due to the scouring action of the river in flood, its piers had to be dug far below the riverbed. Two of them, resting on firm clay 43 meters below the low-water level, were the deepest in the world. The seven steel spans were each 108 meters long and weighed 924 tons. The Jubilee Bridge, built in 1883–87, carried the EIR over the Hooghly River 45 kilometers north of Calcutta. Though shorter than the Dufferin—only 370 meters in all—it required much longer spans to accommodate the heavy navigation on the river. Two of its steel spans were 160 meters between piers. The Landsdowne Bridge over the Indus at Sukkur was notable for having the world's longest flat span, 250 meters in length. It was erected in 1887–89 of steel girders prefabricated in England and placed on abutments of Portland concrete. In contrast to these bridges over large navigable rivers, the Upper Sone Bridge, built in 1900, was made up of short 32-meter spans; however, the width of the floodplain and the shifting nature of the riverbed required ninety-three such spans to cover the 3 kilometers from one side of the valley to the other. Together, these four bridges cost £1.4 million, out of the £10 million spent on all railroad bridges in India in that period.[61]

Not surprisingly, the guaranteed companies were a drain on the Indian Treasury. Only one railroad, the EIR's trunk line from Calcutta to Delhi, made a profit of over 5 percent and did not require a subsidy. Altogether the government's guarantee payments to the railway companies in their first forty years of operation came to a staggering £50 billion.[62]

The government allowed the railways to set their own rates and fares. Its only control over costs, therefore, was its veto power over capital expenditures and maintenance. It used this power to balance its own budget by controlling railway finances. In good years when tax revenues were high, the Finance Department allowed construction to proceed. In years of drought or recession or war, it cut off the funds for the railways' capital budget, even for projects already underway. This made long-range planning impossible. The managers of the railway companies, aging financiers living in London, were little troubled by this system, knowing their stockholders were guaranteed a safe return on their investments. They had no incentive to compete, or innovate, or increase traffic by lowering rates or improving service.[63]

There is no evidence in India of the gross corruption of politicians, fraudulent stock issues, or financial scandals connected with railway booms in Britain and America. Yet there seems to have been a certain amount of petty corruption, as Edward Pierce, son of a BB&CI manager, recalled:

> My father was a very honest man by any standard but he got so many thousands of rupees before a brick was even laid, when he gave the contracts to the right parties. For orders to supply rails or sleepers or cement, contractors presented him with money and gold bangles and my mother was given jewellery. Stations used to be virtually sold. Your stationmaster used to give you a bribe to be placed at that particular station. His salary was negligible but the stations were alloted wagons which were the gift of the stationmaster to give to the merchants who booked them, so his income was enormous. He paid the district traffic superintendent and the company

inspectors who'd go down their district once a month—when a brown envelope was slipped into their hands containing this tip.[64]

Apart from petty corruption, what the guarantee system produced was technical overbuilding by engineers who cared more for quality than profit, and the response thereto by parsimonious politicians, the two-gauge system.

After 1900 the railways ceased being a burden on the treasury. But the profits were deceptive, for they were the result of deferred maintenance, overcrowded third-class accommodations, and slow freight handling. For the New Guarantee period was also a time of organizational muddling. This was perhaps to be expected, since both free enterprise and state operation had been tried and found wanting. New construction was undertaken by private firms, with a government guarantee of 3.5 or 4 percent profits. The older railway companies were gradually bought up by the government: the EIR in 1880, the Eastern Bengal in 1884, the South Indian in 1890, the GIP in 1900. Yet the India Office evidently did not trust the government of India to operate railways, for it signed contracts, usually with the previous owners, to operate their former lines for the government. These management companies were guaranteed a profit of 4 percent, and four-fifths of any profit above that went to the government.

This period was one of mergers and consolidations of railway lines, both private and state-owned. Large networks arose, such as the North Western State Railway in the Punjab and Sind, the Madras and Southern Mahratta in the southern Deccan, and the two giants, the EIR and the GIP. There were not two but several types of railways in India: state-owned and operated, privately owned and operated, state-owned but privately managed, privately owned and state-managed (a few), and owned by the Princely States. Thus in 1902 there were 96 railways operated by different administrations: 24 companies, 4 government agencies, and 5 princely states. Even state acquisition of private railways did not simplify the situation: in 1920 the government owned 73 percent of all the track in India but operated only 21 percent, whereas private companies, which only owned 15 percent, still managed 70 percent (the rest were in the Princely States). Such fragmentation was inefficient, costly, and confusing.[65]

Since the very start of the Indian railways, the guarantee system has generated a lively debate. The need for a guarantee of dividends is seldom questioned, for railroads everywhere received government subsidies of one sort or another. Though possessing much hoarded wealth, India had little liquid capital and a poorly developed capital market. Neither Dwarkanath Tagore nor the Nizam of Hyderabad was able to raise money in India to build even short lines. Therefore the capital had to come from Britain. Over time, the bias toward British capital became self-reinforcing. Shares were sold and loans floated in London. Indians who wished to invest in their own railways had to do so in England; of the 50,000 holders of Indian railway shares in 1868, only 400 were Indian; or, put another way, 99 percent of the capital invested in Indian railways was British, and 1 percent was Indian.[66]

As for the rate of interest guaranteed in the contracts, it was less than other British colonies and foreign countries paid for railway capital raised in Britain at the time.[67] What shareholders of Indian railways stock were investing in, after all, was not so much the profitability of the railways of India, as the ability of the Indian government to collect taxes from its subjects.

The debate does not revolve around the need for British capital, but around the power which such investments conferred. As Ramswarup Tiwari put it,

> As far as the introduction of British capital was concerned, it had enough justification, inasmuch as Indian capitalists were not forthcoming to undertake these projects. British capital was therefore indispensable. But the accepted policy of the time, instead of importing foreign capital, led to the importation of foreign capitalists as well, which resulted in a colossal loss of revenue to the Indian Exchequer.[68]

Since British capitalists would not have invested in Indian railways without a guarantee, the only alternative was state construction and management. In other words, in nineteenth-century India the alternative to foreign capitalists was foreign politicians and bureaucrats. In the end, one's judgment of the guarantees depends on whether or not one accepts the policies of these foreign rulers as being sufficiently for the good of India to justify the subsidies.[69]

World War I and After (1914–1947)

The golden years of the Indian railways ended in 1914, and the strains of World War I brought their weaknesses out into the open.

Indian exports of food, coal, equipment, and troops soared while the army requisitioned much of the rolling stock. The railways were barely able to meet the new demands, and as the war dragged on, they fell further behind. The Finance Department used railway income as a source of revenue. Shipments of vital parts and supplies from Britain were choked off by the submarine campaign and by Britain's more urgent needs. Old equipment wore out, got patched up, and wore out again. Whole lines were dismantled and their rails, locomotives, and rolling stock were shipped to the Mesopotamian theater of war. Railroad workshops, then the largest mechanical industries in India, were converted to the manufacture of military materiel. Employees were drafted or quit to work for the armed forces. Businesses, desperate to get their shipments through, had to resort to bribery on a massive scale.[70]

So deteriorated and overcrowded were the trains by 1918 that they eroded even the legendary Indian tolerance for discomfort. Postwar demonstrations connected with Gandhi's civil disobedience campaigns targeted the railways, now seen as ubiquitous symbols of British misrule. Incidents of violence and vandalism interfered with their operation. Nationalists in the Legislative Assembly repeatedly embarrassed the government with the railway issue.

In 1920 the government of India, chastened by the Amritsar Massacre of the previous year, appointed the East India Railway Committee chaired by Sir William Acworth, a well-known railway economist. Of the nine other members, three were Indian: V. S. Srinivasa Sastri, member of the Council of State, and the businessmen Sir Rajendra Nath Mookerjee of Calcutta and Sir Purshotamdas Thakurdas of Bombay.[71] The report, which the committee issued in 1921, dealt mainly with finances and administration and blamed the railways' troubles on mismanagement and irrational budgeting:

> How much the economic development of India has suffered, not from hesitation to provide for the future—no attempt has been made to do this—but from the utter failure even to keep abreast of the day-to-day requirements of the traffic actually in sight and clamoring to be carried, it is impossible to say.[72]

It showed itself sensitive to the politics of the situation:

> A large section of the Indian public supports the adoption of this system [state management], because it believes that company management does not encourage the development of indigenous industries by sufficiently favourable treatment; that it gives preferential

treatment to import and export goods; that under the present system of company management large profits are made in British interests; and that hitherto the companies have not employed Indians in higher appointments except to a very limited extent, and have not granted them adequate facilities for technical training. . . . There is also in addition . . . a positive feeling caused by the awakening national self-consciousness that Indians should have more control in the management of the railways in their own country. . . . We therefore do not hesitate, though most of us have approached the question with a strong prepossession in favour of private enterprise as a general proposition, to recommend that in India the State should manage directly the railways which it already owns.[73]

Many debates and several committees later, the government of India accepted the Acworth Committee's recommendations. It established a Railway Board in 1922 and separated railway finance from the general budget two years later, finally allowing rational planning and adequate maintenance. The next year it took over the operation of the EIR and GIP. Gradually the others were incorporated into the state system. By 1934 the state owned 74 percent of the tracks, a very slight increase over the early 1920s; but the share of track which it actually operated had risen from 21 to 45 percent.[74] The process continued until by 1944 practically the entire Indian rail network was state-owned and operated.

During the 1920s, more efficient operations and budgeting helped the railways recover from their wartime losses. The Acworth Committee recommended extensive rebuilding and new construction. The railways were quick to respond. They applied the principles of mass production learned in the war to the workshops, cutting overhaul times on locomotives by half, and by four-fifths on rolling stock. They renovated stations, yards, and bridges, pooled their freight cars, and extended their lines. The first electrified railroad, a suburban line in Bombay, began service in 1925; it was followed by the electrification of the Thall and Bhore Ghats in 1929, and of other suburban lines.

These improvements ended abruptly with the onset of the Depression. Between 1929–30 and 1932–33 the number of passengers fell by 25 percent, and freight shipments declined 22 percent. The Pope Committee of 1932 recommended a drastic retrenchment at the expense of construction, maintenance, replacement, and reserves.[75] Echoing Lord Lawrence's complaints of seventy years earlier, the Indian Railway Enquiry Committee of 1937 stated:

We cannot help feeling that in the past 15 years, stations, workshops and marshalling yards have often been built to be the last word in railway technique rather than on a careful calculation of probable requirements, and that prestige has perhaps counted for more than prudence. It is the worst feature of such overgrown schemes that they continue to burden the railways with excessive interest charges involved.[76]

By 1939 the Indian railways were even less prepared for war than they had been in 1914. Once again, the demand for transportation soared while the railways' ability to satisfy it declined. Track and rolling stock were shipped to the Middle East. Workshops made munitions while railroad maintenance was postponed. Locomotive breakdowns on the EIR increased eightfold, from 50 to 400 failures per month out of a total of 1,700 locomotives.[77] Service deteriorated while trains fell into "battered disrepair."[78] Only the financial side of the railways looked good, but their large surpluses were once again funnelled into the government's general revenues.[79] When Pakistan and the Republic of India inherited the railways of British India in 1947, they were in sorry shape after two decades of neglect.

The Locomotive Industry

While the attention of most historians has been focused on the finances, politics, and management of the Indian railways, other aspects—the supply of rails, the manufacture of locomotives and rolling stock, and the railways' personnel policies—also give valuable insights into the process of technology transfer. We will consider the supply of rails in conjunction with the Indian iron and steel industry in chapter 8, and the training of railway personnel in chapter 9, which deals with technical education. The supply of locomotives, however, is strictly a railroad matter; let us turn to it now.

Steam locomotives have this peculiarity, that the facilities required to repair them—foundries, forges, and machine shops—are virtually the same as those needed to manufacture them. In the nineteenth century, nations that built railroads almost immediately began manufacturing their own locomotives: the United States in 1836, Russia in 1845, Canada from 1866 on.

And so did India. The GIP built India's first locomotive in 1865. Later the BB&CI built 444 locomotives in its Ajmer workshops; 217 were built by the EIR at Jamalpur; 27 at Lahore by the

North Western Railway; and a few elsewhere. The BB&CI concentrated mainly on six-wheel meter-gauge locomotives, and the North Western on shunting engines, while the EIR built many standard gauge machines of the 0–6–0, 0–6–4, and 0–8–0 classes. All in all, some 700 locomotives were built in India before Independence.

The successful manufacture of locomotives in India and the considerable export of used locomotives from India to countries in Africa, Southeast Asia, and the Mideast are evidence that India had a potential comparative advantage in this industry. Yet the railway workshops only built 4 percent of the locomotives used in British India. Another 14,420 locomotives—almost 80 percent—were imported from Britain; the rest were German or American.[80] This was not the result of market forces, but of policy decisions.

The main problem in the locomotive industry was not technology but the business cycle, as the demand for new locomotives fluctuated wildly. In boom periods, railways needed new locomotives before they could generate the revenue to pay for them. Tariff barriers alone could not help a country's manufacturers. Only a government policy to encourage local purchases and smooth out the flow of orders could give domestic manufacturers a chance. In Canada, where the government was responsive to local interests, this is how the industry got a foothold.[81]

The government of India also intervened to help the industry, but the British, not the Indian one. The Vulcan Foundry, which had made India's first locomotive, the "Lord Falkland," in 1852, sold 47.5 percent of its production to India. Other major suppliers included Beyer, Peacock & Company and North British Locomotive. In all, India purchased one out of every five locomotives made in Britain.[82]

In the late nineteenth century, the British railways increasingly manufactured their own locomotives. Locomotive builders made up for the loss of the British market by concentrating on exports to underdeveloped countries, India in particular. A boom in business and railway construction in India at the turn of the century created a crisis in the industry. Overwhelmed with orders, the British locomotive builders had to quote distant delivery dates. The railway companies, unable to get locomotives from their usual suppliers soon enough, turned to American or German manufacturers, or began making their own.

The British manufacturers, seeing their market slipping away, implored Secretary of State for India George Hamilton to restrict the flow of non-British locomotives into India. In an era of free trade, the India Office could not simply impose discriminatory duties. Instead, Hamilton appointed a committee composed of representatives of British locomotive manufacturers, British and Indian railways, and the Colonial and India offices to investigate the question. What the committee came up with was a technical solution to a commercial problem. It enshrined the British locomotive designs developed by the British Engineering Standards Association, or BESA, as appropriate for India; and it chose five BESA-approved models (4–4–0, 4–4–2, 0–6–0, 4–6–0, and 2–8–0) as the standard Indian locomotives.

Ostensibly, the BESA standards were designed to reduce the plethora of locomotive types, to simplify maintenance, to flatten the manufacturers' production curves by allowing them to build up stocks of parts in advance of rush periods, and to reduce costs by permitting bulk purchases: all the traditional and reasonable goals of industrial standardization. But in fact they had another purpose: to make the Indian railways buy British. From 1903 to 1914, the company-owned railways ordered most of their locomotives from Britain, despite lower prices and quicker deliveries from other suppliers.[83]

This is not to say that British locomotives were overpriced. Rather, they were better and more expensively built than those of their competitors. Their wheels, set in rigid plate frames, were driven by inside cylinders through cranked axles, typically in an 0–6–0 configuration known in India as the "Scindia class"; it was a system designed for smooth, solid tracks. Their fireboxes were of copper, their boiler tubes of copper or brass, and their slide valves were forged, not cast. Not only were they designed to pull heavy loads under hard conditions, they were expected to run for a long time with little maintenance; on average, British locomotives in India lasted 35 to 40 years, and some as long as 60.

In contrast, North American locomotives used flexible leading trucks, outside cylinders, steel fire boxes and boiler tubes, and cast-iron slide valves; all inexpensive substitutes designed to provide low-cost tractive power over cheap wobbly tracks. Though fast, powerful, and cheap, they were built for "a short life and a gay one." Such locomotives were only purchased when British supplies failed. An American report blamed this state of affairs on the prejudice of the Scots engineer against American equipment: "In his hands it is certain not to come up to requirements."[84]

What we have is a classic confrontation between the British and

American value systems and engineering styles. Nathan Rosenberg has described their difference in these terms:

> These [British consulting] engineers were imbued with a professional tradition which often led to an obsession with technical perfection in a purely engineering sense, and they imposed their own tastes and idiosyncracies upon product design. In America, by contrast, the engineer and engineering skills were more effectively subordinated to business discipline and commercial criteria and did not dominate them. The result was to perpetuate, in Great Britain, a preoccupation with purely technical aspects of the final *product* rather than with the productive *process*.[85]

By being under British rule, India obtained better locomotives, at a higher cost, than it might have. Whether this was technically a good choice is open to debate. Economically and politically, it was a decision made in Britain at India's expense, for it diverted demand away from a potential Indian industry.

It also affected the engineering profession in India. Until the 1920s, the mechanical engineers on the Indian railways were all Europeans; Indians were not welcomed into the profession, as we shall see in Chapter 9. For many capable and ambitious British railway engineers, India was a way station in careers that spanned several continents. Engineers working in India made a number of contributions to railroad technology. William Brunton, in ordering locomotives for the Scinde Railway in 1857, specified "that any piece of any engine shall fit and be applicable to perform the same duty on any other of the same set," one of the first instances of standardization in the locomotive industry.[86] F. J. Cartazzi invented a radial axle-box; C. E. Sandiford built compound locomotives; and A. Caprotti invented the rotary cam valve gear, first used in India.[87]

But India did not retain good engineers for long. Some complained that the pay and pension benefits could not compensate them for the hardships of a tropical exile and, more telling perhaps, that better opportunities beckoned elsewhere. Wrote one anonymous but indignant engineer in 1878:

> The whole world is open to the engineer. In America, in Canada, the Australian colonies, and the Cape, ample work can be found for the man of energy in congenial climates and among congenial people. Let those who turn their thoughts toward India for themselves or their sons remember that it is no longer a country of wealth and luxury [sic]. Life in India is one of hard work, of dis-

appointment, and, too often, of sorrow and trouble. The 'pagoda tree' no longer provides a feast as a recompense for this. It yields only mere sustenance, and this fare can be found in many other and happier lands.[88]

After the turn of the century the BESA standards, which imposed certain fixed locomotive types on India, hampered innovation and creativity in railroad engineering. Locomotive superintendents and chief mechanical engineers, who drew up the specifications for the railways' purchases of materiel, had to get them approved by consulting engineers working at their companies' headquarters in London. An editorial in *Locomotive Magazine* complained in 1908 that standards "preclude the locomotive engineer on the spot from exercizing . . . ingenuity of design. . . . a railway mechanical [engineer] in India rarely, now-a-days, designs an engine, because he might not be permitted to construct it, even were his own shops capable of undertaking the work."[89] Thus the creative side of locomotive design was effectively divorced from the machinery itself, and from the physical and economic environment in which it was to operate.

During World War I, Indian locomotive imports virtually ceased. Instead, as part of the war effort, India exported used railway materiel, including locomotives, to Iraq, where British and Indian units were fighting the Ottoman army. By 1918, the Indian railways were saddled with worn-out equipment badly in need of replacement. The Indian government therefore invited bids for 400 locomotives and boilers per year for the next twelve years. This set off a tug-of-war between various parties interested in locomotive manufacturing. Nationalists in the Legislative Assembly agitated for support to Indian industries. The giant British steel and munitions manufacturer Armstrong Whitworth and Company approached Secretary of State for India Sir Edwin Montagu and Viceroy Baron Chelmsford to request permission to manufacture locomotives in India, with a guarantee of government purchases. In October 1920 Montagu wrote Chelmsford that "the establishment on a sound basis of locomotive works in India appears to be very desirable in the interests of India." When other manufacturers protested, however, the government refused to offer guarantees. Soon thereafter Armstrong Whitworth and Company and another arms manufacturer, William Beardmore and Company, opened factories in Britain to manufacture locomotives for India, and that opportunity was lost.[90]

In September 1921, an Indian government communique estimated that the railways would require 160 new locomotives each year in 1923 and 1924, and 400 a year after that, plus an equal number of spare boilers. In response, the Indian Tariff Board proposed a tariff on locomotives equal to those of Australia (27.5 percent) or Canada (22.5 percent). Lured by these prospects, the British firm of Kerr Stuart and Company decided to build locomotives in India. In 1923–24 it incorporated a subsidiary, the Peninsular Locomotive Company, and built a plant at Jamshedpur, near the Tata steel mills. A year later, however, the Retrenchment Committee led by Lord Inchcape recommended drastic cuts in the Indian railways' reconstruction program. The Railway Board thereupon decreed that it would only require 60 new locomotives a year, and the Tariff Board rejected the request of the Peninsular Locomotive Company for tariff protection, on the grounds that the minimal economic production of locomotives was 200 a year, more than India needed. This was a false excuse, since British manufacturers got by with as few as 10 a year. The government's refusal to grant either tariff protection or guaranteed orders doomed the company; it never made a single locomotive and ended up five years later as a state railway workshop.[91] Thus ended the last attempt to build locomotives in India before Independence.

Since the 1950s, India not only has satisfied its own demand for locomotives, but also has exported them throughout Asia and Africa. But for the BESA standards and the waffling policies of the 1920s, India could have been building locomotives fifty years earlier. It would thereby have obtained a substantial industry and strengthened the empire in time of war, but at the expense of British locomotive manufacturers. The choice between a declining British industry and an emerging India was a difficult one, and difficult choices were resolved by procrastination.

Consequences and Comparisons

In the nineteenth century, railway enthusiasts were convinced that technology would transform the very soul of India. One of them, George W. Macgeorge, consulting engineer to the government of India for railways, wrote in 1894:

A land where the very names of innovation, progress, energy, and the practical arts of life were unknown, or were abhorred, and which appeared sunk in a lethargic sleep too profound for any possibility of awakening . . . under the guiding direction of Providence it is from the British nation that the vast continent of India has received the leaven of a new moral and material regeneration, which can now never cease to operate until it has raised the country to a high level of power and civilisation. The most potent factor in this truly wonderful resurrection of a whole people, so visibly taking place before the eyes of the present generation, is unquestionably the railway system of the country; and there is little reason to doubt that the powerful onward impetus already imparted by railway communication—even if every other instrument of English power were relaxed or removed—would continue to prevail, and that it will ever remain a lasting memorial of the influence of Great Britain on the destinies of India.[92]

Aside from generating purple rhetoric, how much did the railways of India actually accomplish? For their stockholders, they proved to be gilt-edged investments, thanks to the guarantee. To India they brought a vastly increased foreign trade, through the exchange of raw materials for manufactured goods. Politically the railways helped consolidate Britain's hold over India: there were noticeably fewer revolts after 1858 than before.

The railways had unexpected consequences as well. These are intertwined with the many other aspects of that great transformation that turned India from a congeries of traditional states into something new on the subcontinent: modern underdeveloped nation-states.

All predictions to the contrary, the Indian people took to the new railways with enthusiasm. The number of passengers rose from 80 million in 1880 to 200 million in 1904, 500 million in 1920–21, and a billion by 1945–46. One reason was the low fares, perhaps the world's lowest. Third-class travel in India cost about one-quarter what it cost in America, per kilometer, and 96 percent of Indian passengers traveled third class, crammed together on wooden benches. Yet in the eyes of many critics, the third-class fares were still excessive, since the companies made profits on them to balance the losses incurred in lavishing luxurious service on their first-class passengers, most of them Europeans. In terms of railway travel, British India was indeed a land of contrasts.

To the surprise of Europeans, Hindus did not violate their caste rules when they sat next to one another on a train. Yet they would not eat together, nor could the railways provide enough different kinds of meals to satisfy all travelers. As a solution, trains stopped half an hour for lunch and dinner, and fifteen minutes for breakfast, to let passengers jump out and prepare their own food by the side of the tracks.[93]

As in other countries, most passengers were commuters or traveled for business. Yet a survey of 10,000 passengers taken in 1938 showed an astonishing 29 percent traveling for "pleasure," a category which included pilgrimages. The mass pilgrimages for which India is famous resulted as much from train travel as from religious fervor. For Hindus, the railways put on special trains to Benares and other holy places, and for Muslims to Bombay and Karachi, whence ships took them to Arabia. In such ways did the railways contribute to religion in India.[94]

As train travel eroded the barriers of caste, it replaced them, in the eyes of hundreds of millions of passengers, with a simpler division of society: Hindus and Muslims in third class and in the lowest jobs, Europeans in first class and in executive positions, and Anglo-Indians in the middle. In the awakening of religious and ethnic consciousness which shaped the new nations of the subcontinent, the railways played their unwitting part.

Yet another consequence of railways was the postal system. Before the mid-nineteenth century, the postal systems run by the individual presidencies and Indian states were expensive, slow, unreliable, and often corrupt. The great modernizer, Lord Dalhousie, introduced stamps, a uniform rate throughout India, and a central administration. But in increasing the speed and security of the mails, no reform was as effective as the spread of railways. After the trunk lines and the Suez Canal were completed, express trains began carrying the mails from the P&O docks at Bombay to the strategic centers of British India: the "Frontier Mail" to Delhi, the "Punjab Limited" to Peshawar, and, most famous of all, the "Imperial Indian Mail" which crossed the subcontinent to Calcutta in less than forty hours.[95]

Like other reforms, the postal system was designed to facilitate intercourse among the British in India. Like other expatriates in the tropics, they had a special fondness for the mail. Temporary visitors living in an uncomfortable environment among people with alien customs, they needed their homeland's political support and personal attachments. To them the arrival of letters and periodicals from home was a major event of the season, the month, or the week. The content of the mail ruled the professional lives of colonials, and its very existence and predictability gave them the reassurance their surroundings conspired to deny them. Hence their love affair with steamers and railways and their insistent demands for improved service, regardless of cost.

But like so many other Western reforms, the postal system had unexpected side effects. In Dalhousie's day, Europeans assumed that Indians would remain ensconced in their slothful Eastern ways for generations to come. Yet the ability of the Post Office to carry letters for a half-anna, a small fraction of the previous rates, was due to the economies of scale made possible by the enormous demand for communications among Indians. By the turn of the century, every small town had a post office and runners carried the mail between every village and the nearest rail line. Most financial transactions in India—money orders and "value payable post" (CODs)—were carried out by the Post Office. As the historian M. N. Das wrote: "The post office also played an important role in breaking down the static nature of the Indian society. . . . Judged from whatever angle, social, cultural, educative or economic, the half-anna postal system of Dalhousie played a remarkable role in the progress of India."[96]

Freight was another matter. In comparison with alternative forms of transportation, the railways were cheap: in 1930–31 they charged only one-eighth as much as bullock carts had in the 1840s and 1850s. Yet Indian freight rates for grain were 40 to 60 percent higher than American freights at the prevailing exchange rates.[97] And the railway companies were nonchalant about the volume of traffic, which did not affect their dividends.

Actual freight shipping costs were higher than the rates alone would indicate. For certain areas, gauge changes required unloading and reloading, with attendant breakage and theft problems. Slow and uncertain schedules increased costs. In wartime and during business booms, a shortage of freight cars led to bottlenecks, delays, and bribery.

Not only were the rates higher than they could have been, they were also skewed. Rates to and from port cities were lower than those between inland points, and long hauls cost less, per kilometer, than short ones. These biases favored imports and exports over domestic trades. They helped India become a leading exporter of agricultural products, but Indian businessmen and nationalists complained that they hampered the development of Indian industry.[98]

The railways opened the Indian economy to international trade and increased all categories of production. Yet, at Independence, the same proportion of the Indian people worked in agriculture, trade, industry, and services as a century before.[99] In the end, the railways had a far smaller impact on the economy of India than any other network of comparable dimensions had in any other country.

When Macgeorge called railways "a lasting memorial of the influence of Great Britain on the destinies of India," he implied that without British rule, India would hardly have acquired such a rail network. Alternately, when the Indian historian Amba Prasad wrote that "India possesses today a magnificent railway system. But the development has not always been on right lines," he implied that someone else—perhaps Indians—could have done it better.[100] Obviously, this issue will never be resolved, yet a comparison with other railway systems may prove instructive.

The country most often compared to British India was Japan. There the inspiration and the capital for railway building came from within the country. The first line, running 29 kilometers from Tokyo to Yokohama, was a government project, inaugurated in 1872 in the presence of the emperor. In the early years of railway building the Japanese hired a number of Europeans, including a British chief engineer. The number of foreign technicians rose from 19 in 1870 to 113 in 1874, then dropped to 43 in 1879 and to 15 in 1885; after that the Japanese dispensed with foreign advisers. They had viewed the foreigners less as railroad builders than as teachers. As early as 1877, only seven years after the railway era began, the line from Kyoto to Otsu was built without any foreign help.

From the beginning, the railways were designed not only to transport goods and people and to benefit investors, but also to contribute to the further development of Japanese industry. These issues were debated in the Diet, not imposed from overseas. As Daniel Thorner has pointed out,

> The foreign orientation of India's economic life and the wasteful use of her limited resources stand in sharp contrast to the domestic orientation of Japan's economy and the careful husbanding of the limited capital available to the Japanese. . . . the difference in railway policy simply illustrates the difference in the direction and emphasis between a country running its own affairs and a dependency whose affairs were being managed by an external power.[101]

China represents the opposite alternative. China came through the railway age theoretically united and independent, but in practice very divided and subject to foreign interventions. Among these interventions was the building of railways by European companies to serve European interests.[102] They were resisted by the Chinese ruling class, resented by the people, and destroyed, damaged, or neglected in every upheaval. As a result, China had only 19,300 kilometers of track in 1942, about one-fourth as much as India, or one-fifth as much per capita.[103]

Yet a third alternative consisted of the smaller tropical dependencies such as Malaya, Indochina, or the African colonies. As these regions were mostly colonized at the end of the nineteenth century, their railways date from the 1890s and after. A number of them were built purely to export raw materials; the lines radiating from the copper belt of Northern Rhodesia and Katanga toward the sea are but the most obvious cases. Others were built in the hopes of opening up new territories, preempting other countries, or spreading "civilization." The motives that led to the construction of these lines were much the same as had created the Indian network, originally. The difference lies not in the motives but in the results. Unlike India, Africa and Southeast Asia were broken up between various powers, none of which had any reason to create an integrated African or Southeast Asian network. They were colonized much later than India, and before the bits and pieces of railways could be linked up, their railroad-building programs were cut short by World War I, the Depression, and the advent of the automobile. As they were much smaller than India, there was no economic reason for developing the ancillary industries which major rail networks encourage. Today, by a nice irony of history, the countries surrounding the Indian Ocean find it economic to purchase railroad supplies from India.[104]

Japan, China, and the smaller tropical countries: therein lies a spectrum of possible alternative scenarios for India. Had the Europeans never conquered the subcontinent, there may have arisen a modernizing state, or a decadent one, or a plethora of small and weak ones. That leaves one last alternative: a different kind of imperialism, one that would have built a railway network beneficial to India rather than to Britain. This is the most difficult of all scenarios to imagine, for, as Thorner points out, "A British Cabinet that tried to implement such policies would have been turned out of office. . . . After all, the prime concern of British railway policy in India was to make India useful to Britain, not to make Britain useful to India."[105]

Notes

1. A. Fraser-Macdonald, *Our Ocean Railways; or, the Rise, Progress, and Development of Ocean Steam Navigation* (London, 1893).

2. Max Liniger-Goumaz has analyzed this flood of books and articles in "Transsaharien et transafricain: Essai bibliographique," *Genève-Afrique* 7, no. 1 (1968): 70–84. See also Henri Brunschwig, "Note sur les technocrates de l'impérialisme français en Afrique noire," *Revue française d'histoire d'outre-mer* 54 (1967): 171–87. Before World War II, the journal *L'Afrique française* regularly published articles on the transsaharan project.

3. Ralph Richardson, "British Trans-African Railways," *Scottish Geographical Magazine* 26 (1910): 633–38; Leo Weinthal, ed., *The Story of the Cape to Cairo Railways and River Route from 1887 to 1922,* 5 vols. (London, 1923–26), especially G. H. Lepper, "Anticipations of the Cape to Cairo Route (What It May Be Like in 1950)," in vol. 3, pp. 445–50.

4. Edward K. Hawkins, *Roads and Road Transport in an Underdeveloped Country: A Case Study of Uganda* (London, 1962), pp. 24–25.

5. R. Godfernaux, *Les chemins de fer coloniaux français* (Paris, 1911), p. 15.

6. John M. Hurd, "Railways," in *The Cambridge Economic History of India,* vol. 2: *c. 1757–c. 1970,* ed. Dharma Kumar (Cambridge, 1983), p. 740; Vinod Dubey, "Railways," in *The Economic History of India,* ed. V. B. Singh (New Delhi, 1965), p. 336.

7. Franz Joseph Wilhelm Baltzer, *Die Kolonialbahnen, mit besonderer Berücksichtigung Afrikas* (Berlin and Leipzig, 1916), pp. 21–22.

8. J. E. Holstrom, cited in S. Herbert Frankel, *Capital Investment in Africa: Its Course and Effects* (London, 1938), p. 32.

9. Lilian C. A. Knowles, *The Economic Development of the British Overseas Empire, 1763–1914,* vol. 1: *The Empire as a Whole* and *The British Tropics* (London, 1924), p. 144.

10. That is also why it is absurd to measure the impact of railways by imagining counterfactual scenarios in which the same amount of freight would somehow have been transported the same distances by preindustrial means; for an example, see Hurd, pp. 740–41.

11. On the external economies of railways, see Moses Abramovitz, "The Economic Characteristics of Railroads and the Problem of Economic Development," *Far Eastern Quarterly* 14, no. 2 (February 1955): 201–16.

12. Daniel Thorner, "The Pattern of Railway Development in India," *Far Eastern Quarterly* 14, no. 2 (February 1955): 212. See also M. Arokiaswami and T. M. Royappa, *The Modern Economic History of India,* 6th ed. (Madras, 1959), p. 45; Daniel Houston Buchanan, *The Development of Capitalistic Enterprise in India* (New York, 1934), p. 184; and Romesh Chunder Dutt, *The Economic History of India in the Victorian Age,* 3rd ed. (London, 1908), pp. 547–48.

13. Herbert Feis, *Europe: The World's Banker, 1870–1914,* 2nd ed. (New York, 1965), p. 27; David Kenneth Fieldhouse, *Economics and Empire, 1830–1914* (Ithaca, N.Y., 1973), p. 56; Gerald S. Graham, "Imperial Finance, Trade, and Communications, 1895–1914," in *The Cambridge History of the British Empire,* vol. 3: *The Empire-Commonwealth, 1870–1919,* ed. E. A. Benians,

James Butler, and C. E. Carrington (Cambridge, 1959), p. 458; A. J. H. Latham, *The International Economy and the Underdeveloped World, 1865–1914* (London and Totowa, N.J., 1978), pp. 56–60; B. R. Tomlinson, *The Political Economy of the Raj, 1914–1947: The Economics of Decolonization in India* (London, 1979), p. 4.

14. The expression is from Sir Clement Hindley, "Indian Railway Developments," *Asiatic Review* 25 (1929), p. 639.

15. Leland Hamilton Jenks, *The Migration of British Capital to 1875* (New York, 1927), pp. 208–9.

16. Rowland Macdonald Stephenson, *Report upon the Practicability and Advantages of the Introduction of Railways into British India* (London, 1844), quoted in Dipesh Chakrabarty, "The Colonial Context of the Bengal Renaissance: A Note on Early Railway-Thinking in Bengal," *Indian Economic and Social History Review* 11, no. 1 (March 1974): 107–11.

17. Quoted in Daniel Thorner, *Investment in Empire: British Railway and Steam Shipping Enterprise in India, 1825–1849* (Philadelphia, 1950), pp. 47–48.

18. Quoted in W. J. Macpherson, "Investment in Indian Railways, 1845–72," *Economic History Review* 8, 2nd ser. (1955): 182.

19. Thorner, *Investment*, p. 96. See also Macpherson, pp. 183–85.

20. Quoted in Dubey, p. 329.

21. Thorner, *Investment*, pp. 22–23, 177; Macpherson, pp. 182–83.

22. Manindra Nath Das, *Studies in the Economic and Social Development of Modern India: 1848–56* (Calcutta, 1959), pp. 10, 18.

23. Das, chap. 1; Ramswarup Deotadin Tiwari, *Railways in Modern India* (New York, 1941), p. 48.

24. George Walter Macgeorge, *Ways and Works in India: Being an Account of the Public Works in that Country from the Earliest Times up to the Present Day* (Westminster, 1894), pp. 300–303; M. A. Rao, *Indian Railways* (New Delhi, 1975); Jogendra Nath Sahni, *Indian Railways: One Hundred Years, 1853 to 1953* (New Delhi, 1953), pp. 2–3. On the role of army engineers, see Edward W. C. Sandes, *The Military Engineer in India,* 2 vols. (Chatham, 1933–35), 2: 102–22.

25. Macgeorge, pp. 310–12.

26. Ibid., p. 318; Ernst Karl Arnold Schulz, *Geschichte und Entwicklung der ostindischen Eisenbahnen* (Berlin, 1909), pp. 12–13, 163–65; Maurice A. Harrison, *Indian Locomotives of Yesterday* (India, Bangla Desh, and Pakistan), part 1: *Broad Gauge* (Bracknell, England, 1972), p. 4.

27. J. N. Westwood, *Railways of India* (Newton Abbot, England, and North Pomfret, Vt. 1974), p. 16; Dutt, pp. 175–76; Rao, pp. 15–20.

28. Quoted in Horace Bell, *Railway Policy in India* (London, 1894), pp. 3–4.

29. Das, pp. 88–89.

30. Ibid., p. 96.

31. "Minute by the Most Noble the Governor-General; Dated the 20th April, 1853: Railways in India," *Parliamentary Papers* 1852–53 (787), vol. 76, p. 595.

32. Rao, p. 17.

33. Das, pp. 102–3.

34. Thorner, "Pattern," p. 205; Macgeorge, pp. 319–27; Schulz, pp. 32–34; Westwood, pp. 31, 37; Edward Davidson (Capt. Royal Engineers), *The Railways of India: With an Account of their Rise, Progress and Construction, written with the aid of the Records of the India Office* (London, 1868), pp. 110–11; Jenks, p. 222.

35. P. S. A. Berridge, *Couplings to the Khyber: The Story of the North Western Railway* (New York, 1969), p. 29; H. C. Hughes, "The Scinde Railway," *Journal of Transportation History* 5, no. 4 (November 1962): 219–24.

36. (Cambridge, 1939.)

37. Brunton, p. 105.

38. Ibid., p. 109.

39. Ibid., pp. 128–29.

40. Ibid., p. 136.

41. Ibid., pp. 144–45.

42. Andrew C. O'Dell and Peter S. Richards, *Railways and Geography*, 2nd ed. (London, 1971), p. 67.

43. Brunton, p. 107.

44. Davidson, pp. 247–51, 273–81; Macgeorge, pp. 349–58; Sahni, pp. 11, 14, 46–52.

45. Macgeorge, pp. 333–35, 372–83; Sahni, pp. 53–55.

46. Jenks, pp. 219–25.

47. These are the figures given by Tiwari, p. 53. Other estimates vary slightly; see Dubey, pp. 330–31; Schulz, p. 58; Westwood, p. 26; Sahni, p. 51; and J. Johnson, *The Economics of Indian Rail Transport* (London, 1963), p. 13.

48. Harrison, p. 5.

49. Dutt, pp. 355–56.

50. Frank Reginald Harris, *Jamsetji Nusserwanji Tata: A Chronicle of His Life* (London, 1925), p. 152.

51. Many historians have echoed Massey's and Lawrence's charges of "extravagance." See for instance Dutt, p. 353; Rao, pp. 27–28; Tiwari, p. 56; and Amba Prasad, *Indian Railways: A Study in Public Utility Administration* (London, 1960), p. 52.

52. Tiwari, pp. 60–61.

53. Macgeorge, p. 417; Tiwari, p. 61.

54. Schulz, pp. 82, 173–84; Thorner, "Pattern," pp. 208–9. Two even narrower gauges, 762 and 610 millimeters, are used in mountain areas and as local tramways; see Schulz, p. 180; Sahni, p. 116; and Johnson, pp. 406–10.

55. Sir Richard Temple, *Man and Events of My Time in India,* pp. 299–300, quoted in Vasant Kumar Bawa, "Salar Jang and the Nizam's State Railway, 1860–1883," *Indian Economic and Social History Review* 2, no. 4 (October 1965): 307–40.

56. Sahni, p. 22.

57. On the state construction phase, see Prasad, pp. 53–57; Rao, pp. 27–29; Dubey, pp. 331–32; and Thorner, "Pattern," pp. 206–7 and "Great Britain and the Development of India's Railways," *Journal of Economic History* 11, no. 4 (1951): 393–94.

58. Harrison, p. 5.

59. Lt.-Col. W. V. Constable, R.E., *Report on the Working of American*

Railways, dated 23rd July, 1901 (London, 1902), p. 2, cited in Fritz Lehmann, "Railway Workshops, Technology, and Personnel in Colonial India," *Journal of Historical Research* (Ranchi, India) 20, no. 1 (August 1977): 49–61.

60. J. C. Carr and Walter Taplin, *History of the British Steel Industry* (Cambridge, Mass., 1962), pp. 159–60.

61. Macgeorge, pp. 341–43, 380, 390–95, and appendix K, pp. 539–43. As a point of comparison, the Eads Bridge over the Mississippi at St. Louis, built in 1868–74, has a central span 158.5 meters long; see David B. Stein and Sarah Ruth Watson, *Bridges and Their Builders,* 2nd ed. (New York, 1957), p. 172; and John A. Kouwenhoven, "The Designing of the Eads Bridge," *Technology and Culture* 23, no. 4 (October 1982): 564–67.

62. Johnson, pp. 10–18; Thorner, "Great Britain," p. 392.

63. Johnson, p. 12; Thorner, "Great Britain," pp. 396–97.

64. Charles Allen, ed., *Plain Tales from the Raj: Images of British India in the Twentieth Century* (London, 1975), pp. 227–28.

65. Buchanan, p. 184; Dubey, p. 333; Dutt, pp. 548–50; Hurd, pp. 739, 755; Sahni, p. 25; Westwood, p. 60.

66. Westwood, p. 37. See also Arokiaswami and Royappa, pp. 30–31 and Hurd, p. 750.

67. For example, in 1870 Japan obtained a British loan at 9 percent.

68. Tiwari, p. 52.

69. On the guarantee, see Arokiaswami and Royappa, pp. 30–31; Dutt, pp. 174–76, 353; Johnson, p. 63; Macpherson, pp. 177–81; Prasad, pp. 51–52; Thorner, "Great Britain," pp. 389–91 and "Pattern," p. 206; Tiwari, p. 52; and Westwood, pp. 13–15.

70. On the railways in World War I, see Tiwari, p. 80; Thorner, "Pattern," p. 210; and Johnson, pp. 18–19.

71. Rao, pp. 37–38.

72. Thorner, "Great Britain," p. 396.

73. Rao, pp. 38–39.

74. Vera P. Anstey, *The Economic Development of India,* 4th ed. (London and New York, 1952), p. 141.

75. Ibid.; Hindley, pp. 643–50; Sahni, p. 36; Dubey, p. 342; Johnson, pp. 21–27.

76. Sahni, p. 37.

77. Duvur Venkatrama Reddy, *Inside Story of the Indian Railways: Startling Revelations of a Retired Executive* (Madras, 1975), p. 70.

78. Sahni, p. 39.

79. Dubey, p. 343.

80. On the locomotives of India, see Lehmann, "Railway Workshops," pp. 46–61; "Great Britain and the Supply of Railway Locomotives to India," *Indian Economic and Social History Review* 2, no. 4 (October 1965): 297–306; and "Empire and Industry: Locomotive Building Industries in Canada and India, 1850–1939," *Proceedings of the Indian History Congress* (40th Sessions, 1979, Waltair, India), pp. 985–96.

81. Lehmann, "Empire and Industry," pp. 987–95.

82. Harrison, pp. 17ff; Sahni, p. 91.

83. Lehmann, "Great Britain," pp. 298–303, and "Empire and Industry,"

pp. 990–91; Harrison, p. 9; Samuel Berrick Saul, *Studies in British Overseas Trade, 1870–1914* (Liverpool, 1970), p. 200.

84. Lehmann, "Railway Workshops," pp. 60–61; Saul, p. 200; Harrison, pp. 7–13.

85. Nathan Rosenberg, *Perspectives on Technology* (Cambridge, 1976), p. 160.

86. Harrison, p. 7.

87. Lehmann, "Railway Workshops," p. 4; Patrick L. J. C. Ransome-Wallis, *The Concise Encyclopedia of World Railway Locomotives* (New York, 1959), p. 497.

88. "Engineers in India," *Fraser's Magazine 98* (November 1878): 565.

89. Quoted in Lehmann, "Great Britain," pp. 300–301.

90. Lehmann, "Empire and Industry," pp. 992–94.

91. Ibid.; John Keenan, *A Steel Man in India* (New York, 1943), pp. 184–85; P. J. Thomas, *India's Basic Industries* (Bombay, 1948), pp. 206–7.

92. Macgeorge, pp. 292–93.

93. Buchanan, pp. 186–89; Westwood, pp. 23, 38–40, 71–73.

94. Robert Perry-Ellis, "India's Railways," *Asiatic Review* 35 (1939): 792–94.

95. On the Indian Post Office see Das, pp. 161–99; Hindley, p. 643; Sahni, pp. 36, 112–15; Sir Geoffrey R. Clarke, *The Post Office of India and Its Story* (London, 1921), and "Post and Telegraph Work in India," *Asiatic Review* 23 (1927): 79–108.

96. Das, pp. 198–99.

97. Hurd, pp. 740, 752.

98. For this interpretation, see Tiwari.

99. Hurd, pp. 745–61; Buchanan, pp. 185–87. On agriculture, see Michelle Burge McAlpin, "Railroads, Prices, and Peasant Rationality; India, 1860–1900," *Journal of Economic History* 34, no. 3 (September 1974): 663–69.

100. Prasad, p. 46.

101. Thorner, "Pattern," p. 214. See also Nobutaka Ike, "The Pattern of Railway Development in Japan," *Far Eastern Quarterly* 14, no. 2 (February 1955): 217–29.

102. A classic case is the French-built Yunnan-Haiphong railroad; see Michel Bruguière, "Le chemin de fer du Yunnan. Paul Doumer et la politique d'intervention française en Chine (1889–1902)," *Revue d'histoire diplomatique* 77 (1963): 23–61, 129–62, 252–78.

103. E-tu Zen Sun, "The Pattern of Railway Development in China," *Far Eastern Quarterly* 14, no. 2 (February 1955): 179–99.

104. On the railways of colonial Africa, see Baltzer; Godfernaux; Lionel Wiener, *Les chemins de fer coloniaux de l'Afrique* (Brussels and Paris, 1931); André Huybrechts, *Transports et structures de développement au Congo. Etude du progrès économique de 1900 à 1970* (Paris, 1970); Anthony Michael O'Connor, *Railways and Development in Uganda: A Study in Economic Geography* (Nairobi, 1965); and S. E. Katzenellenbogen, *Railways and the Copper Mines of Katanga* (Oxford, 1973). On Malaya, see C. A. Fisher, "The Railway Geography of British Malaya," *Scottish Geographical Magazine* 64, no. 3 (December 1848): 123–36.

105. Thorner, "Great Britain," pp. 400–401.

4

The Imperial
Telecommunications Networks

The lines of communication that hold empires together never seem strong enough to those whose power and security depend on them. The lines that bound European countries to their colonies in the mid-nineteenth century were weak indeed. Before the 1840s it took 5 to 8 months for a letter to travel between Britain and India, and the writer could not expect to receive an answer in less than 2 years. Even after steamships took over the mail service, it still took 6 weeks in each direction. Within India, the mails were just as slow. No wonder the first telegraph lines were greeted with such enthusiasm among imperialists! In 1854, as soon as the telegraph line between Calcutta and Bombay was completed, Governor-General Dalhousie wrote to a friend: "The post takes ten days between the two places. Thus in less than one day the Government made communications which, before the telegraph was, would have occupied a whole *month*—what a political reinforcement is this!"[1] Later in the century, J. Henniker Heaton, member of Parliament, told the Royal Colonial Institute:

> Now it is often gloomily predicted by purblind students of history that this tremendous agglomeration must inevitably break up and dissolve, like its predecessors. "Where," they ask, "are the Greek, the Roman, the Spanish, the Napoleonic Empires? What is there in the British Empire to preserve it from the fate of these?" I venture to reply, that in the postal and telegraphic services the Empire of our Queen possesses a cohesive force which was utterly lacking in former cases. Stronger than death-dealing war-ships, stronger than the might of devoted legions, stronger than wealth and genius of administration, stronger even than the unswerving justice of Queen

Victoria's rule, are the scraps of paper that are borne in myriads over the seas, and the two or three slender wires that connect the scattered parts of her realm.[2]

The causal relations between communications technology and imperial rule went in both directions. Much of the world's telegraph network was erected to satisfy the imperialists' demand for improved communications at any cost. The Indian, Algerian, and Indochinese telegraph systems were political, not commercial, projects. The cables around Africa, in the West Indies, and across the Pacific were subsidized by various governments for imperial reasons. And long-distance radiotelegraphy was partly funded by various imperial wireless chain projects. The web of power that tied the colonial empires together was made of electricity as well as steam and iron.

Submarine Telegraph Cables, 1850–70

To nineteenth-century Europeans, steamships, railways, and telegraphs were matters for great astonishment and self-congratulation. Of all these innovations, none contributed to the shrinking of the world quite so obviously as the submarine telegraph cable. Like other inventions, it was also an instrument of power, so it is not surprising to find it intertwined with the power struggles of the time: private enterprise and governments; the dominance of the Western nations over the non-Western world; and, as the nineteenth century gave way to the twentieth, the growing rivalries between the nations of the West which led to two world wars.

Amidst the complexities of cable history we can distinguish three phases. In the first the world cable network was laid, a technical and economic achievement. Once it was in place, a period of international distrust and rivalry began, putting the lie to those who believed that better communications would lead to world peace and harmony. Finally a new technology—the wireless—lifted the burden of suspicion from the cable networks.

From the very beginnings of electric telegraphy, inventors tried to run their lines underwater. They first insulated their wires with tar or rubber, but this solution proved short-lived. The problem was solved in the 1840s by the fortuitous discovery of gutta-percha, the latex of the *Palaquium* tree that grows in Southeast Asia and the East Indies. A natural plastic which can be molded to any shape, gutta-percha is both a good electrical insulator and impervious to seawater. In 1843 two residents of Singapore, Dr. José d'Almeida and Dr. William Montgomerie, sent samples of it to the Royal Asiatic Society and the Society for the Encouragement of Arts, Manufactures and Commerce in London. Five years later, Lt. Werner von Siemens of the Prussian artillery developed a machine to coat cables with this remarkable substance and laid an insulated cable across Kiel harbor.[3]

Underwater telegraphy now entered a phase of experimentation and development. In 1850 John and Jacob Brett laid an insulated copper wire from Dover to Calais, but it was broken a few hours later by a fisherman's anchor. A year later they tried again with four strands of copper wire coated with gutta-percha and protected by a winding of iron rope; this one held up.[4] Inspired by this success, entrepreneurs laid a cable to Ireland in 1853, then one across the Atlantic in 1858. The technology was not up to their demands, however. The thinness of the cables caused poor reception and slow transmission. After a few months it ceased completely, but not before carrying one useful message: by cancelling an order to ship two regiments from Canada to India, the British government saved £50,000.

The failure of the first Atlantic cable led to serious reassessments. Unlike steam and iron, electricity was beyond the understanding of the engineers of that period. The British Board of Trade and the Atlantic Telegraph Company turned to the scientist William Thomson (later Lord Kelvin) for technical advice. Thomson designed more sensitive sending and receiving instruments and devised methods of sounding the ocean floor. Meanwhile, others were improving cables and laying techniques. By 1866 all was ready, and a new cable was laid across the Atlantic, this time successfully.[5]

The years 1851–70 witnessed two other attempts to link the continents by telegraphic cables. Significantly, they were both along the main axes of the major colonial empires of the time: from Britain to India, and from France to Algeria.

The Indian Rebellion of 1857 turned a strong interest in communications between Britain and India into something approaching panic. Promoters and the Indian government pushed several schemes at once without waiting for the technology to mature. In the process, many mistakes were made and much money was wasted, but the efforts hastened the evolution of telegraphic technology and led directly to the British dominance of that field for the next half-century.

In 1856 two entrepreneurs, Lionel and Francis Gisbourne, obtained a concession from the Turkish and Egyptian governments to stretch a land line from Alexandria to Suez. Meanwhile, a special committee of the House of Commons which met in 1857 endorsed the idea of a cable to India. In 1858 and 1859 the Gisbournes' new company, the Red Sea and India Telegraph Company, laid a cable from Constantinople to Alexandria, then another from Suez to Aden and Karachi; the British government guaranteed a dividend of 4.5 percent or £36,000 per year for fifty years.[6] Unfortunately, cable technology was still primitive. The Red Sea cable was too thin, consisting of a single strand of copper coated with gutta-percha and wound with hemp. It was laid without any slack, so that it was stretched between peaks on the sea bed, which had not been surveyed beforehand. In the warm tropical waters, worms ate through the insulation and barnacles encrusted the cable until it broke under their weight. It failed before it could transmit a single word, and its £800,000 cost was a total loss to the British government.[7]

The second attempt was a throwback to a proven technology, the land line. As there were already land lines through Europe to Constantinople, this left two long stretches to be covered: across the Ottoman Empire from Constantinople to the Persian Gulf, and across Persia and Beluchistan to India. Work started at both ends. In 1858 the Ottoman government allowed British army personnel to install a telegraph line between Constantinople and Baghdad. Soon after, in 1862, the Bombay government established the Indo-European Telegraph Department under the direction of Col. Patrick Stewart of the Royal Engineers. Maj. F. J. Goldsmid and Lt. Col. Lewis Pelley were sent on a mission to put up a land line along the Arabian Sea between Karachi and Gwadur. Finally in 1864 a submarine cable was laid between Gwadur and Fao, at the head of the Persian Gulf. This cable was armored and weighed 2.5 tons per kilometer, four times more than the ill-fated Red Sea cable. At Fao it met a new Turkish land line from Baghdad. At last, on January 27, 1865, Britain and India were linked by telegraph.[8]

Though it worked, it was a poor connection. The "Turkish line" was expensive, charging £5 for a twenty-word message. It was slow: only once in 1866 did a telegram get through in less than 24 hours; others took a month or more, and the average was over 6 days. And it was erratic. In the winter the wires broke under the weight of the snow. In other areas, they were a tempting prey for nomadic tribesmen. Transmission was plagued by poorly trained operators. By the Indo-Ottoman Telegraphic Convention of 1864, the Ottoman government had agreed to provide clerks "possessing a knowledge of the English language sufficient for the perfect performance of that service"; yet many telegrams, relayed from clerk to clerk across the length of the Ottoman Empire, arrived in gibberish. But worst of all, from the British point of view, was the fact that the telegraph departments of France, Bavaria, and the Austrian Empire, through which the lines passed, gave their own messages priority over foreign ones, or held up British telegrams out of spite or curiosity. The defects of the Turkish line, wrote one indignant Briton, "are as much the result of faulty management in Europe, as of the barbarous character of the countries of the East." In 1866 Parliament appointed a Select Committee on East India Communications to look into the matter.[9]

The result was the laying of other telegraph lines to supplement the Turkish one. One of them was a land line, which the Siemens Company offered to build across Prussia, Russia, and Persia for the exclusive use of British-Indian communications. Having obtained the agreement of the governments concerned in 1867, Siemens founded the Indo-European Telegraph Company and completed the new line by January 31, 1870. The competition had good results: within two years the time it took a message to travel between India and Britain fell to 6 hours and 7 minutes on the Russian line, and to 1 day, 6 hours and 20 minutes on the Turkish line.[10]

Still, the British government knew that a land line, no matter how well managed, was only a temporary expedient. The idea of a submarine cable to India, dormant since 1859, was revived in 1866 after the success of the new Atlantic cable. In 1869 the British Indian Submarine Telegraph Company, chaired by the Manchester industrialist John Pender, commissioned the Telegraph Construction and Maintenance Company, cable manufacturers, to lay a cable from Suez to Karachi and from Alexandria to Malta. Meanwhile, the Falmouth, Gibraltar and Malta Telegraph Company was laying the rest of the line to Britain. On July 14, 1870, Britain and India were joined by an entirely British-owned and controlled cable, except for a short land line across the Isthmus of Suez.[11] Britain now had what politicians and publicists had long demanded: rapid, safe, secret communications with India, and several backup systems in case of trouble. It was the spinal cord of the British Empire.

Cable Technology to 1914

Cable technology can be considered under three headings: the cable itself, cable laying, and transmission. The first cables were single strands of copper similar to aerial telegraph wire, thinly coated with rubber or gutta-percha. It did not take long to show that cables had to be designed for specific underwater environments. Several factors entered into the calculations. The first was that electric impulses got smoothed out, hence hard to read, as they traveled along the cables. To counteract this capacity effect and allow reasonably rapid signals, the thickness of the copper conductor had to be made proportional to the square of its length. A cable between France and Algeria might need 22 to 40 kilograms of copper per kilometer, whereas a North Atlantic cable required between 80 and 160.

Next came the coating of gutta-percha, which weighed almost as much as the conductor and cost considerably more. Then the cable had to be armored, that is, wrapped in several protective layers: brass tape against boring insects, tarred hemp to cushion against blows, and steel wire around the outside. Across long stretches of deep ocean floor covered with alluvial deposits, the cable was out of harm's way. On continental shelves and on jagged seabeds like the Caribbean, Mediterranean, or Red seas, cables had to be much heavier and better armored; near the coasts with their tidal currents, ships' anchors, and trawling tackle, cables needed to be thicker still. Thus ocean floor cables weighed a ton or 2 per kilometer, but near a coast cables weighing up to 12 tons per kilometer were commonly used.

Not only were submarine cables thick, heavy, and complex, they also had to be perfect, for the slightest impurity in the copper could slow transmission speeds drastically, and the tiniest flaw in the insulation, such as an air bubble or a worm hole, could quickly render the cable useless. All these factors conspired to make cables expensive—up to £200 per kilometer—and thus out of reach of all but well-capitalized companies and prosperous governments.

Cable manufacture was one of the high-technology industries of the late nineteenth century, dominated by a few firms. The first was the Telegraph Construction and Maintenance Company (Telcon), which made two-thirds of all the world's cables before 1900. Then came three other British firms: Siemens Brothers (London), India Rubber, Gutta Percha and Telegraph Works, and W. T. Henley Telegraph Works. Only at the end of the century did non-British firms appear: Société industrielle des téléphones, Establissements Grammont, and Norddeutsche Seekabelwerke.[12]

Though costly, cables turned out to be more durable than anyone had thought. They were expected to last twenty-five years; in fact, many of them performed for up to a century. By the twentieth century, many cables were technically obsolete yet still worked well. Burdened with an enormous capital investment, the industry became extraordinarily conservative.[13]

Laying a cable was even more complex than manufacturing it. The ocean floor had to be sounded to find a path that avoided peaks and cliffs. Cable ships had to be large enough to carry hundreds of kilometers of cable coiled up in a water-filled tank, and maneuverable enough to navigate an exact and steady course even in the midst of storms and currents. They carried special brakes and pulleys to pay out the cable at a steady rate. As the ship advanced, it kept in telegraphic contact with its point of departure via the cable, and electrical measurements were taken continuously to spot any possible weaknesses. Cable laying involved many delicate operations such as meeting other cable ships at sea, joining cable ends, and landing the heavy shore cable in shallow waters. Some of the early long-distance cables to America and India were laid by the *Great Eastern,* at the time the only ship which could carry the 4,000 tons of cable needed to cross an ocean. After 1880, Telcon, Siemens, and other manufacturers bought specially designed cable ships. Their cost was high enough to keep small firms out of the business. The advantage of being first was self-reinforcing. Even countries with cable programs of their own had to rely for a time on British manufacturers and cable ships. Of the 41 cable ships in the world in 1904, 28 were British, 5 were French, and the rest divided among 6 other countries.[14]

Though the cables themselves hardly changed between 1866 and 1900, advances in transmission techniques improved their efficiency many times over. To sharpen signals blurred by the capacity effect of long cables, William Thomson devised a system of curb transmission which sent a reverse pulse immediately after the main pulse. He also invented the siphon recorder, which printed a picture of the signal on a paper tape that was read by the telegraph operators. Duplexing, introduced in the 1870s, permitted a cable to be used simultaneously in both directions, doubling its capacity. Other devices to improve transmission included the automatic transmitter, the signal regenerator, and the cable relay, all of which greatly reduced the work of skilled telegraph operators.[15]

The result was to increase the speed and capacity of cables. In the sixties and seventies, cables transmitted, on the average, 9 to 13 words per minute; by 1900 transatlantic cables could handle 50 words per minute. In parallel with these electromechanical improvements, codes were developed which crammed ever more meaning into fewer and fewer words; by the turn of the century, common business codes could express an average of 28 clear words per code word.[16] In the 1920s, spurred on by the competition of the wireless and the need to replace old cables with improved ones, cable companies reached speeds of 400 to 500 words per minute.[17]

While these figures mattered to the cable companies, the public was more interested in another measure of speed: how long it took to transmit a telegram over a given route. This measure of speed depended on demand and competition as much as on technology. Nowhere was demand as strong and competition as fierce as on the North Atlantic route. By the 1890s, telegrams between the London and New York Stock Exchanges took 3 minutes from sender to receiver. Elsewhere, a more lackadaisical attitude prevailed: London to Bombay took 35 minutes, to Hong Kong 80, and to Sydney 100.[18]

Submarine Cables and the British Empire, 1870–1914

The first cable to India, that panic-driven, ill-starred enterprise, cooled for many years the British government's eagerness to subsidize cables. The successful cable of 1870, in contrast, marked the beginning of a surge of private cable companies which not only linked together the scattered pieces of the British Empire, but reinforced Britain's position as the foremost naval, commercial, and financial power in the world.

In 1868 the British government had nationalized all domestic telegraph companies. The compensation paid their stockholders enabled them to invest in several new cable ventures, including the Falmouth, Gibraltar and Malta and the British Indian Submarine Telegraph Companies, which coalesced in November 1872 to form the Eastern Telegraph Company. The new firm quickly obtained almost half the telegraph traffic between India and Europe. The other half went via Germany, Russia, and Persia on the land lines of the

Indo-European Telegraph Company, or via the lines of the Indo-European Telegraph Department, which linked up with those of the Ottoman and Austro-Hungarian empires. If the land lines survived despite their higher rates and worse service, it was thanks to the "joint-purse" or cartel agreement of 1878, designed to maintain the largest possible number of alternate routes for strategic reasons. In later years, for both strategic and commercial reasons, telegraphic links between India and Britain were increased still further; by 1908 there were, besides the Persian Gulf cable, three cables through the Mediterranean and Red seas, a direct land line to Teheran, and from there one line through Turkey and two through Russia. Beyond these six were yet other, strategically reassuring routes: around Africa, via Hong Kong and Siberia, and across the Pacific, North America, and the Atlantic. Under no circumstances could Britain ever be cut off from India.[19]

From 1873 to 1906 the profits of the Eastern Telegraph Company were excellent, varying between 6.69 and 9.6 percent. John Pender, chairman of both Eastern and Telcon, seized the opportunity to expand. An affiliate, the Eastern Extension Australasia and China Telegraph Company, laid cables from Madras to Penang and Singapore, then on to Java, Australia, and New Zealand on the one hand, and to Saigon and Hong Kong on the other. Meanwhile, other Eastern affiliates laid cables to Brazil, Argentina, and the west coast of South America.

By World War I, the Eastern and Associated Companies had become one of the world's most powerful multinational conglomerates. Their capital, valued at nearly £13 million, was controlled by the Pender family and the Marquess of Tweeddale. Their over 180,000 kilometers of cables throughout the South Atlantic, the Mediterranean, the Indian Ocean, and the Eastern Pacific represented two-thirds of all British cables, or two-fifths of all cables in the world. Through their London headquarters came much of the world's commercial cable traffic, and a good part of the rest was funneled through Eastern's secondary nodes at Aden, Cape Town, Singapore, Bombay, and Hong Kong. Eastern was one of the pillars of British commercial and strategic power.[20]

Aside from Eastern, there were several lesser British networks. The India Rubber, Gutta Percha and Telegraph Works founded the West African Telegraph Company in 1885 to serve the French and Portuguese towns of Saint-Louis, Dakar, Grand Bassam, Cotonou, Libreville, São Thomé, Principe, and Loanda. It also launched the

Table 4.1 Submarine Cables, British and Non-British, 1887–1908

| Year | British | | Non-British | | |
	Kilometers	%	Kilometers	%	Total
1887	149,269	70	64,948	30	214,217
1894	172,465	63	100,137	37	272,602
1901	225,795	63	132,342	37	358,137
1904	248,265	60	163,766	40	412,031
1908	265,971	56	207,137	44	473,108

Sources: Maxíme de Margerie, *Le réseau auglais des câbles sous-marins* (Paris, n.d. [1909?]), p. 36. Pierre Jouhannaud, *Les câbles sous-marins, leur protection en temps de paix et en temps de guerre* (Paris, 1904), pp. 31–32.

South American Cable Company, a financial failure later sold to France. Others included two private cable companies to North America—the Anglo-American and the Direct United States Tele-graph—and some smaller companies in the Caribbean.

Despite frantic cable laying by other major powers after the turn of the century, Britain maintained a comfortable lead in the cable business right up until World War I (Table 4.1). In 1901, fully 91 percent of British cables were privately owned. This reflects not so much the British preference for free enterprise as the fact that British companies, having arrived on the scene first, had established them-selves on the profitable commercial routes.

Until 1878, Eastern received no subsidies or dividend guaran-tees, but they did get considerable official help in the form of land-ing rights, ocean-bed surveys, diplomatic support, and government cable traffic. This alliance benefited business, especially British busi-ness. Manufacturers, banks, import and export firms, shipping lines, and tramp steamers used the cables, making London the center of world trade and finance.

Yet imperialism made demands which business could not af-ford to meet. After cabling to some parts of the empire became a habit, a demand arose for cables to the lesser colonies, regardless of profit considerations. This demand often came from soldiers and administrators at moments of colonial expansion and warfare. The British annexation of the South African Republic in 1877 precipi-tated South African demands for a telegraphic connection to Europe. At first there was talk of a Cape-to-Cairo land line, but the difficul-ties of installing it through unexplored parts of Africa delayed the project. In 1878 James Pender, John Pender's son, persuaded the

Cape Colony, Natal, Mozambique, and Zanzibar to accept a cable instead. The Zulu War of 1879 turned the matter into an imperial necessity, and the British government agreed to pay £55,000 per annum to the Eastern and South African Telegraph Company to operate a cable from Durban to Zanzibar and Aden, linking it to Eastern's trunk line.[21]

Troubles on the periphery of empire led to several other subsidized cables. During the British invasion of Egypt in 1882, Eastern laid a cable from Suez to Suakin on the Red Sea coast of the Sudan. At the time of the Ashanti War of 1873–74, the War Office lobbied unsuccessfully for a cable to the Gold Coast. The matter came up again, however, when Britain proclaimed a protectorate over Nigeria in 1885–86. Though there was already a West African coastal cable, the British government decided it could not tolerate having its telegrams pass through French and Spanish territory and therefore offered Eastern £19,000 a year to operate a cable from the Gambia to Sierra Leone, the Gold Coast, and Nigeria. For similar reasons, after the Boxer Rebellion of 1900 a cable was laid between Shanghai and the British base at Weihaiwei, avoiding the land line through China.[22]

It is tempting to believe that by putting colonies into rapid contact with London, cables gave the Colonial and India offices a tighter rein over their distant subordinates, thus substituting centralized control for the little subimperialisms of the periphery. The evidence, however, points the other way. For information about crises on the frontiers of the empire, London still depended on the men on the spot, who distorted the facts to suit their own ambitions. In West Africa cables magnified the rivalries between British, French, and German agents and frustrated London's attempts to reduce the number of troops by increasing their mobility. In the Jameson Raid of 1895–96 in South Africa, writes Robert Kubicek, "The telegraph intensified involvement but denied [Colonial Secretary] Chamberlain its corollary, control." A few years later, in 1899, Chamberlain failed to restrain the British high commissioner in South Africa, Lord Milner, from exacerbating relations with the Transvaal: "The telegraph did not extend the [Colonial] office's opportunities for learning of or controlling events in South Africa. In fact, its operation reinforced the office's passive characteristics, and the activist propensities of the man of the spot."[23]

At the end of the nineteenth century, when Britain was linked to almost all its colonies through either commercial or subsidized cables, another sort of imperialism arose, which pitted the great powers against one another. The long peace of the nineteenth century gave way to a prewar era, an atmosphere of mutual suspicions fueled by small incidents and imaginary scenarios. In the strategic thinking of the time, cables figured prominently. Yet in the underwater world of cable strategy, the potential enemies were different from the enemies on the surface.

An example of the paranoia of the time is an article entitled "Our Telegraphic Isolation," written in 1896 by Percy Hurd. In it he called the British cable network "an excellent fair-weather system, but it is little more." The British West Indies, he noted, needed "help to free them from their telegraphic dependence upon the United States" and "the commercial domination of New York." Cables to India were vulnerable in the Mediterranean and Red seas, as was the line to Australia where it crossed India from Bombay to Madras. What was needed was "a system of telegraphic communication completely under British control."[24]

This was not an isolated voice. To prevent foreign powers from stopping, censoring, or even just reading British telegrams, the Colonial Defence Committee, the Admiralty, and the War Office had long insisted on the need for duplicate routes. Thus the Halifax and Bermuda cable was laid in 1889 and prolonged to Jamaica in 1897 in order to avoid the United States. During the Boer War the Colonial Defence Committee felt it was not enough that South Africa could cable to Britain via the east and west coasts of Africa. In 1901 it obtained a deep-sea cable from Britain to South Africa via Cape Verde, Ascension, and Saint Helena. From South Africa, another cable was laid to Australia via Mauritius, Rodrigues, and Cocos Islands, one free from even the slender risks of the Britain-Australia cable's being cut in Egypt or in India.[25]

By this time the military and the Colonial Defence Committee had developed, in the words of P. M. Kennedy, "a virtual fetish" for "all-red" routes, so named after the color of the British Empire on maps of that period. As the Inter-Departmental Committee on Cable Communications put it in 1902,

We regard it as desirable that every important colony or naval base should possess one cable to this country which touches on British territory or on the territory of some friendly neutral. We think that, after this, there should be as many alternative cables as possible, but that these should be allowed to follow the normal routes suggested by commercial considerations.[26]

All-red cables that duplicated existing commercial cables, however, were of little interest to the Eastern and Associated Companies. Until the 1890s, Eastern and the government had worked together in harmony because the government paid ever larger subsidies for strategic cables; the South Africa–Australia cable, for instance, cost the government £1,750,000.

When Joseph Chamberlain took over the Colonial Office in 1895, he began actively backing the demands of Australia and New Zealand for yet another cable to Britain, one going through Canada. It was to be the final link in the all-red route around the globe. Eastern fought the idea on the sensible grounds that there was little likelihood of cable traffic's developing between Canada and Australia or New Zealand. The government's motives for this last cable were purely strategic: it would both lie far from French naval bases and be, in the postmaster general's words, "a means of escaping from the many difficulties and embarrassments which cluster round the eastern world and the eastern route."[27] Since Eastern would have none of it, the governments involved formed a Pacific Cable Board and divided the £2 million cost; Britain and Canada each paid five-eighteenths, Australia six-eighteenths, and New Zealand two-eighteenths. It was the longest cable in the world, from Vancouver to Norfolk Island and from there to Australia and New Zealand. In December 1902 the world was at last girdled by an all-red cable.[28]

In 1911 a cable subcommittee of the Committee of Imperial Defence concluded that the all-red system was virtually complete, and the communications of the British Empire were, for all practical purposes, secure. Yet voices of suspicion could still be heard. In a paper read before the Royal Colonial Institute in April 1911, the telegraph engineer Charles Bright pointed to unsuspected weaknesses. Around South America, British interests were being threatened by "the great efforts now being made by the United States to advance their trade, and hinder that of other countries." Around Australia, "at any moment the cable might be cut. . . . we should have no means of communications. . . . we might have hostile cruisers at our door." Between Britain and Canada, all telegrams went through commercial cables which were either American-owned or held in alliance with American trusts. Hence, "not a single one of

these Atlantic cables . . . could at any time be relied upon as a
genuinely All-British strategic connection in the event of trouble
with the United States or with other countries interested in disturbing
our Imperial communications." To avert this danger, Bright proposed
a government-owned direct Britain-Canada cable, and a new land
line across Canada "at some distance from the American frontier."[29]
Only with hindsight can we dismiss such fears as foolish. From the
perspective of the time, the British Empire, despite its great power,
did indeed look vulnerable.

What conclusions can we draw about cables and the British
Empire? There is no doubt that cables contributed mightily to the
growth in world trade, to the division of labor within the empire,
and to Britain's economic position as the commercial center of its
empire and of the world. The political gains are harder to measure.
Thanks to cables, Britain remained in touch with all its colonies,
naval bases, allies, and neutrals, and was able to repress all rebel-
lions in its territories. Though cables made Britain stronger, they did
not make the British Empire more secure, however, for they incited
the jealousies of others and contributed to the international paranoia
of our century.

The French Cable Network, 1856–95

In the early days of cable, the French were as eager as the British
to use the new technology. Like the British, their first efforts pointed
across the Atlantic, and toward their empire. The first attempt to
link France and North Africa by cable took place in 1856. The
French government, which held the monopoly of internal telegraphic
communications, extended that privilege to Algeria. So eager was
it to communicate across the Mediterranean that it did not wait for
the technology to mature, nor for a French cable industry to arise.
Instead, it hired John Watkins Brett, who had laid the Channel
cable a few years before. Brett's cable to Algeria via Corsica and
Sardinia failed within two years. The French government then
turned in 1860 to Glass, Elliot and Company, promoters of the Atlantic cable. Their first cable broke off in Toulon. Another attempt
in 1861, from Port-Vendre to Algiers via Menorca, lasted only a
year. After those failures France, like Britain, fell back on an easier
technology: a land line through Spain to Cartagena, a short cable
laid from Cartagena to Oran in 1864, and another land line from

Oran to Algiers. Finally in 1870 the Falmouth, Gibraltar and Malta Telegraph Company (soon to become the Eastern Telegraph Company) laid a cable from Marseille to Bône in Algeria, which worked well for many years. Yet the French government was not content with just a connection to a foreign company's cable. Furthermore, the influx of settlers, capital, and military forces into Algeria after the Franco-Prussian War required more than one link. Hence, as traffic increased, the government laid its own cables between France and North Africa: to Algiers in 1871, 1879, and 1880; to Oran in 1892; to Bizerte and Tunis in 1893; from Oran to Tangier in 1901; and yet another from Marseille to Algiers in 1913.[30]

Unlike the North African cables, the French Atlantic cables were privately owned and operated. International telecommunications were left to private enterprises because governments, which willingly granted landing rights to foreign companies, would have refused such a concession to another government. The first French Atlantic cable was laid in 1869 for the Société du câble transatlantique francais by Siemens Brothers of London. In 1872 it was sold to a British firm, the Anglo-American Telegraph Company, which employed only British telegraph clerks.

The second French Atlantic cable was also British-made and laid by a British ship in 1879, but it remained in the hands of a French firm, the Compagnie française du télégraphe de Paris à New-York. Competition with British firms, however, made it unprofitable. It went bankrupt in 1895 and its assets were merged with those of another small company, the Société française des télégraphes sous-marins, to create the Compagnie française des câbles télégraphiques, a government-subsidized and controlled organization. The latter company laid a third cable in 1898 directly from Brest to Cape Cod; it was the longest and heaviest cable made up to that point, the product of a French manufacturer, the Société industrielle des téléphones.[31]

While French enterprise went into the Atlantic and Mediterranean cables, it did little else before the mid-1890s. Other parts of the world were either already linked to France by way of the British network or were too poor and remote to interest French capitalists and the parsimonious French government. In many of their colonies, the French hired British firms to provide telecommunications. Thus the French West African colonies were first linked to France in 1884–85 by the Spanish National Submarine Telegraph Company and the West African Telegraph Company, both affiliates of the India Rubber, Gutta-Percha and Telegraph Works. Indochina was first connected to the world network in 1871, when the China Submarine Telegraph Company (soon to be Eastern Extension Australasia and China Telegraph Company) agreed to land its Singapore–Hong Kong cable at Cap Saint-Jacques near Saigon. In 1883–84, the French government negotiated another contract with Eastern Extension, to lay and maintain a cable from Saigon to Haiphong in the new colony of Tonkin, and from there to Hong Kong.[32]

French companies laid and operated a few short cables between French colonies and the British network. Thus in 1890–91 the Société française des télégraphes sous-marins connected Haiti, Martinique, French and Dutch Guyana, and northern Brazil. Between the Caribbean and North America or Europe, however, messages still had to go through British or American lines. In 1893 the same company laid a cable from New Caledonia to Bundaberg on the Australian coast. The impetus and part of the funds came from the governments of Queensland and New South Wales, which hoped it would be the start of a transpacific line to San Francisco via Fiji, Samoa, and Hawaii. Loud protests by the governments of Great Britain and Victoria quashed the project, leaving an unprofitable short line to serve the very remote French outpost at Noumea. In 1895, during the French conquest of Madagascar, the same company laid another short cable across the Straits of Mozambique, to connect with the Eastern and South African Telegraph Company's line from Durban to Aden.[33]

By 1895 British cables linked France to almost all its colonies except North Africa. The cost was high; according to one author it cost the French government 2,337,000 francs (£93,480) a year to subsidize foreign companies for the Tenerife-Senegal, West African, Saigon-Haiphong, and Obock-Perim cables.[34] Yet this solution was a lot less expensive than to lay a competing network at 3,000 to 5,000 francs a kilometer.

This at least is how the French Chamber of Deputies felt. It is often asserted that Britain was an "absent-minded" or "reluctant" imperialist, while France was eager for conquests in the tropics to make up for its humiliating defeat of 1871. Spiritually that may be so, but financially quite the opposite is true. The British government was willing to spend a great deal on the technical armature of imperialism—the Royal Navy and its bases, and subsidies to shipping lines and cable companies—while British capitalists invested heavily in shipping, telegraphs, colonial railways, and plantations. In contrast, the French government, representing the views of the French taxpayers, wanted as cheap an empire as possible. This is reflected in the French colonial cable projects of the late nineteenth century.

Many cable projects were presented, only to be rejected for financial reasons. In 1866 the Ministry of the Navy and Colonies discussed a cable project between China and Europe via Indochina and India, but nothing came of the scheme.[35] More modest cable projects were presented in the seventies and eighties. In 1878 the Ministry of Posts and Telegraphs contacted the India Rubber and Siemens Brothers companies regarding cables from the Cape Verde Islands to Senegal and from New Caledonia to Sydney. In 1886 a project to lay a cable from Reunion to Madagascar, Djibouti, and Tunis was presented to the Chamber of Deputies, but never got out of the budget committee. Another one, a year later, to link the French West Indies with New York was defeated on the floor of the Chamber.[36]

A more ambitious plan led to the Azores affair of 1892–93. In February 1892 the Portuguese government granted exclusive landing rights in the Azores to the Telegraph Construction and Maintenance Company. Through some intermediaries, French Foreign Minister Ribot persuaded the Portuguese parliament to reject this agreement, and to grant the concession to the Société française des télégraphes sous-marins instead. The latter company then contracted with the French Ministry of Posts and Telegraphs to lay a cable from Brest to Haiti via Lisbon and the Azores. A new minister of commerce and industry, Jules Siegfried, found the dividend guarantee (2.3 million francs per year) too high, and the budget committee of the French Chamber of Deputies refused to ratify the contract. As a result, Portugal granted the concession to the European and Azores Telegraph Company, an affiliate of Eastern, which sold it to the German Post Office, to the dismay of French patriots.[37]

British Abuses and French Reactions, 1884–1914

In a letter to Maxime Hélène, author of a book extolling the great construction projects of the century, Ferdinand de Lesseps wrote: "All these enterprises of universal interest—some contemplated, others under construction or projected—have an identical purpose: to bring peoples closer together and thereby to bring about an era in which men, by knowing one another, will finally stop fighting."[38] Such an idea—peace on earth through better communications—is too familiar and persistent, even in our own much more cynical times, to be dismissed as naive. In the case of telegraph cables, however, this illusion was slowly eroded by imperialism and great-power rivalries.

In France all cable projects required some sort of government subsidy, and until 1898 they were ignored by a government caught between the need to maintain a huge army facing Germany and the natural reluctance of citizens to pay more taxes. To shake the politicians out of their parsimonious ways required a serious threat to the prestige of France and to the security of its empire. The threat was provided by certain British misbehaviors, amplified by outraged French journalists.

One irritant to the French was information. News of commercial importance—commodity prices, contracts, ships' arrivals and departures, etc.—passed through London before reaching Paris. British newspapers and the Reuters news agency received reports of world conditions sooner and in more detail than their French counterparts. This helped British trade at the expense of the French.[39]

News also flowed in the other direction, and with it flowed culture, a subject to which the French were particularly sensitive. A French agent in Central America complained:

> General news dispatches arrive every day from the United States and are immediately published by the local papers. They contain hardly anything but news about the great republic to the north and England. Only exceptionally do they bring any information concerning France. . . . By a natural process one becomes interested in people who are mentioned at every moment, by and by one lives their life, and one naturally enters into long-term relations with them. As for those who are never mentioned, they are quite soon forgotten.

To which a French journalist commented: "This does not just involve the commerce and industry [of France]; its influence is also at stake, and the diffusion of its ideas, and its good name in the entire universe."[40]

These were minor irritants compared to foreign cable control over political intelligence. Britain exercised this control all along its cables. Because the British government required its cable companies to employ only British citizens as telegraph clerks, there were British agents in the French naval bases of Dakar and Saigon, watching the ships and reading the telegrams. In London the Colonial Defence Committee listened in on the foreign cable traffic at the Central Telegraph Office.[41] And at other cable nodes—Bathurst, Suez, Aden, Hong Kong—other agents read the cables. In Maxime de Margerie's words, "electricity was the ally of English diplomacy."[42]

Little incidents began to accumulate. In 1885, when Admiral Courbet was defeated at Langson in Tonkin, the British embassy in Paris knew about it before the French government. At the time of the French ultimatum to Siam in 1893, a French government telegram to Admiral Humann was held up by the Eastern Telegraph Company until the British cabinet would read and approve it. During the Sino-Japanese War of 1894–95, dispatches to the French Foreign Ministry were delayed in London for 12 hours. After the death of Sultan Mulay Hassan of Morocco on June 11, 1894, the British consul at Tangier monopolized the country's only cable for an entire night. A year later, in September 1895, the British government knew of the French capture of Tananarive in Madagascar three days before the French government.[43] Until that point the imperialist powers, though suspicious of one another, avoided direct confrontations. In 1898, however, confrontations flared, war seemed imminent, and the fate of cables in war became a serious worry.

In the early days of cables, France and the United States had proposed that cables be considered neutral in wartime, but Britain categorically refused. The International Cable Convention of 1885 contained a clause inserted at Britain's insistence, recognizing the right of belligerents to cut one another's cables.[44] Cutting cables, however, was no easy matter. Near the shore, where they were most vulnerable, they were likely to be protected by powerful guns. At sea, the operation required an exact knowledge of the cable's location and a cable ship with grappling equipment. Since British firms had laid all British and most non-British cables in the world and owned 24 of the world's 30 cable ships, the Royal Navy had an overwhelming advantage in this area.

In the spring of 1898 the United States went to war against Spain. Spanish communications with Cuba via the United States were censored. American ships attempted to cut the cables to Cuba from Haiti and Jamaica, but only succeeded in breaking the French one from Haiti. Meanwhile in the Far East, Admiral Dewey seized the Hong Kong–Manila cable. To French writers it seemed an ominous lesson: in the event of a Franco-British war, the French Empire would be in a worse situation than the Spanish colonies had been, for the British were far more adept at cable cutting than the Americans.[45]

The fear of a Franco-British war was not so far-fetched in 1898. In July of that year, Captain Marchand reached the banks of the Nile in the Sudan, after a two-year trek from Gabon. Three months later General Kitchener, who had defeated the Sudanese dervishes at Omdurman in early September, marched up to Fashoda to prevent Marchand from establishing a French claim to a part of the Nile. In the ensuing crisis, Kitchener remained in touch with Britain by way of a cable laid in the Nile to Khartoum, while Marchand was cut off from France.

During those tense days, when war seemed likely, the cable to Senegal suddenly went dead. The governor of Senegal mobilized the native reserves and loaded the cannon overlooking Dakar harbor. Five days later, the cable was repaired. Was it simple coincidence?[46]

A year later, at the outbreak of the Boer War, Britain again showed its muscle. On November 17, 1899, the postmaster general informed the International Telegraph Bureau in Bern that all coded telegrams were forbidden south of Aden, and that all clear-language telegrams were subject to censorship. To other European countries it was galling that their communications with their East African colonies were subject to the veto of a British clerk in the Aden cable office.[47]

French reaction was entirely political; that is, it took a crisis to bring about a change. Complaints about British behavior began after the Moroccan affair of 1894, when Harry Alis wrote in the *Journal des débats*: "The newspapers wonder what can be, under these circumstances, the security of the interests of other nations if England, which holds all sources of information, can thus suspend, at will, communications which are not its own." He suggested that "only France has enough colonies and capital so that a telegraphic competition with England is both necessary and possible."[48] At the same time the Société des études coloniales et maritimes, a colonialist lobby, resolved that "it is necessary for a great country like France to have independent national telegraphic communications with its colonies."[49] A few others added their opinions: Henri Bousquet, writing in 1895, worried about a possible war with Britain; and in 1896 J. Depelley, director of the Compagnie française des câbles télégraphiques, spoke before the Union coloniale française on the subject of British cable power and the dangers to the French colonial empire.[50] But their complaints were ignored until 1898. Then came Fashoda, followed by the British announcement of censorship at Aden, and the crisis broke. On November 27, 1899 the deputy Henrique introduced a resolution on cables in the Chamber of Deputies. On December 8, Minister of Colonies Decrais announced that the government was preparing a cable bill. A week later the Council of Ministers accepted a plan for a French imperial cable network to cost 100 million francs.[51]

In early 1900 articles appeared in various journals, lobbying for a French world cable network.[52] Commercial and colonial societies—the Conseillers du commerce extérieur de la France, the Chamber of Commerce of Lyon, and the Société d'économie industrielle et commerciale, a trade lobby, added their voices to the chorus.[53]

Finally on May 1, Minister of Posts and Telegraphs Millerand delivered the government's report on cables. In it he described "the insufficiency of our submarine telegraph network and the numerous inconveniences which come from our having to use almost exclusively foreign lines." But he warned that "to insure in satisfactory conditions telegraphic communications between the metropolis and French colonies in West Africa, the Indian Ocean and Indochina, other important sacrifices are still necessary."[54] The projects that were discussed in the Chamber of Deputies did indeed involve a sacrifice, on the order of 130 million francs. And that was the dilemma: a French network would be prohibitively expensive and, unlike the British one, commercially unprofitable. In the many discussions and pamphlets on cables in the year 1900, one finds, here and there, the seed of a clever solution to this dilemma: that France should join with the United States, the Netherlands, even Germany, in laying a second world cable network. Such was the depth of anti-British feeling.[55]

In the next few years numerous cable projects were discussed, ranging from the most modest to the truly unrealistic. Bills were passed and sums appropriated for cables.[56] Parliamentary speeches and articles in the press petered out and almost vanished after 1904, however, when France and Britain signed the Entente Cordiale. Yet the crisis left a legacy: the beginnings of an independent French cable network.

The French imperial cable network began in 1896–98, when the Compagnie française des câbles télégraphiques laid a cable from New York to Puerto Plata (Dominican Republic), followed in 1899 by a new Brest-New York cable. France was now linked by French cables to its Caribbean possessions.[57]

The West African and Indochinese connections took longer to establish. On July 25, 1901 the Chamber of Deputies authorized the purchase of the West African Telegraph Company's cables touching at French colonies. These consisted of two parts: the Dakar-Bathurst-Conakry cable, and the Grand Bassam–Accra–Cotonou–São Thomé–Libreville cable. The company also agreed to cut out the stops at Bathurst, Accra, and São Thomé, leaving only French possessions along the cables. Between Conakry (French Guinea) and Cotonou (Dahomey), the French still had to use British cables or the very unreliable land lines through the Soudan.[58] Dakar, then becoming the major French naval base and commercial entrepôt in West Africa, was linked to France by a new cable laid in 1905 by the Société industrielle des téléphones. The French West African network was completed in 1912 by a Conakry–Monrovia–Grand Bassam cable laid in cooperation with the German firm Deutsche Südamerikanische Telegraphengesellschaft, and a Libreville–Cap Lopez–Pointe-Noire cable laid in 1913. Henceforth France could communicate directly by French cables with all its possessions on the Atlantic coast of Africa.[59]

While the French government was buying up the cables of the West African Telegraph Company, it also entered into negotiations to acquire the transatlantic cable from Dakar to Pernambuco in Brazil. However, it could not simply buy that cable from its owner, the South American Telegraph Company, for the Brazilian government would have refused to let a foreign power hold landing rights on its shores. The French government circumvented this legal problem by secretly purchasing all the shares of the South American Telegraph Company from its holding company, the India Rubber group. Was South American now a British or a French firm? Was it private or government-owned? The situation was not clarified until 1914, when the company was transferred to France and renamed Compagnie des câbles sud-américains.[60] This clever deal, which casts some doubts on the fervent rhetoric of panicky patriots, was not to be the last instance of political ambiguity in the telecommunications industry.

A third focus of French activity in the early years of this century was Indochina. That colony had long been linked to France by the Eastern Extension's cables. This left the colony exposed to two perils: a break in the cables (as happened on the Haiphong–Hong Kong portion in 1897), and British hostility. Therefore in 1901 the Chamber of Deputies appropriated funds for an alternate cable, to run from Tourane (near Hue) to Amoy in South China, the southernmost point on the network of the Danish-owned Great Northern Telegraph Company, which crossed Siberia, Russia, and Scandinavia. The Tourane-Amoy cable, made in France, was laid in 1902 by a French cable ship.[61]

The Great Northern route, though free from British eyes, was vulnerable to such irritants as the Russo-Japanese War and the Russian Revolution of 1905. In 1906, therefore, France tried a third route: from Saigon to Pontianak in Borneo, linking up with the cables of the Dutch East Indies government to Celebes. From there, telegrams traveled via the cables of the Deutsch-Niederlandische Telegraphengesellschaft (a German-Dutch consortium) to Guam, and the American-owned Commercial Pacific Company to the United States.[62] This cable also had political drawbacks. Like the Tourane-Amoy cable, it got very little traffic and gave trouble technically as well. Both were interrupted in late 1913 and, the war intervening, never repaired.[63]

By the outbreak of World War I, France was still far from having a world network like Britain's. But at least it had its own cables to North America, the Caribbean, and Brazil, and to its colonies in North, West, and Equatorial Africa. Elsewhere, France relied on the kindness of strangers.

The Indian Telegraphs

The diffusion of the inland telegraph followed the same pattern as those other nineteenth-century imperial linkages, the postal system, steamship lines, and railroads. It first went to India, then to the white-settler colonies, and later, very gradually, it spread to parts of Africa. It would be wrong, however, to think of the process of diffusion as mechanical and somehow preordained. In the case of the telegraph in particular, the process of diffusion was molded and hastened by visionary individuals.

In India, the telegraph was created by William B. O'Shaughnessy, an assistant surgeon in the Bengal Army. He was a man of wide-ranging interests and abilities; in addition to his medical duties, he also directed the mint, taught chemistry, and experimented with electric motors, batteries, and lightning rods. His interest in electricity led him to develop an electric telegraph.

O'Shaughnessy began his telegraphic experiments in early 1839, the year in which Samuel Morse and Wheatstone and Cooke laid their first telegraph lines. He set up a 22-kilometer telegraph circuit near Calcutta, 18 of which consisted of iron wire hung on trees. In the process, he came up with two remarkable discoveries: that a body of water was as good a conductor as a wire, and that a wire insulated with pitch and tarred yarn could cross a body of water without appreciable loss of current. He published his findings in the *Journal of the Asiatic Society of Bengal*.[64] Showing that it could be done was not enough, however; once dismantled, the telegraph was forgotten for a decade, until high officials saw a need for it.

By the late 1840s the telegraph had proved itself in England and America, and lobbyists were eager to see it spread to India. In September 1849 the Court of Directors of the East India Company asked Governor-General Dalhousie to investigate the question. Dalhousie turned to O'Shaughnessy, who laid a new telegraph line from Alipore near Calcutta to Diamond Harbour, some 43 kilometers to the south. Part of it was insulated and laid in an underground circuit, while another section was hung on bamboo poles. By March 1851 the line was working, and in December it was opened to the public. A year later it was extended another 85 kilometers to Kedgeree on the Bay of Bengal, to give advance notice of the arrival of ships from England, a question of obsessive interest to the European community of Bengal.[65]

Dalhousie was enthusiastic. An eager imperialist who had sent armies into Burma and the Punjab, he was frustrated by the slowness of communications, both within his realm and with the India Office in London. On April 14, 1852 he wrote the Court of Directors demanding permission and funds to build a telegraph network:

> Everything, all the world over, moves faster now-a-days than it used to do, except the transactions of Indian business. What with the number of functionaries, boards, references, correspondences, and several Governments in India, what with the distance, the reference for further information made from England, the fresh correspondences arising from that reference, and the consultation of the several authorities in England, the progress of any great public measure, even when all are equally disposed to promote it, is often discouragingly slow.[66]

Dalhousie planned the trunk lines of the network to run from Bombay to Madras, from Bombay to Agra and Calcutta, and from Agra to Delhi and Peshawar. He sent O'Shaughnessy to London to appear before the directors and hasten their approval. To O'Shaughnessy's surprise, the directors had already approved the project. He therefore ordered everything he needed for a 6,000-kilometer-long telegraph network. He also toured telegraph installations in England and on the Continent, recruited sixty soldiers as telegraph installers, and published an instruction manual.

In late 1853 O'Shaughnessy was back in India with his men and equipment. Construction began at once. Conditions were more difficult than in Britain, as insects, storms, monkeys, elephants, and humans conspired to pull down the wires and frustrate the telegraphers. In places, 4.9-meter-high granite posts were used in place of wood. O'Shaughnessy's instruments were crude compared to Cooke and Wheatstone's, but they could be made and serviced by Indian craftsmen or by soldiers with little training. Despite environmental and technical problems, work proceeded fast, for O'Shaughnessy had a generous budget, administrative talents equal to his inventive spirit, and Dalhousie's full backing. Within five months the line from Calcutta to Agra was finished, and on February 1, 1855 the trunk lines were opened to the public. Dalhousie wrote his friend Couper: "The communication between Calcutta and Madras direct by land, a month ago, took twelve days—yesterday a communication was made, round by Bombay, in two hours. Again, I ask, are we such slow coaches out here?"[67]

During the Rebellion of 1857–58, the military used the telegraph to good effect, laying new lines between Madras and Calcutta and in strategic areas. John Lawrence, chief commissioner of the Punjab and later viceroy of India, was moved to declare "the telegraph saved India" (for the British that is). Government officials did not need convincing, and the telegraph system was given the highest priority. At first it was a separate department reporting directly to the governor-general, thereby avoiding both the stifling bureaucracies of India and the confusion of competing private enterprises. In 1883 it was merged with the Post Office Department, which allowed telegrams to be handled by thousands of small post offices which forwarded them by mail to and from the nearest telegraph office. Access to the network was thus extended to the entire population.

Telegraph rates were much lower than in Europe or America.

At first fixed at 1 rupee per sixteen-word message for every 644 kilometers, the 1-rupee fee was later made uniform throughout India regardless of distance. The telegraph system thereby served the Indian middle-class and business communities as well as the administration. Statistics give an impression of the system's penetration into Indian life. In December 1856 the network had 46 offices and 6,840 kilometers of line. Immediately after the Rebellion, there was a rush to put up telegraph lines; by 1865 the network was 28,164 kilometers long, of which 4,828 belonged to the railroads. By 1900 there were 4,949 offices and 85,150 kilometers of lines, and by 1947 the system was 188,600 kilometers long.[68]

French Colonial Posts and Telegraphs

In communications as in other technical fields, India was the most advanced of the Afro-Asian colonies. Within the British Empire there were places which were not connected to the outside world until the twentieth century. On balance, however, the British possessions were well served, because most of them had coastlines and were accessible to steamers and cables.

French possessions contained a larger proportion of deserts and land-locked areas, and they were less populated and less valuable economically. France had acquired them later and was less willing to invest in colonies than Britain was. Given pre-1914 technology, the French Empire therefore offered far greater obstacles to communications. If communications in the French Empire contrasted sharply with those of India, it was a consequence of geography as much as policy.

At one end of the spectrum were the two richest French colonies, Algeria and Indochina. French forces invaded Algeria in 1830 but did not crush all resistance until 1857. For military reasons, the army needed swift communications before technology had advanced enough to provide them. Hence the haste in laying submarine cables which we saw earlier. On land, the military government set up a 552-kilometer-long optical semaphore telegraph from Algiers to Oran and Tlemcen as early as 1842. By 1854 the optical system had grown to 1,498 kilometers, to which were added 249 kilometers of electric telegraph lines. By 1861, Algeria was well served, with telegraph offices in 38 of the major towns, and 3,179 kilometers of lines. Until then the telegraphs and mails were handled by the army.

After 1860 a separate Post Office was set up to handle these services. By 1920 the number of post and telegraph offices had risen to 720, and the telegraph network stretched to every village. Algeria was as well served as India, and almost as well as France.[69]

Indochina was conquered much later than Algeria: Cochinchina in 1858–62; Tonkin, Annam, and Cambodia in 1873–84; Laos in 1893. Everywhere the telegraph and postal services followed the flag; as A. Berbain, a high postal official, explained: "The postal and telegraph services thus had to be created from scratch; their development occurred alongside the occupation of this immense territory, not in order to help a possible economic growth, but following purely strategic or political aims."[70] Until 1900 coverage was very uneven, reflecting the varying durations of French occupation: in 1896–97, Cochinchina had 64 offices, Tonkin 51, Annam 21, Cambodia 16, and Laos 8. After 1901, the telegraph and postal services were merged and the network filled out. The number of telegraph offices rose from 15 in 1864 to 222 in 1901. By the early twenties Indochina had 425 offices, more than all of French sub-Saharan Africa and Madagascar together. Telegraph lines increased in proportion: from 400 kilometers in 1864 to 11,942 in 1902 and 31,155 in 1921. Though the Indochinese network resembled that of Algeria in length of lines and number of offices, it carried only 2 million messages compared to Algeria's 13 million, because of Algeria's much larger European population.[71]

At the other end of the spectrum were the equatorial colonies of Gabon, Congo, Ubangi-Shari, and Chad. French penetration began in Gabon in the 1840s. By 1880 France had claimed the right bank of the Congo and Ubangi rivers. The pacification of Chad was not completed until the first decade of the twentieth century. L'Afrique equatoriale française (AEF), as it was called after 1910, was almost five times the size of France, but it had few advantages of any sort: few people, little agriculture, hardly any mineral wealth, and enormous distances across swamps, mountains, rain forests, and deserts. Because its conquest cost France far less money and manpower than any other colony, the French government saw little reason to invest in it. Colonials called it the Cinderella of the French Empire.

As in Indochina, communications in AEF followed the flag, but with long delays. The West African Telegraph Company's cable reached Libreville in 1885 and cost France 75,000 francs per year.[72] While Libreville had good mail and cable service, beyond the town things got more difficult. In 1909, when French troops were stationed throughout the colony, it still took 6 or 7 months to get a response to a letter from Brazzaville to Fort-Lamy, the capital of Chad. And in 1912 the official yearbook of the colony noted that a letter from Bordeaux took 23 days to reach Brazzaville by ship and train, another 13 to Bangui by river steamer, and then 2 to 3 months from Bangui to Fort-Lamy by canoe and on foot, depending on the season.[73]

Speeding up the mails would have required railroads. Though there were plans for railroads, their realization still lay far in the future. Hence the eagerness for telegraph lines. In 1890, ten years after he had claimed the Moyen-Congo for France, Savorgnan de Brazza urged an extension of the West African cable to Loango, the point on the Atlantic nearest Brazzaville and the start of the porters' path into the interior. Rather than agree to pay the high price of a cable, the government chose to save money by putting up a 827-kilometer-long land line along the coast. Though cheaper, it took ten years to survey and build and was not opened to traffic until 1899. Even then, it was frequently down because of faulty construction and poor materials. The same was true of the Loango-Brazzaville line, a 461-kilometer-long stretch, half of it across jungle-covered mountains. Begun in 1894, it had to be abandoned because the aluminum wire shipped from France had corroded. It was begun again in 1899 with heavy iron wire, and in the end it cost half a million francs.[74] Yet it still suffered breakdowns, as an inspector complained in 1901:

> With a large budget, it was installed in a solid way that would ensure its regular functioning. However, it has just experienced, for lack of maintenance, interruptions of ten to twelve days, caused by the collapse of wooden poles which in this moist climate should be replaced every year. . . . The present slowness of transmissions is seriously prejudicial: it scares away the international clientele and exposes the Colony to damages.[75]

Another inspector, three years later, echoed the same complaint: in the rainy season, the line was down an average of 10 days a month. To remedy it would require replacing the wooden poles with iron ones, an enormous expense in a country where most transport was on the heads of porters.[76] As a result, many urgent messages were sent via Leopoldville and the Congo Free State, an effective if somewhat humiliating solution.[77]

North of Brazzaville, telegraph construction got even more difficult. Large areas were almost constantly flooded, distances were enormous, and the population among the sparsest in Africa. The administration avoided the worst stretch, between Brazzaville and Liranga at the confluence of the Congo and Ubangi rivers, by a simple expedient: it used the Belgian line across the river, which was crossed at Brazzaville and at Liranga by an optical telegraph or by clerks paddling a canoe. The Belgians showed little enthusiasm for this service and delayed the French telegrams to let their own pass through first.

By 1912 AEF had 2,112 kilometers of telegraph lines but only 32 telegraph offices; half of them were in Gabon, and none at all in the largest territory, Chad.[78] Chad was partly inhabited by warlike desert people who kept numbers of French troops occupied. Though the army cried out for a telegraph, the isolation of Chad made it one of the last places in Africa to be connected to the outside world. Only in 1910 was money appropriated to build a land line from Fort-Crampel, in Ubangi-Shari, north to Fort-Lamy, the capital of Chad. Seven-meter-high poles had to be carried by porters or oxen through a region that had neither suitable trees nor roads. Wire and other materials had to come from France via Nigeria by riverboat, canoe, and porters' heads. A telegraph mission, led by Captain Lancrenon, surveyed the route and cut a 20 meter-wide path through the bush over a distance of 860 kilometers. Lancrenon's team had to train local farmers to become carpenters, bricklayers, and blacksmiths. At last, in October 1912, two years after he had started, Lancrenon connected Fort-Lamy to the outside world.

While in Chad, Lancrenon's team was supposed to set up two other telegraph lines. The distances, difficult terrain, and exorbitant transportation costs brought forth some ingenious technical solutions. One line was to go from Miltou, on the Bangui–Fort-Lamy trunk line, north to the small outpost of Ati, on the edge of the desert. So little traffic was expected on this line that even a single wire hung on poles would have been too costly. Instead, Lancrenon set up ten heliographs, optical telegraphs using mirrors which could flash signals between posts 40 or 50 kilometers apart, but only on clear sunny days.[79]

North of Fort-Lamy another line was to go to N'Guigmi, the easternmost military outpost in French West Africa, in what is now Niger. There it was to connect to the telegraph to Dakar and the cable to France. In 1912, however, just as Lancrenon was surveying the path for the line, he was ordered to stop and set up a new device in its place: the wireless.[80]

Colonial Wireless Networks to 1918

In the interwar years a technology came to maturity which had been a dream at the time the colonial empires were established: the wireless. Not only did it revolutionize communications, it also changed the patterns of power which had supported the new imperialism.

In the late nineteenth century, radio waves, until then a scientific phenomenon, were turned to practical use. In 1895 Guglielmo Marconi succeeded in transmitting over a distance of 2 to 3 kilometers. A year later, he moved to Britain and took out his first patent. Technical successes quickly followed: in 1899 he transmitted across the Channel, and in 1901 across the Atlantic from Cornwall to Newfoundland.

In the first decade of this century, wireless was mainly used for ship-to-shore communications. Elsewhere, the telegraph was considered more secure, reliable, and economical. At the time, wireless still had many deficiencies which hindered its application to long-distance communications. One was its high cost. The accepted wisdom of the time was that longer distances required longer waves, in a 500-to-1 ratio; hence to transmit over a distance of 5,000 kilometers required 10,000-meter waves. To produce such long waves of sufficient power required enormous transmitters which consumed over 100 kilowatts of power and needed a whole field of antennas supported on 100-meter-high masts. The cost of such a transmitter was estimated to be around £60,000, of which over half went for the antenna. Even then, wireless was subject to natural static and frequent breakdowns. Hence the cautious attitude of the business community and the British government toward Marconi's bolder plans.[81]

Yet long-distance wireless telegraphy could not be ignored, for two reasons: because technological advance had a momentum of its own and, more significantly, because inventors in other countries were eager to improve the wireless and challenge Britain's predominance in communications.

As early as 1906 Marconi had offered to build a network of wireless stations 1,600 kilometers apart throughout the British Empire, but the Colonial Office rejected his suggestion as too radical. The other governments of the empire were more interested. In 1911 Marconi's Wireless Company already owned and operated stations in Ireland and Canada. At the Imperial Conference that year Sir Joseph Ward, premier of New Zealand, proposed an "Imperial Wireless Chain" stretching from Britain to New Zealand. The conference passed a motion which read: "That the great importance of wireless telegraphy for social, commercial, and defensive purposes renders it desirable that a chain of British State–owned wireless stations be established within the Empire."[82] By now, the pressure of competition was being sharply felt: "In view of the fact that various foreign countries were already commencing to erect long-distance installations and that it was therefore incumbent on the Imperial Government to act with promptitude, it was considered essential that the system be established at the earliest possible moment."[83] In 1912 Marconi's Wireless Company and the Post Office agreed to set up a chain of stations from Britain to South Africa and to Singapore; Australia and New Zealand were to set up their own. Only two stations were built before the outbreak of war put a halt to the scheme: one at Leafield near Oxford, and the other, hastily completed in late 1914, at Abu Zabal near Cairo.[84]

As the British were well aware, other colonial powers perceived the wireless not just as a technical alternative to cables, but as a national alternative to British cables.[85] The German company Telefunken had built the world's most powerful station at Nauen near Berlin. However, the only German colony it communicated with was Togo. Smaller stations with a more limited reach were being built in China, Cameroon, Southwest Africa, German East Africa, New Guinea, and the Pacific islands of Yap, Nauru, and Samoa.[86] In Belgium, King Albert asked Brussels University professor Robert Goldschmidt to set up stations in Belgium and the Belgian Congo. With French help, Goldschmidt had constructed the first station of the chain, at Laeken near Brussels, when the war broke out. Similarly, Italy had a station in Eritrea, and Portugal had one in Mozambique and one in Angola, built with Marconi equipment.[87]

France, with its large empire and poor cable communications, was especially interested in radiotelegraphy. Before World War I there was no systematic state-supported research program, nor were there any organic links between universities and industry. Wireless research was the domain of private enthusiasts like the scientific instrument maker Eugène Ducrétêt and army captain Gustave Ferrié. Soon, however, the navy became interested, and in 1904 it began equipping all its ships with wireless sets.[88]

At that point, experiments began in the colonies. The tropics were a particularly hostile environment, for electrical storms produced static which interfered with wireless reception. In 1901 postal and telegraph inspector Magne, who had built the Loango-Brazzaville telegraph line, tried transmitting across the estuary of the Gabon River. After innumerable difficulties, he was able to register a modest success, communicating across 35 kilometers of water. A year later, the eruption of Mount Pelée in Martinique broke the cable linking that island to Guadeloupe and the rest of the world. Magne was sent to Guadeloupe and Captain Ferrié to Martinique, where they were able to reestablish communications.[89]

In 1904 Captain Péri, head of the army telegraph service of Indochina, built three tiny 1-kilowatt transmitters and a year later succeeded in putting transmitters on carts. These were designed for frontier outposts, where Chinese "bandits" were prone to cut telegraph wires.[90] Two years after that, Captain Ferrié was able to communicate by wireless between Paris and the French forces in Morocco.[91] The experimental stage was now drawing to a close, and France was preparing to create a more permanent wireless network in its colonies.

Until 1910, French wireless equipment was handcrafted for each application. Radio waves were produced by sparks from a condenser recharged by ordinary alternating current at 50 or 60 cycles per second; the resulting waves were easily confused with natural static. In 1910 Emile Girardeau founded the Société française radio-électrique, or SFR, to compete with Marconi's company and Telefunken. His plan was to create a continuous spark which would emit waves at 437 cycles or more, the so-called "musical frequency" waves.

His first order came from the French Congo. In 1909 Martial Merlin, governor of the colony, ordered wireless equipment from SFR to link Pointe-Noire to Brazzaville, where the telegraph had proved so erratic.[92] The two stations, built in late 1911, transmitted 36,423 words in four weeks, just under the promised 10,000 words a week. Not only did they duplicate the land telegraph lines between the two towns, they also allowed the colony of the Congo to communicate with the Belgian Congo, Cameroon, and ships at sea. Wrote inspector Tixier: "The establishment of the Brazzaville and

Pointe-Noire stations has largely fulfilled our expectations; it demonstrated that one could obtain from the wireless in the equatorial zone, despite intense electrical phenomena and despite the forest, a service similar to what it provides in temperate countries, though only by largely increasing the energy involved."[93]

After that, orders poured in from all the colonies for wireless stations to communicate with ships, islands, and isolated outposts like Fort-Lamy. In the years 1910–14 several dozen small stations were erected in the French colonies on the orders of the ministries of Colonies, Navy, and War, or of the colonial governments themselves.[94]

Yet a smattering of small stations does not constitute an imperial wireless chain. In 1911–12 France, like Britain, decided to create such a chain. The initiative came from Colonial Minister Adolphe Messimy and from Albert Sarraut, governor-general of Indochina. Messimy's plan called for ten powerful stations 4,000 to 5,000 kilometers apart: Paris, Timbuctoo, Dakar, Panama, Fort-Lamy, Djibouti, Madagascar, and Indochina. Others would fill in the network later. He estimated their cost at a million francs (£40,000) apiece, the price of 200 kilometers of cable. His motives were the characteristically French ones of culture and prestige: "[There is] no place on earth, except the northern Pacific, where we cannot create a French center for electrical transmission, and thus for the diffusion of our thought as well as our economic activity. Will we neglect this opportunity to be important in the world?"[95]

Yet Messimy was not in office long enough to see his plan implemented. When Albert Sarraut left France in 1911 to become governor-general of Indochina, he decided to build a wireless station in Saigon powerful enough to communicate directly with France, and a medium-size one at Hanoi to serve the colony and neighboring countries.[96] He had to agree that Indochina would bear the entire cost, with no help from France. After some delays caused by a power struggle between the Ministry of Colonies and the Ministry of Posts and Telegraphs, both of which aspired to control all wireless transmissions in the French Empire, the equipment was ordered in October 1913 and finally completed nine months later. In July 1914, just as it was being loaded on board a ship in Marseille, war broke out, and it was unloaded again. Thus the French Empire, like the British, almost got a wireless chain before World War I, but not quite.[97]

Within hours of the outbreak of war in 1914, British cable ships cut the German cable between Emden and New York and landed the ends in Cornwall and Nova Scotia. In their colonies, the Germans destroyed their wireless stations to prevent them from falling into Allied hands.[98] Meanwhile, the Belgians destroyed their new wireless station at Laeken just before the German army arrived, and the Central Powers seized the British land lines to India via Russia and the Ottoman Empire.

During the war, the French and British imperial communications escaped damage but suffered from neglect. The powerful transmitter Sarraut had ordered for Saigon was hastily assembled at Lyon-la-Doua and served to communicate with Russia.[99] All communications with India, Indochina, and beyond now went through the Eastern Telegraph Companies' cables, and regular telegrams suffered up to three days' delay. Both the Royal Navy and the French War Ministry set up small wireless stations throughout their empires, but only to serve military needs.[100]

British Wireless after 1918

If the war destroyed or hampered colonial communications networks, it also gave a great push to the new technology of radio-telegraphy. Schemes for global wireless networks which had seemed very daring in 1912 appeared not only sensible but necessary after 1918. Technological advances brought within reach what had seemed impossible before the war: direct communications over distances of 10,000 kilometers or more. Two innovations were especially important. One was the high-frequency alternator developed during the war and first produced commercially by General Electric in the United States and by SFR in France in 1918. Soon thereafter, in 1919–20, came the thermionic valves, developed by de Forrest and Fleming and manufactured by Marconi. These devices were more energy-efficient and produced purer radio waves than the older spark transmitters they displaced.

These innovations stimulated the rivalries that pitted Britain, until then the ruler of world telecommunications, against France and the United States.[101] After World War I, Britain faced a difficult commercial and political dilemma, for it could afford neither to fall behind in radiotelegraphy, nor to jeopardize its cable system. The result was seven years of procrastination and compromise.

In 1919 Marconi's Wireless Company offered to build and operate four high-powered transmitters in Britain, India, South Africa, and Australia. An Imperial Wireless Telegraphy Committee rejected the idea of privately owned imperial stations and proposed instead that the Post Office operate a chain of smaller stations located at 3,600-kilometer intervals. Australia and South Africa objected, and the Imperial Conference of 1921 damned the proposal with faint praise, for it would have required too many retransmissions. The impasse lasted a couple of years, while Britain's rivals moved ahead. By 1923, the United States and France each had over 3,000 kilowatts of transmission power, while the British Empire only had 700, barely ahead of Germany.[102] That year, Marconi obtained permission from the South African and Australian governments to build stations for them, while the British Post Office was to build its own station at Rugby. India and Africa would come later.[103]

The confusion of the years 1920–26 was as much technical as commercial. As a result of the researches of de Forrest, Fleming, Marconi, and others, the arc transmitter, with its open continuous spark, was replaced by thermionic valve transmitters. While these were more reliable and their transmissions were easier to receive, they still operated in the longwave band, where the range was proportional to the size, power, and cost of the equipment. Even a small 2-kilowatt station, with a range of a few hundred kilometers, weighed 12 to 15 tons and took months to set up. At the other extreme was a giant station like Rugby. Operating on the 20,000-meter band, it was designed to be heard in Canada, India, South Africa, and Australia. To achieve these ranges, it required 1,800 kilowatts of power, two large buildings, and an antenna over 3 kilometers long resting on sixteen 250-meter-high masts. The entire station cost around half a million pounds. Small wonder some experts considered the wireless to be as costly as a duplicate cable system, and far less secure or reliable.[104]

Just as Britain was finally getting started on its Imperial Wireless Chain, along came an innovation which threw the whole field of international telecommunications into another upheaval: shortwave.

Off and on, Marconi and other radio pioneers had experimented with waves of less than 200 meters. In 1918 the Italian navy had asked Marconi to work on a wireless system for ships that would not be received beyond the horizon but the results were mediocre. Shortwaves had a way of "disappearing" after a few kilometers, only to "reappear" thousands of kilometers away, in an unpredictable fashion. In fact, so erratic were the results that the British Post Office allowed amateurs to experiment with the "useless" bandwidth.

Two postwar advances brought shortwave into the "useful" realm. One was the thermionic valve, which allowed a precise tuning of the wavelength; the other was the directional antenna developed by Charles S. Franklin, an engineer at Marconi's. Incredible distances could now be achieved with very little electrical energy. In 1923 Marconi was able to send messages to the Cape Verde Islands, 4,130 kilometers away, with a 1-kilowatt transmitter. A year later, regular transmissions from a 12-kilowatt station were heard in Australia, Canada, the United States, and South Africa.[105]

The emergence of this new technology broke the logjam of confusion and hesitation. In July 1924 the British government decided to build a shortwave station for commercial communications within the empire and to allow private companies to compete in international communications. Since shortwave only worked at certain hours of the night, however, and only in one direction at a time, the huge and costly Rugby station still served a military need, for it alone could transmit to the entire Royal Navy at once.

Over the next three years shortwave stations were built throughout the British Empire. Commercial service to Canada began in October 1926, followed a few months later by service to Australia, South Africa, and India.[106] The greatest virtue of shortwave was its low cost. Whereas an intercontinental longwave station cost around £500,000, a shortwave station with a similar range could be put up for £40,000. It was also capable of transmitting 100 to 250 words per minute, versus 35 to 100 words per minute for cables. Hence it was cheaper to operate; a radiotelegram to India, for example, cost 25 percent less than a cable.[107] Suddenly the world cable network was in jeopardy, and with it, one of the pillars of British commercial and diplomatic power.

Until the advent of shortwave, the rivalry between cables and wireless had been "a ding-dong struggle, with technical improvements on one side or the other giving temporary advantages, but not for long."[108] Cable technology had in fact undergone some remarkable improvements since the advent of the radio.[109] In the 1880s, British physicist Oliver Heaviside showed that it was theoretically possible to compensate for the distorting effects of capacitance by adding inductance along the cable. Loaded cables wrapped in highly magnetic metals could handle ten times as many words per minute as unloaded cables. Other improvements included tapered loading to allow messages in both directions at the same time; regenerators, which automatically sharpened signals; and various substitutes for gutta-percha, which reduced capacitance still further. By the late twenties, cables could handle five channels in each direction simultaneously, and a total of 500 words per minute.[110] But only new cables could do so, while an enormous investment in old cables lay on the bottom of the ocean, becoming yearly more obsolete. In the early twenties, while the wireless was still very costly and demand for telecommunications was booming, the cable companies did well, meeting the increased demand by laying new cables. By 1927 the world's cable network had reached 650,000 kilometers, of which 250,000 belonged to the Eastern and Associated Companies.[111]

But if the cable network was getting longer, its share of world telecommunications traffic was shrinking. By 1927 it had lost half its business to the new shortwave beam stations. The British government reacted swiftly. In 1928 it called an Imperial Wireless and Cable Conference attended by representatives from Britain, Canada, Australia, New Zealand, South Africa, the Irish Free State, and India, and one from the "Colonies and Protectorates." The conference considered the following dilemma:

> The Beam services could always afford to undercut the cable rates, and if competition were unrestricted could render the cable systems unremunerative. Having regard to these considerations it was suggested to us that those responsible for the Cable Companies might be pressed, unless a satisfactory means of obviating the effect of acute competition could be provided, to liquidate their undertakings at once and distribute their large reserves among the shareholders, rather than to remain in operation and dissipate their resources.

This was no mere commercial problem, but a crisis of national, even imperial, proportions:

> In this connection information has been laid before us which points to an attempt on the part of certain foreign interests to secure an increased share in the control and operation of world communications. . . . It is obvious that, if the Eastern Telegraph and Associated Companies went into voluntary liquidation and wished to dispose of their assets, the opportunities presented to foreign interests to strengthen their position would be considerable.[112]

In other words, Eastern blackmailed the British Empire.

After reviewing the alternatives, the conference recommended the merger of all empire cable and wireless communications systems. This was not the idea of Marconi, whose aversion to the cable companies was said to be "akin to that attributed to the devil for holy water."[113] But more was at stake than the inventor's feelings. On March 14, 1928, J. Dennison Pender, chairman of Eastern, and Lord Inverforth, chairman of Marconi's Wireless Company, wrote to the conference announcing their merger. The conference recommended that a new company be created to regroup all the international communications of the empire. Under government supervision, it would serve the interests of Britain and the empire, not just those of the two major companies. What faith there had once been in free enterprise was replaced by a belief in semiprivate monopolies large enough to do battle with other nations' monopolistic firms. The French view of the world market had triumphed in Britain as well.

Parliament accepted the conference proposals and in 1929 it created the Imperial and International Communications, Ltd. This firm acquired all the communications assets of the Eastern and Associated Companies, other British private cables in the Atlantic and the West Indies, the government's Atlantic cables, those of the Pacific Cable Board, and the wireless stations of the Post Office and Marconi. In 1931 it also obtained the Persian Gulf cables from the now-defunct Indo-European Telegraph Department. Three years later it changed its name to Cable and Wireless, Ltd.[114]

French Wireless after 1918

In the postwar international wireless race, the French had less capital but a stronger motivation than the British, for they owned only 5 percent of the world's cables, through which passed only 2.2 percent of the world's cable traffic. To France, the wireless offered the means to bypass the British cable network. Yet within France this suggestion caused much wrangling. The Ministry of Posts, Telegraphs, and Telephones (PTT) claimed a monopoly over all French communications but opposed the new systems; as one high postal official told Emile Girardeau, "What! you believe in that wireless?"[115] On the other side were the military (especially Colonel Ferrié), private industry (the Société française radio-électrique), the Ministry of Colonies, and their parliamentary friends. In July 1919 the government decreed that France would build a network of large "intercolonial" stations, while the colonies themselves would pay for smaller ones.[116]

The public-private issues was thornier. Right after the war, two French companies had the potential to enter the long-distance wireless field. One was Girardeau's SFR, which had the most advanced technology but was short of funds. The other was the Compagnie générale de télégraphie sans fil—CSF for short—which belonged jointly to the Compagnie française des câbles télégraphiques, Marconi's Wireless, and a group of French banks; this company was well capitalized, but had no manufacturing capacity in France. A year after the war ended the two firms merged under the name CSF, with Girardeau as its head. France now had a company capable of taking on the international competition. In September 1919 CSF, Marconi's Wireless, and RCA agreed to divide the world among them, with RCA and CFS sharing the exclusive right to communicate between the United States and France.

Thus armed, CSF could now confront the French government. In 1920, after some intense parliamentary debates, the PTT was granted the right to communicate with the colonies, while CSF obtained the privilege of international telecommunications. In 1921 CSF launched affiliates to handle its communications business. Radio-France built the world's most powerful transmitter at Sainte-Assise near Paris, while Radio-Orient put up another in newly conquered Beirut. Soon after, French banks bought out Marconi's share of CSF, and the government tightened its control over it, making it a quasi-public company.[117]

Since 1911 there had been plans to equip Indochina with a wireless station strong enough to communicate directly with France. Despite much talk, these plans were further from realization in 1920 than they had been in 1914. Governor Maurice Long, tired of the procrastinations of the French government, contracted with CSF to build and operate a station in Indochina. To communicate with France over 10,500 kilometers of land mass, the Centre radio-électrique de Saigon had to have 1,250 kilowatts of power and an antenna covering 72 hectares. It began receiving in August 1922 and transmitting in January 1924. In its first year it captured half the cable traffic between France and Indochina. The colony, then in the midst of an economic boom, also equipped itself with several lesser stations: two 25-kilowatt transmitters in Saigon and Hanoi, and fourteen smaller ones in outposts along the Chinese border and in medium-sized towns. By 1927, Indochina was well provided with wireless as well as wired telegraphy.[118]

Soon after Indochina and CSF set up the first links in the French imperial wireless chain, the PTT started its own network. In 1920 it built a transmitter at Bordeaux and receivers in several colonial cities. In the early 1920s powerful stations were erected in

Tananarive, Brazzaville, Djibouti, and Bamako. To these intercolonial stations, the various colonies added their own smaller stations. By 1926 the French colonies had sixty-seven longwave stations; ironically, Algeria was among the last to get one, because it already had excellent cable service.[119]

In 1925 Major Metz of the French army engineers predicted: "France will have in 1926 or 1927 at the latest, direct communications with its main colonies and communications involving only one retransmission with all its secondary colonies. It will be at the head of the European nations in establishing its imperial network."[120] While shortwave presented Britain with a commercial crisis of the first magnitude, for France it was a political opportunity. The first shortwave station was built at Sainte-Assise in 1925 and began transmitting to Indochina in January 1926; transmissions from Indochina started in late 1927. As a result, traffic jumped from 4,484 words a day in 1924 to 11,007 in 1928.

The impact of shortwave was felt even more strongly in Africa, for France's African colonies were poor and underpopulated and had only a very few longwave stations in the mid-1920s. Shortwave stations of 2.5 kilowatts able to communicate with France were set up in Dakar, Cotonou, Bamako, and Brazzaville in 1928, and numerous small 10-watt stations with a range of 2,000 to 3,500 kilometers were placed in all the other towns in French Africa and on all French islands around the world. These transmitters cost one-thirtieth as much to build, and one-tenth as much to maintain and operate, as longwave stations of the same range. Even portable 8-watt stations, with a kite for an antenna and an African turning a crank as a source of power, could reach 500 kilometers in daylight and 1,200 at night. Such equipment brought telecommunications to every small town and isolated outpost in the French Empire.[121]

Because it was so much cheaper than longwave, shortwave made possible new forms of communication. One was the intercontinental radiotelephone, inaugurated in 1928 with a conversation between Albert Sarraut, then minister of colonies, and Pierre Pasquier, governor of Indochina. Another was broadcasting to the colonies, begun in May 1931 by Radio-Coloniale, a year and a half before the start of the BBC's Empire Service.[122]

With the wireless, and especially with shortwave, France had finally overcome a handicap dating back to the nineteenth century: its dependence on British cables. André Touzet, professor of law at the University of Hanoi, called the new wireless communications "the end of a servitude, one could in a certain measure say the end of an isolation."[123]

Notes

1. Letter to George Couper (December 9, 1854) quoted in Mel Gorman, "Sir William O'Shaughnessy, Lord Dalhousie, and the Establishment of the Telegraph System in India," *Technology and Culture* 12 (1971): 597.

2. J. Henniker Heaton, "The Postal and Telegraphic Communications of the Empire," *Proceedings of the Royal Colonial Institute* 19 (1887–88): 172.

3. The *Palaquium* tree is also known as *Isonandra* and as *Dichopsis*. Chemically, gutta-percha is trans-polyisoprene-1,4; though still used to cover golf balls, in its electrical uses it has given way to a synthetic substitute, polyethylene. The most thorough study of gutta-percha is Dr. Eugen Friedrich August Obach's *Cantor Lectures on Gutta-Percha* (London, 1898), delivered to the Society for the Encouragement of Arts, Manufactures, and Commerce, London, November 29, 1897.

4. On the beginnings of submarine telegraphy, see Willoughby Smith, *The Rise and Extension of Submarine Telegraphy* (London, 1891); Bernard S. Finn, *Submarine Telegraphy: The Grand Victorian Technology* (London, 1973), p. 17; Jeffry Kieve, *The Electric Telegraph: A Social and Economic History* (Newton Abbott, England, 1973), pp. 101–4; and Catherine Bertho, *Télégraphes et téléphones: de Valmy au microprocesseur* (Paris, 1981), p. 118.

5. Bern Dibner, *The Atlantic Cable* (Norwalk, Conn., 1959), pp. 7–32; Finn, pp. 17–23.

6. Maxime de Margerie, *Le réseau anglais de câbles sous-marins* (Paris, n.d. [1909?]), p. 100.

7. Alfred Gay, *Les cables sous-marins,* 2 vols. (Paris, 1902–1903), 2: 157–59; Halford Lancaster Hoskins, *British Routes to India* (London, 1928), pp. 374–78.

8. "Indo-European Telegraph Department, 1865–1931," pp. i–iii, in India Office Records (London), L/PWD/7. Henry Archibald Mallock, *Report on the Indo-European Telegraph Department, being a History of the Department from 1863 to 1888 and a Description of the Country through which the Line Passes,* 2nd ed. (Calcutta, 1890), p. 5; Christina Phelps Harris, "The Persian Gulf Submarine Telegraph of 1864," *Geographical Journal* 135, pt. 2 (June 1969): 170–71. See also Hoskins, pp. 376–80.

9. J. C. Parkinson, *The Ocean Telegraph to India: A Narrative and a Diary* (Edinburgh, 1870), pp. 280–91. See also Hoskins, pp. 382–88.

10. Mallock, pp. 6–9; "Indo-European Telegraph Department," p. ii; Hoskins, pp. 396–97.

11. Parkinson's *Ocean Telegraph to India* narrates the laying of the Bombay-Aden-Suez cable by the *Great Eastern.*

12. Frank James Brown, *The Cable and Wireless Communications of the World; a Survey of Present-Day means of International Communication by Cable and Wireless, Containing Chapters on Cable and Wireless Finance* (London, 1927), pp. 25–30; Thomas Lenschau, *Das Weltkabelnetz* (Halle,

1903), pp. 28–29; John Dick Scott, *Siemens Brothers, 1858–1958: An Essay in the History of Industry* (London, 1958), p. 124.

13. Brown, pp. 37–38; Gerald Reginald Mansel Garratt, *One Hundred Years of Submarine Cables* (London, 1950), p. 40.

14. H. Casevitz, "La télégraphie sous-marine en France," *Société des ingénieurs civils de France. Mémoires et comptes-rendus des travaux* 53, no. 7 (August 1900): 365–73; P. M. Kennedy, "Imperial Cable Communications and Strategy, 1870–1914," *English Historical Review* 86 (1971): 740.

15. Charles Bright, "Imperial Telegraphs," pt. 1, *United Empire* 2 (August 1911): 541–52 and pt. 2, ibid. (September 1911): 622–33; Brown, pp. 51–58, 75–76; Garratt, pp. 35–36; Finn, pp. 33–35.

16. de Margerie, p. 79.

17. Brown, pp. 51–55; Garratt, pp. 31–36 and 53; Scott, p. 124.

18. Brown, p. 55; Hoskins, p. 395.

19. "Indo-European Telegraph Department," pp. i–iv; Harris, p. 191; Mallock, pp. 3–11; de Margerie, pp. 26–27.

20. The two official histories of the Eastern and Associated Companies are K. C. Baglehole, *A Century of Service: A Brief History of Cable and Wireless Ltd., 1868–1968* (Welwyn Garden City and London, 1969), pp. 1–18, and Hugh Barty-King, *Girdle Round the Earth: The Story of Cable and Wireless and its Predecessors to Mark the Group's Jubilee, 1929–1979* (London, 1979), pp. 3–203. For complete lists of the world's cables, see Union télégraphique internationale, *Nomenclature des câbles formant le réseau sous-marin du globe* (Bern, 1877, 1903, 1910, and 1934).

21. Leo Weinthal, ed., *The Story of the Cape to Cairo Railway and River Route from 1887 to 1922*, 5 vols. (London, 1923–26), 1: 211–17; Lois Alward Raphael, *The Cape-to-Cairo Dream, a Study in British Imperialism* (New York, 1936), pp. 49–65; Kennedy, p. 741; de Margerie, pp. 82–86.

22. Charles Lesage, *La rivalité anglo-germanique. Les câbles sous-marins allemands* (Paris, 1915), pp. 110–12; de Margerie, pp. 82–87; Kennedy, pp. 737–41.

23. Robert V. Kubicek, *The Administration of Imperialism: Joseph Chamberlain at the Colonial Office* (Durham, N.C., 1969), pp. 30–33, 97, 109. For a similar opinion, see Kennedy, p. 751, and C. M. Woodhouse, "The Missing Telegrams and the Jameson Raid," *History Today* 12 (June 1962): 395–404 and (July 1962): 506–14.

24. Percy A. Hurd, "Our Telegraphic Isolation," *Contemporary Review* 69 (1896): 899–908.

25. Kennedy, pp. 729–38; de Margerie, p. 89.

26. Kennedy, pp. 738–39.

27. Heaton, "Postal and Telegraphic Communications," p. 187.

28. George Johnson, *The All-Red Line. The Annals and Aims of the Pacific Cable Project* (Ottawa, 1903).

29. Bright, "Imperial Telegraphs" and "Inter-Imperial Telegraphy," *Quarterly Review* 220 (1914): 134–51.

30. Gay, 2:161; Bertho, pp. 118–24; de Margerie, pp. 11–12, 18, 27; Léon Jacob, *Les intérêts français et les relations télégraphiques internationales* (Paris, 1912), p. 10.

31. de Margerie, pp. 22–23, 180; Jacob, pp. 10–12, 20; Bertho, p. 120;

Lesage, pp. 6, 47–48; Harry Alis (pseudonym for Jules-Hyppolite Percher), article in *Journal des débats,* July 16, 1894 and "Les câbles sous-marins" in *Nos africains* (Paris, 1894), pp. 535–50; Charles Cazalet, "Les câbles sous-marins nationaux," *Revue économique de Bordeaux* 12, no. 71 (March 1900), pp. 47–48; Philippe Bata, "Le réseau de câbles télégraphiques sous-marins français des origines à 1914" (Mémoire de maîtrise, Université Paris–I Sorbonne, 1981), pp. 117–18. On French cable factories see Cazalet, p. 47; Camille Guy, *Les colonies françaises* (Paris, 1900) 3: 562; and Bertho, pp. 116–17, 124.

32. Archives Nationales Section Outre-Mer, Paris (henceforth ANSOM), Colonies Série Moderne 313 Indochine W31 (1) and (4). The maintenance contract was needed because France could not afford to keep a cable ship in the Far East for such a short cable.

33. ANSOM, Affaires Politiques 2554 (10): "Création du câble trans-pacifique 1892–94"; Jacques Haussmann, "La question des câbles," *Revue de Paris* 7, no. 6 (March 15, 1900): 256–58; Cazalet, p. 47; Lesage, p. 226; Alis, "Les câbles sous-marins" p. 547.

34. J. Depelley, *Les câbles sous-marins et la défense de nos colonies. Conférence faite sous le patronage de l'Union coloniale française* (Paris, 1896), p. 29 and map.

35. *Journal officiel de l'Empire français,* September 17, 1869, p. 1; ANSOM, Colonies Série Moderne 313 Indochine W30 (1).

36. Alis in *Journal des débats,* July 23, 1894, p. 1, and "Les câbles sous-marins," p. 545; Cazalet, pp. 48–49; Bertho, p. 121; Depelley, *Les câbles sous-marins,* p. 30.

37. Lesage, pp. 94–97; Haussman, pp. 261–69; Depelley, *Les câbles sous-marins,* pp. 30–31.

38. Letter dated November 6, 1882 in Maxime Hélène (pseudonym of Maxime Vuillaume), *Les travaux publics au XIXe siècle. Les nouvelles routes du globe* (Paris, 1883), p. 7.

39. Kennedy, p. 748; Lenschau, pp. 45–46; Jacob, pp. 21–22.

40. Haussmann, pp. 274–76.

41. Cazalet, p. 50; Kennedy, p. 748.

42. de Margerie, p. 37.

43. Cazalet, p. 43; de Margerie, pp. 37–38; Haussman, 252–53; Charles Lemire, *Le défense nationale. La France et les câbles sous-marins avec nos possessions et les pays étrangers* (Paris, 1900), p. 8; R. Hennig, "Die deutsche Seekabelpolitik zur Befreiung vom englischen Weltmonopol," *Meereskunde* 6, no. 4 (1912): 7–8.

44. The legal aspects of cables are treated extensively in Pierre Jouhannaud, *Les câbles sous-marins français, leur protection en temps de paix et en temps de guerre* (Paris, 1904).

45. Cazalet, p. 45; Jacob, pp. 25–26; J. Depelley, "Les câbles télégraphiques en temps de guerre," *Revue des deux mondes* (January 1, 1900): 185–91.

46. Kennedy, p. 728; Haussmann, pp. 251–53; de Margerie, pp. 37–38.

47. Lenschau, p. 14; Jacob, pp. 23–24; Depelley, "Les câbles télégraphiques," p. 184; Lemire, p. 7.

48. Harry Alis in *Journal des débats,* July 8 and 14, 1894.

49. ANSOM, Affaires Politiques 2554 (4): "Organisation des liaisons télégraphiques des colonies 1890/97."

50. Henri Bousquet, *La question des câbles sous-marins en France* (Paris, 1895); Depelley, *Les câbles sous-marins*.

51. "Rapport du 1er mai 1900 au Président de la République française," *Journal officiel* (May 12, 1900): 2984–3013.

52. Depelley, "Les câbles télégraphiques," pp. 181–96; Haussmann, "La question des câbles"; Cazalet, pp. 41–51; Casevitz, pp. 365–73.

53. See for example Lemire, *La défense nationale*. Lemire was vice-president of the colonial branch of the Conseillers du commerce exterieur and of the Syndicat de la presse coloniale and member of the Société française de colonisation, in other words an important colonial lobbyist.

54. "Rapport du 1er mai 1900," p. 2996.

55. Haussmann, p. 270; Lemire, p. 47.

56. According to Philippe Bata, the French government paid 6 to 8 million francs a year for cables and maintenance between 1900 and 1914; "Le réseau," p. 135.

57. ANSOM, Affaires Politiques 2554 (5): "Projet de ligne télégraphique reliant Puerto Plata à New York 1892"; Lesage, p. 86; de Margerie, p. 181.

58. de Margerie, pp. 38–39; Lesage, pp. 137–39, 224–26; *Les postes et télégraphes en Afrique occidentale* (Corbeil, 1907), pp. 99–100.

59. Lesage, pp. 86, 224–26; de Margerie, p. 27; France, Ministère des Colonies, *Colonies françaises. Guide postal et télégraphique* (Paris, 1911), p. 15; C. Bouérat, "Les débuts du service des Postes et Télégraphes en Côte d'Ivoire (1880–1905)," *Bulletin de la Société internationale d'histoire postale* 19–20 (1972): 70.

60. Lesage, pp. 137–48.

61. Léon Mascart, "Le câble sous-marin Tourane-Amoy," *Revue générale des sciences* 13, no. 1 (January 15, 1902): 27–35; Mascart was the captain of the cable ship *François-Arago* which laid the cable.

62. Lesage, pp. 214–15, 226; Jacob, p. 19; de Margerie, p. 30; A. Berbain, *Note sur le service postal, télégraphique et téléphonique de l'Indochine* (Hanoi-Haiphong, 1923), p. 6; Jorma Ahvenainen, *The Far Eastern Telegraphs: The History of Telegraphic Communications between the Far East, Europe, and America before the First World War* (Helsinki, 1981), pp. 172–78.

63. Archives of the Ministry of Posts and Telecommunications, Paris: F90 bis 4460: "Liaisons téléphoniques par câbles sous-marins."

64. William Brooke O'Shaughnessy, "Memorandum Relative to Experiments on the Communication of Telegraphic Signals by Induced Electricity," *Journal of the Asiatic Society of Bengal* (September 1839): 716.

65. William Brooke O'Shaughnessy, *The Electric Telegraph in British India: A Manual of Instructions for the Subordinate Officers, Artificers, and Signallers Employed in the Department* (London, 1853), p. iv; Krishnalal Jethalal Shridharani, *Story of the Indian Telegraphs; A Century of Progress* (New Delhi, 1956), pp. 4–8; Manindra Nath Das, *Studies in the Economic and Social Development of Modern India: 1848–56* (Calcutta, 1959), pp. 111–18; Gorman, pp. 584–85.

66. O'Shaughnessy, *Electric Telegraph*, pp. xi–xii.

67. Das, pp. 146–47. On this phase of Indian telegraphs see O'Shaugh-

nessy, *Electric Telegraph*, pp. v–xii; Das, pp. 111–56; Gorman, pp. 584–95; Shridharani, pp. 9–27; and Edward W. C. Sandes, *The Military Engineer in India*, 2 vols. (Chatham, 1933– 35), 2: 279–300.

68. India, Telegraph Department, *Summary of the Principal Measures Carried out in the Government Telegraph Department during the Administration of Sir John Lawrence, Bart., G.C.B., G.M.S.I., and D.C.L., Viceroy and Governor General of India, 1864–68* (Calcutta, 1869); Sir Geoffrey R. Clarke, "Post and Telegraph Work in India," *Asiatic Review* 23 (1927): 86–89; Shridharani, pp. 21–28, 38, 54–58.

69. *Exposé du développement des services postaux, télégraphiques et téléphoniques en Algérie depuis la conquête* (Algiers, 1930); France, Ministry of Posts and Telegraphs, *Carte du réseau télégraphique de l'Algérie et de la Tunisie* (1870, 1879, 1885) in the library of the Ministry of Posts and Tele- communications, Paris.

70. Berbain, p. 2.

71. On the Indochinese telegraphs, see Berbain; Lucien Cazaux, "Le service des Postes et Télégraphes en Cochinchine depuis 1861 à 1880," *Bulletin de la société des etudes indochinoises de Saïgon* (1926): 185–207; France, Conseil supérieur de l'Indochine, *Note sur la situation et le fonctionnement du service des Postes et Télégraphes en 1902* (n.p., n.d.), pp. 1–9; and France, Ministère des Colonies, *Colonies françaises. Guide postal et télégraphique* (Paris, 1911), pp. 125–53.

72. ANSOM, Gabon-Congo XII (22): Télégraphes 1888–1898; Service d'inspection des Colonies: Mission Verrier, 1893, Congo français, no. 28: Postes et télégraphes.

73. Martial Merlin, "L'oeuvre des récentes missions en Afrique équa- toriale française (Conférence)," *Bulletin de la Société de géographie commer- ciale de Paris* 36 (1914): 264; Afrique équatoriale française, *Annuaire 1912*, pp. 117–21.

74. ANSOM, Gabon-Congo XII (23): Câble de Loango à Libreville 1888–98; Gabon-Congo XII (24b): Letter from Inspector Magne, August 25, 1899.

75. ANSOM, Service d'Inspection des Colonies: Mission Bouchaut, 1901, Congo français, no. 77: Postes et télégraphes.

76. ANSOM, Service d'Inspection des Colonies, Mission Arnaud-Revel, 1903–1904, Congo français no. 38: Postes et télégraphes (Letter from Inspec- tor Arnaud to the Minister of Colonies, April 21, 1904).

77. ANSOM, Affaires Politiques 3122: Mission Arnaud-Revel, 1903–04, Rapport no. 42: Postes et télégraphes (Letter from Inspector Revel, January 27, 1904).

78. Afrique équatoriale française, *Annuaire 1912*, pp. 122–26.

79. This was probably the "Héliographe de campagne modèle 1909," a military device; see France, Ministère de la Guerre, *Ecole des Transmissions*, vol. 2: *Télégraphie et téléphonie sans fil. Télégraphie optique et signalisation optique* (Paris, 1934), pp. 583–85.

80. Merlin, pp. 264–66; P. Lancrenon, *Les travaux de la mission télé- graphique du Tchad (1910–1913)* (Paris, 1914); Colonel Largeau, "La situa- tion du territoire militaire du Tchad au début de 1912," *Afrique française. Renseignements coloniaux* (January 1913), pp. 81–82.

81. For a succinct criticism of wireless by a telegraph engineer, see Bright, "Inter-Imperial Telegraphy."

82. "Minutes of Proceedings of the Imperial Conference, 1911," *Parliamentary Papers* 1911 (Cd. 5745), vol. 54, pp. 307–15.

83. "Marconi's Wireless Company Limited. Copy of Agreement between Marconi's Wireless Telegraph Company, Limited, Commendatore Guglielmo Marconi, and the Postmaster General, with Regard to the Establishment of a Chain of Imperial Wireless Stations; together with a Copy of the Treasury Minute Thereon," *Parliamentary Papers* 1912–13 (265), vol. 44, p. 2.

84. W. J. Baker, *A History of the Marconi Company* (London, 1970), pp. 137–60.

85. "Advisory Committee on Wireless Telegraphy. Report of the Committee appointed by the Postmaster General to Consider and Report on the Merits of the Existing Systems of Long Distance Wireless Telegraphy, and in Particular as to their Capacity for Continuous Communication over the Distances Required by the Imperial Chain," *Parliamentary Papers* 1913 (Cd. 6781), vol. 33, pp. 4–5.

86. Hennig, pp. 32–33; Kennedy, pp. 748–49; Lesage, pp. 210–11; Baker, pp. 158–59.

87. Robert Goldschmidt, "Les relations télégraphiques entre la Belgique et le Congo," in *Notes sur la question des transports en Afrique précédées d'un rapport au Roi,* ed. Count R. de Briey (Paris, 1918), pp. 559–77; Emile Girardeau, *Souvenirs de longue vie* (Paris, 1968), p. 61; René Duval, *Histoire de la radio en France* (Paris, 1980), pp. 24–25; Weinthal, 3: 439; Baker, p. 137.

88. Catherine Bertho, "La recherche publique en télécommunication, 1880–1941," *Télécommunications: Revue française des télécommunications* (October 1983), p. 2; Duval, pp. 19–21; Maurice Guierre, *Les ondes et les hommes, histoire de la radio* (Paris, 1951), p. 59.

89. L. Magne, "Note sur les expériences faites au Congo français et sur une installation de poste aux Antilles (avec cartes, croquis et planches hors texte)," *Revue coloniale* n.s. 11 (March–April 1903): 501–42 and 12 (May–June 1903): 654–93.

90. Lieutenant-Colonel Cluzan, "Les télégraphistes coloniaux, pionniers des télécommunications Outre-Mer," *Tropiques* 393 (March 1957): 3–8; André Touzet, *Le réseau radiotélégraphique indochinois* (Hanoi, 1918), pp. 4–5; Indochina, Gouvernement général, *La télégraphie sans fil en Indochine* (Hanoi-Haiphong, 1921), p. 4; J. de Galembert, *Les administrateurs et les services publics indochinois,* 2nd ed. (Hanoi, 1931), pp. 516–17.

91. Guierre, pp. 55–57.

92. Girardeau, *Souvenirs,* pp. 57–61; Duval, pp. 25–26; Guierre, pp. 87–91; Société française radio-électrique, *Vingt-cinq ans de TSF* (Paris, 1935), pp. 9–10, 40.

93. ANSOM, Travaux Publics 152, no. 15: Rapport de mission Fillon, Moyen-Congo 1912–13; no. 92: Service de la Télégraphie sans Fil, Rapport de M. Tixier, Inspecteur Adjoint de la Colonie.

94. Jacob, p. 4; Société française radio-électrique, p. 12; Girardeau, *Souvenirs,* p. 72; Cluzan, pp. 4–5; Jean d'Arbaumont, *Historique des télégraphistes coloniaux* (Paris, 1955), p. 159; ANSOM, Colonies Série Moderne:

Indochine NF890: "Note sur le service du Secrétariat et du Contreseign" (January 18, 1912).

95. A. Messimy, "Le réseau mondial français de télégraphie sans fil," *Revue de Paris* 19, no. 13 (July 1, 1912): 34–44.

96. ANSOM, Colonies Série Moderne: Indochine NF890: "Note sur le réseau impérial de T.S.F." (December 27, 1911).

97. Touzet, pp. 3–6. Indochina, *La télégraphie sans fil,* pp. 3–6. Société française radio-électrique, p. 40; Girardeau, *Souvenirs,* p. 76; ANSOM, Colonies Série Moderne: Indochine NF890: "Note du Secrétariat et Service du Contreseign au Service de l'Indochine (3e Section)" (January 10, 1912).

98. Howard Robinson, *Carrying British Mails Overseas* (London, 1964), p. 227; J. Saxon Mills, *The Press and Communications of the Empire* (London, 1924), pp. 76–77; Baker, p. 160.

99. Girardeau, *Souvenirs,* p. 77; Duval, p. 25; Société française radio-électrique, p. 47.

100. Weinthal, 3:437–39; Baker, 161; Guierre, p. 108.

101. Emile Girardeau, *Comment furent créées et organisées les radio-communications internationales* (Paris, 1951), p. 2; Bertho, "La recherche publique," p. 8; Guierre, p. 124; Pascal Griset, "La naissance de Radio-France," *Télécommunications* 49 (October, 1983), p. 86.

102. Bright, "The Empire's Telegraph and Trade," *Fortnightly Review* 63 (1923): 463.

103. Baker, pp. 206–11.

104. "Report of the Wireless Telegraphy Commission, Viscount Milner, Chairman," *Parliamentary Papers* 1922 (Cmd. 1572), vol. 10, pp. 729 ff. and "Second Report" *Parliamentary Papers* 1926 (Cmd. 2781), vol. 15, pp. 969 ff.; Mills, pp. 178–89; E. V. Appleton, *Empire Communication* (London, 1933), p. 8.

105. Baker, pp. 172–73, 212–20; Appleton, pp. 7–8.

106. Baker, pp. 214, 224; Shridharani, pp. 126–27; Clarke, p. 91; "Imperial Wireless and Cable Conference, 1928 Report," *Parliamentary Papers* 1928 (Cmd. 3163), vol. 10, p. 7.

107. Shridharani, pp. 126–27; "Imperial Wireless and Cable Conference," p. 7; Brown, pp. 105–6; Baker, p. 229.

108. Baker, p. 229.

109. This phenomenon of defensive innovation is not unusual in the history of technology; for instance, the sailing ship improved more in the half century after the arrival of steamers than in any previous century.

110. Appleton, p. 5; Brown, pp. 76–82; Gerald R. M. Garratt, *One Hundred Years of Submarine Cables* (London, 1950), pp. 40–52; G. L. Lawford and L. R. Nicholson, *The Telcon Story, 1850–1950* (London, 1950), pp. 94–104; John Dick Scott, *Siemens Brothers, 1858–1958. An Essay in the History of Industry* (London, 1958), pp. 129–33, 207.

111. Brown, pp. 5, 17.

112. "Imperial Wireless and Cable Conference," pp. 9–10.

113. Baker, p. 223.

114. "Imperial Wireless and Cable Conference," pp. 5–21. "Agreements made in pursuance of the recommendations of the Imperial Wireless and Cable Conference, 1928," *Parliamentary Papers* 1928–29 (Cmd. 3338), vol.

16, pp. 531 ff.; Garratt, pp. 42–43; Brown, pp. 73–78; Baker, pp. 229–32; "Indo-European Telegraph Department," p. iv; Harris, pp. 181–82.

115. Girardeau, *Souvenirs*, p. 100.

116. "Le réseau colonial de télégraphie sans fil," *L'Afrique française* 36, no. 5 (May 1926): 274; Griset, p. 85; Duval, p. 27.

117. Société française radio-électrique, p. 11; Guierre, pp. 121–38; Griset, pp. 85–88; Duval, pp. 28–30; Girardeau, *Souvenirs*, pp. 102–5 and *Comment furent créées*, pp. 6–16.

118. L. Gallin, "Renseignements statistiques sur le développement des communications radiotélégraphiques en Indochine," *Bulletin économique de l'Indochine* 32, no. 199 (1929): 369–99; Indochina, *La télégraphie sans fil*, pp. 9–10; Galembert, pp. 518–20; Berbain, p. 7; Touzet, pp. 7–21; Indochina, Gouvernement général, *Annuaire statistique de l'Indochine* (*1937–1938*) (Hanoi, 1939); "Organisation des services de postes, télégraphes et téléphones," *Indochine. Documents officiels*, ed. M. S. Lévi, 2 vols. (Paris, 1931), 2: 60–69.

119. Documents concerning the Brazzaville and Bamako stations can be found in the archives of the Ministry of Posts and Telecommunications in Paris: F90 bis 1690: "Poste intercolonial de Bamako," and 1694–1695: "Poste intercolonial de Brazzaville." On the French colonial network in general, see also: "Le réseau colonial," pp. 274–75; Albert Sarraut, *La mise en valeur des colonies françaises* (Paris, 1923), pp. 237, 334–35; Maigret, *Les communications par T.S.F. entre la France et les colonies* (Paris, 1933), pp. 3–7; *Les postes, télégraphes, téléphones et la télégraphie sans fil en Afrique occidentale française* (Paris, 1931), pp. 13–15; and *Exposé du développement des services postaux, télégraphiques et téléphoniques en Algérie depuis la conquête* (Algiers, 1930), p. 59.

120. Quoted in "Le réseau colonial," p. 276.

121. Henri Staut, "La radiotélégraphie coloniale et les ondes courtes," and "L'application des ondes courtes à la radiotélégraphie commerciale en A.O.F.," *Bulletin du Comité d'études historiques et scientifiques de l'A.O.F.* 7 (1926): 517–53, and 9 (1928): 12–29; Gallin, pp. 369–99; Cluzan, pp. 3–8; Touzet, p. 13; Société française radio-électrique, pp. 77–81.

122. Girardeau, *Souvenirs*, p. 185; Maigret, p. 8.

123. Touzet, p. 21.

5

Cities, Sanitation, and Segregation

In tropical Asia and the Middle East, cities predated the arrival of the Europeans by several millennia. Indeed, Europe learned this form of social organization from the more ancient seats of civilization. In the nineteenth century, Europeans returned the favor by introducing into the age-old urbanizing process several new functions and technical systems which prepared the way for today's sprawling tropical megalopolises.

From the beginning, cities have served as centers of administration, religion, culture, trade, and crafts. As such, their primary relations were with their own hinterlands; they absorbed foodstuffs and other rural products, and dispensed urban goods and services. Cities have also, from earliest times, maintained relations with other cities, often over long distances. Some served as economic or political centers. The rest were nodes in long-distance networks that distributed the goods, services, power, and culture originating in more distant cities. Gradually, with many setbacks, empires and trading networks arose and the nodal functions of cities grew in importance. Preindustrial empires can all be described as networks of trade and power linking cities.

Nineteenth-century European expansion accelerated this process radically. Not only did trade follow the flag, so did economic diversification and specialization. Cities were a necessity for the new transportation and communications networks which the Europeans built. Shipping, river transport, railways, and telecommunication lines were attracted to existing cities. Europeans found some old cities—Algiers, Cairo, and Delhi, for example—suitably located as centers of imperial rule and world trade. But more often, they had to create new ones as the terminal and branching points of their networks. While a few new cities began as mining camps (Elisabethville) or resorts (Simla), most were ports of call like Aden, Cape Town, and Singapore, or transshipment points like Leopoldville for the Congo, Hong Kong for China, Calcutta and Bombay for India, and Lagos for Nigeria. Dozens of such cities radiated European power and culture into their hinterlands, redirecting the economies and cultures of the tropics outward, toward the sea and the distant West.

Thus the new technologies associated with the colonizers help explain the founding and growth of colonial cities. They also figure in their settlement and land-use patterns. Certain uses of land were directly related to the functions of the city. Harbors, docks, and warehouses occupied large areas of port cities, as did the stations, yards, and workshops of railroad centers. Capitals like New Delhi and Algiers had substantial administrative districts. Other cities had areas devoted to mills, smelters, or commercial buildings.

The modernization of colonial cities went beyond the productive areas and included residential, recreational, and shopping districts as well, which embodied the very latest concepts in urban planning. The period we are studying was the greatest era of city building the Western world had ever experienced. Old European towns burst out of their walls, while in America whole cities sprang up in the empty landscape. Social elites and urban planners agreed on the need to direct their growth and make the cities livable, even beautiful. The modernization of colonial cities was therefore a transfer to the tropics of Western aesthetics as well as technology and economics.

Municipal planning and public works in colonial cities were also motivated by the growing interest in sanitation and the need to organize a disparate population. Afro-Asian cities grew by stratified immigration. At the top were the European officials and businessmen, whose lifestyles ranged from that of the European bourgeoisie to the ostentatious splendors of Oriental despots. At the bottom were former peasants, most of them poor and uneducated. And in between arose a middle class of sorts. Some were the original inhabitants of the area, or the most assimilated of the native peoples; others were immigrants from far away who brought their talents as merchants, clerks, railway workers, and craftsmen. Thus colonial cities were multiethnic, and the ethnic stratification was compounded by economic differences. It is not surprising, then, to find these different groups living separately, for both cultural and economic reasons. Ghettos and ethnic neighborhoods are not modern inventions, but were multiplied under colonial rule.

What was new in late nineteenth- and early twentieth-century colonial cities was the reinforcement of these tendencies by scientific and technological innovations. Here, among Europeans and those few who could afford to emulate them, a growing concern for hygiene provided the motive for segregation, while new technologies of water supply, sewage disposal, and building materials provided the means. And since these technologies were costly, they were reserved for the better-off.

The imported urban technologies were almost identical to those which were then transforming the towns of the West. In Europe and America too, water supply and sewage disposal were first installed in the wealthier neighborhoods and gradually spread to the poorer ones. In the cities of the Western world, sanitation systems caught up with the growth of population; by and by even poor neighborhoods received clean water, toilets, sewers, garbage removal, and a measure of rat and mosquito control. In the cities of the tropical colonies, however, per capita incomes were much lower, and the swelling of urban populations soon outran the installation of new systems. Hence, the poor fell further behind, and colonial cities gathered increasing numbers of people with little access to those sanitary improvements which European city planners and administrators were so proud to have introduced.

In the Western world after the mid-nineteenth century, municipal health officials and sanitation engineers strove to separate the germs from the people. In the tropical cities, when the officials could not achieve this objective, they substituted another: to separate the people with germs from those without. This is not to say that segregation was a deliberate policy (though it was in some places), nor that urban technologies were a mere excuse to separate the races; but rather that the interaction between European technologies and non-Western economies reinforced segregated residential patterns, wittingly or unwittingly.[1]

Walter Elkan and Roger van Zwanenberg described this process in the formation of Kampala (Uganda), a town divided between the authority of the British administration and the Kabaka of Buganda. The British municipal authorities levied taxes; built roads, sewers, and water mains; and collected garbage. In the Buganda-administered parts of Kampala there were no taxes and no municipal services or public utilities.

A natural consequence of this division of authority was, of course, that life in the municipality was healthy but expensive, whereas in the Kibuga, which made no attempt to provide any sort of services, it was insanitary and cheap. Given the racial distribution of income, this led naturally to a racial distribution of residence. Africans came to live in slums in the Kibuga whilst Asians and Europeans settled in the more salubrious municipality. . . . One solution that was strongly advocated in the years before the First World War was racial segregation, but in Kampala it was neither statutorily enacted nor implemented as a deliberate act of policy because there was in fact no need to do so. All that was necessary was that the government should lay down high enough building standards in the municipality; the price mechanism could be relied upon to do the rest. . . . Unfortunately an elementary understanding of the economic consequences of administrative action was never regarded as a necessary qualification for colonial administrators, and the development of segregation and slum dwelling in Kampala was just one of many examples of inadvertent, undesirable consequences resulting from acts of administration that were intended for other purposes.[2]

In this chapter we will consider the introduction of two technological systems considered essential in any modern Western city—water supply and sewerage—and see how they reinforced social prejudices and economic disparities in three important colonial cities: Hong Kong, Calcutta, and Dakar.

Hong Kong Water

With hindsight, Hong Kong now seems a miraculous blend of Chinese entrepreneurship and British enlightened despotism. Its economic success is also the result of its location. As a small off-shore island, it became the intersection between China and a maritime world long dominated by Britain. Yet the same geography which brought Hong Kong such a fortune in trade also made it, from an engineering point of view, the most difficult of all major cities to supply with fresh water.

The South China coast receives considerable rainfall, over 2 meters a year. As happens throughout the monsoon belt, it comes in powerful storms during a short season; three-quarters of Hong Kong's rain falls between May and September. The storm waters wash off the rocky slopes of the island, and little trickles down into the soil. Such a problem, common to dry climates, has a familiar solution:

catchment basins, reservoirs, and pipes or aqueducts. These are, however, expensive and slow to build, and Hong Kong, unlike Rome, did not have centuries to create an adequate water-supply system. For Hong Kong's geographical dilemma was compounded by its demographic one: periodically flooded by refugees, Hong Kong never quite caught up, before World War II, with the needs of its swelling population.

China ceded Hong Kong island to Great Britain by the treaty of Nanking in 1842, at the end of the Opium War. Two towns sprang up on the island: the shipping port of Victoria, and the smaller fishing port of Aberdeen. In 1861, after the Second Opium War, Britain acquired the town of Kowloon on the mainland. Finally, in 1898, the entire Kowloon Peninsula, called New Territories, was leased for ninety-nine years.

The population of the colony grew quickly (see Table 5.1). Some of this growth—4.8 percent per year on the average—was due to the colony's role as an entrepôt harbor. But much of the increase consisted of destitute refugees from the troubles of the mainland: the T'ai P'ing Rebellion (1850–64), the Boxer Rebellion (1900–1901), the Revolution of 1911 and the civil wars that followed it, and the Japanese invasion of southern China after 1938. In the twentieth century, about half of the colony's population lived on the island, and the other half on the peninsula or on boats.[3]

In the first decade of Hong Kong, the inhabitants relied on natural springs and private wells. The first five municipal wells were not dug until 1851. They quickly proved inadequate and probably contributed to the cholera epidemic of 1857. A new governor, Sir Hercules Robinson, arrived in 1859 in the midst of a severe drought. He offered a reward for the best scheme to avert future shortages. S. B. Rawlings, a clerk in the Royal Engineers, submitted the winning plan: a 9,000-cubic-meter reservoir at Pokfulam on the slopes of Victoria Peak, and a 25-centimeter pipe to carry the water to Victoria.[4] By the time it was completed in 1864, the Pokfulam Reservoir was only sufficient for part of the island's population. Mainly the res-

Table 5.1 Population of Hong Kong, 1841–1941

1841	15,000	1901	283,975
1861	119,321	1921	625,166
1881	160,402	1941	1,639,337

ervoir served to supply ships in the harbor and the western half of
the city, where Europeans lived. As Governor Robinson wrote to the
Colonial Office, "My constant thought has been . . . how best to
keep [the Chinese] to themselves and preserve the European and
American community from the injury and inconvenience of intermix-
ture with them."[5]

Under Governor Arthur Kennedy (1872–77), an extension of
the Pokfulam Reservoir helped alleviate the water shortage. His successor, Sir John Pope Hennessy (1877–82), showed little concern for
the situation. Hennessy believed that water was needed only for the
harbor, the European community, and to put out fires; the Chinese
dry-earth sanitation system, by which scavengers collected night soil
for sale to farmers, made additional water supplies unnecessary:
"The Chinese inhabitants maintain that the attempts now and then
made by successive Surveyor-Generals and Colonial Surgeons to
force what is called 'western sanitary science' upon them, are not
based on sound principles."[6]

In Britain, meanwhile, public health had developed into a major political issue. In 1854 Dr. John Snow had discovered the connection between polluted water and cholera. The epidemic of 1865–
66 led to the Public Health Act of 1866 and the Public Health
Commission, first appointed in 1869. After some delay, these ideas
spread to the colonies. A dispute between Governor Hennessy and
the chief military officer in Hong Kong, General Donovan, prompted
the Colonial Office to send a former Royal Engineer, Osbert Chadwick, to investigate the sanitary condition of the colony. In 1882
Chadwick issued a report containing several recommendations: a
new building ordinance, improved methods of removing night soil
and garbage, new drainage and water-supply systems, and a sanitary
board.[7] Such recommendations were in line with contemporary European practice, with one exception: not even Chadwick could suggest a full-fledged sewerage system of the sort Western cities were
then building. The water-carriage method of sewage disposal, a necessary complement of the water closet, would have consumed more
water than fell on the island.

Yet the Chadwick Report provided the impetus for a major water-supply project, the Tytam Reservoir. High in the middle of the
island, this reservoir held 1.4 million cubic meters and was linked to
Victoria by a 2,400-meter-long tunnel and a 5,000-meter-long con-
duit. In 1897 it was enlarged by 340,000 cubic meters, then in 1907
by another 891,000. In 1917 the largest and last reservoir on the is-

land was completed: Tytam Tank, with a capacity of 6.45 million cubic meters. Two geographers wrote:

> Thus, in one big effort, the Island's water storage was nearly trebled. Little wonder that this was a moment of self-congratulation and a complacent view of the future. . . . Climate, in the shape of long droughts, and an ever increasing population together saw to it that this triumph was short-lived.[8]

By 1918 the reservoirs and their catchment areas covered every possible drainage and storage site, one-third of the island's surface. Nowhere on earth was there such a highly developed water-catchment system. And still it did not suffice.[9]

Meanwhile the New Territories, acquired in 1898, were being transformed from a part of rural China into a part of the Hong Kong metropolitan area. The town of Kowloon grew as fast as Victoria across the harbor, from 14,200 inhabitants in 1891 to 67,497 in 1911. The municipality sank three wells and built a reservoir, but this quickly proved inadequate; by 1911 it provided less than 7 liters per person per day. In the 1920s the colony, facing shortages on both the island and the mainland, began building large reservoirs in the New Territories and laying pipelines across the harbor to Victoria. Despite the new reservoirs, the population once again outraced the water-storage capacity. In the drought of 1929 the colony was down to 14,000 cubic meters per day from the reservoirs, and another 5,000 brought in by tankers; only 20 liters per person per day.

Once again the colony responded with a major project. The Jubilee Dam in the New Territories was, when fully completed in 1936, the highest in the British Commonwealth and held almost 13 million cubic meters, as much as all the colony's other reservoirs put together. Yet already by 1939 the population had reached a million, and a 24-hour supply could only be assured in the rainy season.[10]

Unlike cities located near rivers or over aquifers, Hong Kong has always measured its water supplies in quantity stored, not in amounts dispensed. Nonetheless it is interesting to compare Hong Kong's water consumption with that of other cities in the same period. In the late nineteenth century, British city dwellers used between 109 and 236 liters per person per day, while Londoners averaged 182. European cities likewise provided between 100 and 200 liters per person per day. Indian cities were uneven in their water consumption. In the 1890s, urban supplies ranged from 45 liters on up: Madras supplied 82 liters and Bombay 182. Calcuttans consumed over 300 liters of water a day in 1913, but only 238 liters in 1931; they also got water from tanks, wells, and the River Hooghly. American city dwellers were the world champion water consumers, using twice as much as Londoners. One expert asserted that New Yorkers consumed 1,136 liters per day, almost as much as the ancient Romans, who needed 1,257 liters.[11]

In contrast, the Hong Kong water allowance in 1885 was 18 liters per day. By 1939, wrote the geographer S. G. Davis, "the daily per capita consumption for Hong Kong was estimated at 16½ gallons [75 liters] . . . probably the highest in its history."[12] At other times, consumption was far less than that, as in the dry years 1859, 1902, 1929, and 1938, when water had to be brought from the mainland by boat and was only available one hour a day. In 1929, as the drought had struck the neighboring provinces of China as well, water was shipped in from Manila and Singapore. The vast construction projects alleviated the shortages, but only briefly. In 1938 the Japanese attacked Canton and thousands of refugees fled to Hong Kong. By 1941 the population had doubled, half a million people were sleeping in the streets, and water was again in short supply. When the Japanese attacked Hong Kong on December 7, 1941, they cut the water supply to the island. As Davis wrote, "Arguments are put forward, with strong claims to support them, that the breakdown of the water supply was sufficient in itself to have forced the capitulation of the Colony."[13]

Calcutta Sewage and Sanitation

Hindsight can be a very misleading guide. Today, Hong Kong is the epitome of successful development, while the word Calcutta conjures up in Western minds a picture of urban decay, disease, and human misery. Yet in the colonial era the two cities had much in common. Both were great harbors serving a populous hinterland; both were cosmopolitan, British-ruled; and both were crowded with poor and often homeless people. However, Calcutta had a distinct advantage over Hong Kong: its hinterland was at peace and its newcomers came attracted by the prospect of work, not driven from their homes by war and terror.

If their societies were in some ways similar, their geographies differed completely. Whereas Hong Kong clung to a small rocky island with too little water, Calcutta's problem was getting rid of water and waterborne pollution.

The concept of pollution is a slippery one. In today's industrial economies it brings to mind toxic chemicals in the environment. Earlier in this century, it meant contamination by bacteria, for example such waterborne diseases as typhoid, dysentery, and cholera. Before the germ theory of disease became accepted in the late nineteenth century, pollution referred to the filth and stench of sewage and decayed organic matter.

In India, however, there existed a competing definition of pollution, a religious one associated with the caste system, which complicated the task of Western sanitary engineers. As Maj. William Clemensha, author of *Sewage Disposal in the Tropics,* noted:

> It may be well to point out, for the benefit of those who are not conversant with the customs of the East, that no caste man will have anything to do with a latrine when it has once been used, on the ground that it constitutes a pollution. . . . people of the nature of mill coolies . . . object to touching the [toilet] handle, because the sweeper touches it, and this alone constitutes pollution.[14]

While a toilet became polluted if used by someone of the wrong caste, a river did not. Thus, said Major Clemensha,

> In India things are different. It must be understood that there is a certain percentage of the population in this country who, when they are thirsty, will drink practically any water they come across. . . . the people themselves are extremely careless about the pollution of water; bathing, washing of clothes, not to mention other, still more objectionable practices, are very common in village tanks, from which the people draw their daily supply. . . . an inspection of the [river]bank in the morning shows very clearly that the foreshore is used as a latrine by the very people who rely on the river for a drinking supply.[15]

A few years later another sanitary engineer, George B. Williams, was less squeamish in describing Clemensha's "objectionable practices," when he referred to

> the predilection which many of the inhabitants of India show for defaecating on the banks of rivers or tanks. Along the banks of the Hooghly the grossest pollution takes place daily, whilst a few yards away numerous people may be seen bathing, drinking water, or even drawing it for domestic use.[16]

To Williams, Indian ideas about pollution reflected a "lack of sense of proportion in such matters shewn by the educated inhabitants of the country and also by sentimental reasons," while "the ordinary nuisances to be met with in the towns and their supporting country so far transcend anything that the ordinary inhabitant of present day England can imagine."[17] But Williams, like Clemensha, Brunton, Macgeorge, and other British engineers in India, were missionaries among the heathen, and their religion was the gospel of progress through machinery. They were well aware that their "present day England" was a very different place from that of earlier times. In sanitary terms, Indian cities resembled some English ones earlier in the century; for example, in one district of Manchester known as "little Ireland" two privies served 250 persons; in another, 7,000 persons used thirty-three "necessaries," community-size chamber pots that were carried and emptied by hand.[18]

Indians, like Europeans before them, were bringing their rural habits into the crowded cities. Urban densities magnify the health hazards caused by pollution (in the Western sense of the word). In the West, as Charles Rosenberg pointed out, "the cholera pandemics were transitory phenomena, destined to occupy the world stage for only a short time—the period during which public health and medical science were catching up with urbanization and the transportation revolution."[19] In India, where cholera and other waterborne diseases lingered on, it was poverty as much as Hinduism which delayed the catching up process.

Calcutta receives as much rainfall as Hong Kong, 2,000 millimeters per year on the average. Yet its problem is not water but drainage, for the city is less than 10 meters above sea level, between the Hooghly River to the west and the swampy Salt Lake to the east. In the early nineteenth century, Calcutta, like other Indian towns, relied on traditional water-supply and waste-disposal systems, with little help from Western technology. Until 1820 Calcuttans got their water from the Great Tank in front of Fort Williams and from several smaller tanks (man-made ponds). After 1820, a small pumping station lifted Hooghly River water into open aqueducts which carried it to adjoining neighborhoods. River water tasted fresh from October to March. The rest of the year the Hooghly flows too slowly to fight the incoming tides, and a mixture of fresh water, sea water, and sewage sloshed back and forth past the city. Calcuttans then preferred to capture rainwater and store it in jars.[20]

Sewage disposal in the city was an adaptation of rural ways. Those who lived near the Hooghly used the riverbanks. Elsewhere, urine, waste, and storm waters flowed down open drains. Houses had privies or shared a privy with several others. Periodically *mehtars* collected the night soil in carts and disposed of it in the river, in neighborhood tanks, or in shallow trenches on the edge of town. The hand removal system had several drawbacks, besides the health hazards. It was costly in labor, equipment, and maintenance and vulnerable to labor unrest. Sensitive Westerners like Clemensha complained of "the horrible smell caused by the passing of a night-soil cart."[21]

That was in the better neighborhoods, around Fort William where the British lived, or in the adjoining residential quarters of the well-to-do and middle-class Indians. In the poorer sections, excreta were "scattered over the adjoining spot and left there to remain for ever to be dried by the sun;" or as John Strachey wrote in 1864: "The most important streets and thoroughfares of the northern division of Calcutta form to all intents and purposes a series of huge public latrines, the abominable condition of which cannot adequately be described."[22]

Improvements in sanitation came to Calcutta from two closely intertwined sources: the public health movement and municipal water supply. Interest in public health, which grew in the West in the nineteenth century, was a belated response to the problems of industrial cities. It was greatly furthered by an unwitting contribution from India: cholera. This disease was endemic to India, especially Bengal. Because of its short incubation period, however, its westward spread had to await improvements in transportation. When it reached Europe and America, it came as epidemics in 1831–32, 1848–49, 1853–54, 1866, and 1873. Each time it struck, it stimulated scientific research and public health policies. Thus the 1831–32 pandemic led the reformer Edwin Chadwick to set up a Bureau of Medical Statistics in the office of the Poor Law Commission and to publish his decisive *Report . . . on an Inquiry into the Sanitary Condition of the Labouring Population of Great Britain* (1842). In it, Chadwick and his associates announced a radically new attitude. Disease was no longer a personal matter to be dealt with by one's physician, but a social and environmental problem to be solved through technology:

The great preventives, drainage, street and house cleansing by means of supplies of water and improved sewerage, and especially the introduction of cheaper and more efficient modes of removing all noxious refuse from the towns, are operations for which aid must be sought from the science of the Civil Engineer, not from the physician, who has done his work when he has pointed out the disease that results from the neglect of proper administrative measures, and has alleviated the sufferings of the victims.[23]

In Britain, Chadwick's ideas began to be implemented with the Public Health Act of 1848, stimulated by another outbreak of cholera. Yet, as George Rosen noted, "the program of the sanitary reformers was based to a large extent on a structure of erroneous theories, and, while they hit upon the right solution, it was mostly for the wrong reasons."[24] The reformers believed in the miasmatic theory, according to which diseases arose in decaying organic matter. While it was later shown that diseases are not caused by dirt and stench, their incidence is sharply reduced by cleanliness, and especially by clean water and proper sewerage.

The causal connections between sewage, water, and cholera was demonstrated in the 1880s. Yet their epidemiological correlation had already been suggested in 1849 by Dr. John Snow in his pamphlet, *On the Mode of Communication of Cholera,* and statistically proven in 1854 in his studies of the incidence of cholera in South London.[25]

The public responded positively to these findings because they coincided with a fastidiousness that was new to Western society. One of the less heralded achievements of industrialization was to lower the cost of cleanliness through technological innovations: mass-produced cottons and soap; metal pipes, boilers, and valves; and porcelain sinks, tubs, and flush toilets appeared in middle-class homes by the midcentury. And they all required fresh water and sewerage to remove the wastes. Middle-class European or American city dwellers could now, in the comfort of their own homes, be as clean as the ancient Romans or medieval Arabs. When out in the streets, they were not only personally disgusted but also socially concerned when faced with the filthy habits of their less fortunate fellow-citizens.

These ideas and innovations spread to the European community of Calcutta, with some delays. In 1847 the Fever Hospital and Municipal Improvements Committee issued a report recommending new hospitals and dispensaries, town planning, water supplies, and general reforms. Such a scattershot approach, reflecting the confusion of the age on the subject of public health, was bound to fail.[26]

Then in the 1850s, things began to move swiftly, for several reasons: the connections between sewage, water, and cholera were becoming accepted; the trade of Calcutta was prospering and with it the European community; and that dynamic technocrat, Dalhousie, was governor-general. In 1854 the Corporation, or municipal government, of Calcutta hired the civil engineer W. Clark to recommend improvements in the sanitation of the city. He proposed a dual system: underground water pipes to carry water to hydrants around the city, and a network of sewage pipes with a pumping station to carry the waste waters out to the Salt Lake. The Corporation approved this plan in 1859. In 1868 the first part of Clark's sewerage scheme began operating; it was located, of course, in the European part of the city. The idea of extending the system to the poorer northern sections was thereupon proposed. Clark and the European justices agreed to it, but the Indian justices, "alarmed at the heavy expenditure which the complete scheme must involve," hesitated.[27]

Calcutta's first waterworks began to function in 1870. Steam engines lifted the Hooghly water into sand filters, from where it flowed to 500 public hydrants and 2,316 houses (out of a total of 16,000). The system provided over 20,000 cubic meters a day, enough, it was estimated, for 400,000 people. Yet it was soon overtaxed. In 1880 a Water Supply Extension Committee proposed to expand the filtered supply to 55,000 cubic meters a day, and to add a second, unfiltered system of 18,000 cubic meters for street cleaning, fire fighting, and flushing the sewers. This scheme was based on the assumption that, as S. W. Goode explained,

> The population of Calcutta and its suburbs had reached its high-water mark, and that special conditions, viz. the existing pressure upon a circumscribed area, the increasing use of machinery or labour-saving appliances, and the excessive preponderance of the male element in the population, would so greatly affect the law of natural progression that it would not suffice to maintain the population in *statu quo,* much less add to its numbers.[28]

Demographic forecasting was not yet a science, even among the urban planners who needed it most. In 1888 another, larger waterworks was built 30 kilometers north of the city, and by 1893 it too proved insufficient. The years 1900–1914 saw the Calcutta water system reach its zenith. Under the direction of waterworks engineer W. B. McCabe, the city built a large reservoir and modernized its equipment. By 1913 it was pumping 280,000 cubic meters a day, 58 percent of it filtered.[29]

In the period 1870–1914, the city of Calcutta was able to expand its sewerage and drainage system in parallel with its water supply.[30] The network of sewers was extended to the northern quarters in the 1880s and later to the surrounding suburbs. Mehtars were instructed to empty their night-soil carts into the sewers. Municipal latrines and bathing platforms and facilities at mills and railway stations were built and connected to the sewers, as were privies in the wealthier neighborhoods. A large pumping station lifted the waste waters into a drain which led to the Bidyadhari River, which flowed into the Bay of Bengal.[31] Like many other cities, Calcutta disposed of its effluents in a nearby body of water, on the then-popular theory that "the solution to pollution is dilution."

Unfortunately Calcutta waste waters carried much more than human excreta. Torrential downpours, the hosing of streets, the practice of cleaning utensils with sand, and the use of silt-laden Hooghly water caused the sluggish Bidyadhari to silt up.[32] Hence the engineers' interest in alternative methods of sewage disposal.

One of these was the ancient system of sewage farming, updated to use water-carried sewage. In dry areas of India such as the Punjab, sewage system effluents served both to irrigate and fertilize nearby farmlands. In moister areas, sewage farming required sedimentation tanks and means of transporting sludge. In the early years of this century, experiments were conducted at Dacca, Tittighar, and Calcutta but proved too costly to sustain.[33]

As a result of the discovery of waterborne disease bacteria, public authorities in Europe and America encouraged research into new methods of sewage treatment. In Britain the Royal Commission on Sewage Disposal issued nine reports between 1898 and 1915 which influenced urban engineers in India.[34] The first method was the septic tank, in which suspended solids are liquefied by anaerobic microorganisms, leaving a clear effluent and little sludge. It was first introduced in Britain in 1896 and in India ten years later. Though relatively inexpensive, it did not result in the destruction of all pathogenic microorganisms. Thus the enthusiasm with which Major Clemensha greeted this method in 1910 was not repeated by Williams in 1924.

Sewage filters, which mixed sewage with air to encourage the growth of aerobic bacteria, were designed to break down the sewage into its chemical components. It was the subject of several experiments, in particular by the city of Darjeeling and the Calcutta Sewage Disposal Committee in 1915–16. The Darjeeling experiment failed, while the one in Calcutta was not implemented because it was too costly.[35]

The last and most modern method of sewage disposal was the activated sludge system, introduced in Manchester in 1913. It combined filtration, agitation with compressed air, and the admixture of humus containing aerobic bacteria; at its best, it produced water and fertilizer. After World War I, treatment plants of this type were built at Jamshedpur and at the Sibpur Engineering College near Calcutta. Though technically advanced, this system was very expensive and required "skilled expert supervision, which it is impossible to expect that sewage workers in India will obtain." Only at Jamshedpur, a steel-mill city with a large number of skilled experts, was such a system economically viable. Elsewhere, and in Calcutta especially, taxes were too low to build and maintain such an advanced system.[36]

The benefits of water and sanitation in Calcutta were spread as unevenly as the distribution of income. The wealthy central neighborhoods enjoyed parks, shade trees, and clean streets. Households connected to the mains and sewers benefited the most. In the poorer districts, people had to wait in line at a neighborhood water tap, or, as often as not, get their water, unfiltered, from a hydrant, an open tank, or the River Hooghly itself. Their wastes were collected in carts or fell into open drains or the very rivers and tanks from which they obtained their water. Yet the poor benefited also from the water-supply and sewerage systems of the city. Tanks and hydrants allowed them to indulge their fondness for bathing, which visitors to the city often noted. The health of Calcuttans improved as well. The death rate fell from 54 per thousand in 1900 (a year of plague and cholera) to 29.2 per thousand in 1913–14; how much of this decline was due to the water system is not clear, but it was probably a significant share.[37]

After 1911, the capital was moved to New Delhi and political and economic troubles racked Bengal. The population of the city continued to swell faster than the resources of the municipal government. All in all, Calcuttans probably had a better water and sanitation system just before World War I than ever before, or since.

Dakar and the Plague

Dakar occupies one of the world's most favored maritime locations, at the westernmost tip of Africa, on the sea-lanes between Europe,

West Africa, and South America. On land, however, it is an awkward site, for the sandy, swampy Cap Vert Peninsula receives little rain and has a difficult access to the hinterlands of Senegal and the Niger. The French, who had come to Senegal in the early seventeenth century, ignored the peninsula for over two centuries, preferring the off-shore island of Gorée and the river town of Saint-Louis.

It was steamers that created Dakar. In 1856 Admiral Hamelin, minister of the navy, decided to establish a coaling station on the route to South America, to rival the Portuguese town of Saint Vincent in the Cape Verde Islands and the Spanish Santa Cruz de Tenerife in the Canaries. For this, Gorée was too small and the bar off Saint-Louis too shallow for steamers. Hence Captain Protêt, commander of the French West African fleet, was ordered to claim the peninsula for France.

The Messageries Imperiales line, for whose benefit this was ostensibly done, refused to refuel there until two jetties were completed in 1866. Even after that, Dakar remained a failure for decades. Ships drawing over 5 meters could not enter the shallow harbor and had to unload with lighters. The growing peanut trade went out through nearby Rufisque. In the early 1880s Dakar was a little town of fewer than 2,000 inhabitants, with twelve shops, four wine merchants, and one baker. Not until the railroad from Saint-Louis and the Senegal River valley reached Dakar in 1885 did the sleepy backwater turn into a viable port. By 1891 it had almost 9,000 inhabitants.[38]

The Fashoda incident of 1898 (see Chapter 4), which had only a temporary impact on international diplomacy, was the turning point for Dakar. In response to the British threat, France decided to build a harbor in West Africa for its cruisers, submarines, and torpedo boats. The project took ten years and cost 21 million francs (£840,000). Deep dredging, over 2 kilometers of new breakwaters, and a dry dock made it a harbor fit for cruisers. By 1908 Dakar was the finest naval base between Cape Town and Gibraltar.

While construction of the naval base was underway, the French government took other measures. In 1902 the capital of French West Africa (Afrique occidentale française, or AOF) was moved to Dakar. A full-scale commercial harbor with piers, railroad sidings, electric cranes, water pipes, and warehouses was built alongside the naval base, at a cost of 10 million francs (£400,000). As a port of call, Dakar was finally able to compete with Saint Vincent and Santa Cruz, and as a shipping harbor for the Senegal-Niger trade, with Rufisque. Between 1906 and 1910 the tonnage of ships stopping at Dakar more than doubled, and new harbor extensions were soon required.

Along with the increase in business and administration came a population explosion; from 1891 to 1909 Dakar grew from 8,737 to 24,831 inhabitants, an increase of almost 6 percent a year. Lured by jobs on the docks and railway yards and in the warehouses and offices of the booming city, men came from Senegal and neighboring countries, and also from North Africa, Europe, and the Levant. Dakar was becoming one of the world's cosmopolitan ports.[39]

The demand for water increased more than proportionately. Until 1898 Africans had obtained water by digging shallow wells among the dunes, and Europeans were supplied by a boat which brought water from a creek down the coast every two or three days. Officials knew that if the town was to grow, and if its port was to supply the fleet and attract commercial shipping, it would have to provide good water in large quantities at low prices. A water-supply system was seen as a commercial and military necessity, not as a sanitary measure.[40]

The first municipal wells were dug in 1899 at Hann, 6 kilometers from the city. By 1910 this source was supplying an average of 60 to 80 liters per person per day. Unfortunately it came from an aquifer lying under a village, a cemetery for yellow fever victims, and an experimental farm with its animals and manure. As a result it contained both salt and coliform bacteria, a sign of sewage pollution. In 1907 the city began digging deep wells at M'Bao, 18 kilometers away. This project, completed in 1912, provided 100 liters per person per day.

That amount is just sufficient for a tropical city.[41] Its distribution, however, was very skewed. As a report to the minister of colonies noted in 1918, one-third of the water was used by the harbor and military installations. Within the civilian population,

> Three thousand inhabitants of the European quarter are well supplied, for the eight hundred subscribers with private taps absorb almost the entire 1,962 cubic meters distributed daily on the average to the population. They even indulge in a frenzied wastefulness, which causes a budgetary loss on the basis of a cost almost ten times the selling price.[42]

The poorer, predominantly African quarters were supplied by hydrants which were supervised by guards. There, the scarcity of

water and the distances between homes and hydrants forced people
to adopt unhygienic habits. Some Africans dug wells to water their
vegetable gardens, even though the water was contaminated. Others
kept water in their homes in jugs and barrels, perfect breeding places
for the *Aedes aeqypti* mosquito, vector of yellow fever. Periodically,
health inspectors came through, fining those who had larva-infested
containers. People who did not understand the differences between
mosquitoes noted that little was done to drain nearby mosquito-
infested swamps and ponds. The water-distribution system, less a
sanitary measure than might be expected, was itself a cause of disease
and resentment.[43]

Sanitation policy lagged behind the water supply. An outbreak
of yellow fever in 1900 and complaints from the contractor for the
naval base that work had to be interrupted due to the high death
rate among Europeans prodded the government to issue some health
laws and decrees. Only in 1905, after yet another yellow fever panic,
was a municipal hygiene service established and funds voted for
drainage, sewerage, and garbage removal. From 1905 to 1908 the
municipality laid sewers, and by 1910, 150 European houses were
connected to them. A few public toilets were erected in African
neighborhoods, but they were soon overwhelmed, largely because
there was not enough water to flush them. Most Africans continued
as before to relieve themselves and empty their pans in vacant lots
or on the beaches.[44]

Nonetheless the newfound activism of the municipal and colonial governments led two public health doctors to exclaim in 1908:
"The sanitation of Dakar has become a complete and remarkable
work, honoring those who were in any way involved in it, and an
important event in the sanitary history of colonial cities."[45]

That judgment turned out to be premature. Health officials
worried, and with cause, that the city was still vulnerable to epidemics of yellow fever and cholera. What appeared instead was a
disease they had not foreseen, the plague, even though it had been
prowling around the world for twenty years. The pandemic originated in Hong Kong in 1894, causing clashes between Chinese residents and British troops who entered their houses and seized the
corpses.[46] It reached Madagascar in 1898, Cape Town and Port
Elizabeth in 1900, Tangier in 1904, Accra in 1908, and Casamance
(southern Senegal) in 1912.

In Dakar the first victim died in early April 1914, yet the disease was not diagnosed and announced until May 10. By then it was

spreading fast: from 4 or 5 deaths a day in May to fifteen a day in August, a month in which 376 people died of the disease. Then it diminished until January 1915 when the plague was officially declared to be over. It had claimed over 1,400 lives in Dakar.[47]

Official reaction to the plague reveals more than medical ideas at work, for the disease is, as Maynard Swanson put it, "both a biological fact and a social metaphor."[48] Their first reaction was to bury the dead with lime and burn or disinfect their homes and belongings. Ships were quarantined, and vaccine ordered from France. In late May the authorities decided to build an isolation camp outside of town for the sick and those suspected of carrying the disease. Quickly the idea spread in the administration and in the white community that *all* Africans were suspect and should be evacuated. On July 7, the Dakar Sanitary Commission demanded "the transfer of the native population which takes pleasure in a deep-rooted and incurable filthiness, to a place far from the city, and the destruction or demolition of all shacks and huts, as the only measure able to stop the spread of the current epidemic."[49] Two weeks later the governor general decreed: "Considering that the present sanitary situation of the city of Dakar, and the permanent danger of epidemics which threaten this city because of the anti-hygienic conditions in which the native population lives, requires urgent and extraordinary measures of special protection." Therefore, he authorized the removal of Africans from Dakar.[50]

The idea of racial segregation on grounds of health was part of the same scientific spirit that discovered the microbial causes of disease. In Accra, writes David Patterson, "ironically, this policy was first suggested by an African physician, Dr. J. F. Easmon, in 1893. His recommendation led to the construction of European bungalows in Victoriaborg. Segregation was official policy by 1901."[51] In 1899 Dr. Ronald Ross, who identified the *Anopheles* mosquito as the vector of malaria, spent three weeks in Freetown, Sierra Leone, verifying his findings. He had spent the early years of his career in India, where towns were divided into a native city, a cantonment for the army, and "civil lines" for European civilians.[52] He was shocked to find Africans and Europeans so intermingled in Freetown. Since malaria was endemic among Africans and mosquitoes only fed after sunset, he suggested that Europeans build their homes in an isolated suburb and keep Africans away at night.[53] Similarly, the British physicians J. W. W. Stephens and S. R. Christophers, who traveled to West Africa and India in 1900 to study malaria, recommended separating Europeans from natives, especially children who were carriers of the *plasmodium* of malaria.[54] As the historian Albert Wirz noted,

> The historical tragedy is that at that very moment when a scientific revolution permitted for the first time a rational attack and prevention against malaria, in the social arena a racist ideology of dominance replaced the previous Enlightenment ideology. In the end only the mosquitoes profited from it.[55]

Yet segregation was not implemented everywhere with the same fervor. In Accra and Freetown, some Europeans migrated to their new suburbs, but many did not; the result was less to improve the health of Europeans than to stimulate racial snobbery among whites and resentment among educated Africans.[56]

In Duala, Cameroon, German administrators devised a plan in 1910 to relocate the entire Duala population inland, leaving the harbor areas to the four hundred whites, with a kilometer-wide "free zone" separating the two. A health reason—to prevent malaria—was put forward as a rationale for the plan, but on this question opinions were divided. The plan, and especially the massive expropriations it required, led to a bitter black-white confrontation. Though never fully implemented before the Germans lost Cameroon in World War I, the segregated housing and the unequal water and sewerage services that accompanied it remained as a legacy for the French to uphold.[57]

The most acute cases of segregation-mania involved the plague. Epidemics, like wars, give awesome powers to the authorities, and of all epidemics, none brought forth the thrill of panic quite so powerfully among Europeans as their ancestral scourge, the Black Death. The divisional engineer of Nairobi, J. H. Patterson, recalled: "A case or two of the plague broke out [in the bazaar], so I gave the natives and Indians who inhabited it an hour's notice to clear out, and on my own responsibility promptly burned the whole place to the ground." After that, though de jure segregation was specifically forbidden by the Colonial Office, "the rigid enforcement of sanitary, police, and building regulations without any racial discrimination by the Colonial and municipal authority will (in any case) suffice."[58] When the plague broke out in South Africa, the Cape Town authorities moved six or seven thousand Africans to a makeshift camp several miles from the city. In Port Elizabeth Africans were simply driven out and had to put up their own housing outside of town.[59]

The "segregation syndrome" was thus widespread among whites in Africa, in Dakar as elsewhere. In 1908 Doctors Ribot and Lafon had recommended

> the absolute division of Dakar into two neighborhoods: the European town and the native town. Blacks, even the most civilized, cannot submit to certain European habits any more than Europeans can adopt certain native customs. The result is a bother for both when their homes are near one another.[60]

Similar statements, usually buttressed by arguments of sanitation, had appeared at the time of the yellow fever scares of 1901 and 1905. The municipal government had never accepted these ideas but did enforce codes that allowed only masonry buildings in the predominantly European part of town and tolerated wooden shacks and straw huts in the African sections. These regulations, aimed at buildings, did not prevent a mixture of people and even of houses in many areas.[61]

With the approval of Governor-General William Ponty, the city began enforcing its segregation policy in July. Brick buildings harboring plague victims were disinfected, but straw huts and wooden shacks were incinerated—1,595 in all. Of the 20,000 Africans living in the city, 5,000 were forcibly evacuated. A piece of land separated from the city by an 800-meter-wide corridor was designated as a "segregation village" and given the name Medina. In September and October 2,900 evacuees were given lots on which to build themselves new homes. No building or health codes prevented the construction of shacks and huts, nor was there any piped water, sewerage, or drainage.[62] In October, as the evacuation was coming to a close, the lieutenant governor of Senegal wrote to the municipal council:

> It was necessary to create near Dakar a large segregation camp to which we could send, after isolation and disinfection, the native population of Dakar. There is danger and mutual annoyance in letting two groups cohabit which have such completely distinct views on their way of life. Let us allow them, if need be let us make them, have two different installations conforming to their tastes: on the one side the European town with all the requirements of modern hygiene, on the other the native town with all the freedom to build

out of wood or straw, to play the tom-toms all night, and to pound millet from four in the morning on.[63]

How playing the tom-toms and pounding millet gave people the plague, he did not explain. If the desire to segregate existed in Dakar as elsewhere in Africa, the rhetoric of excuses was different. Unlike Cape Town, where political leaders talked of "raw Kaffirs" and "barbarians," the politicians of Dakar spoke of different people's different lifestyles. This in turn reflected the very unusual political situation of Dakar.

Dakar, Gorée, Saint-Louis, and Rufisque were the "four communes" of Senegal, in which anyone residing five years or more became a French citizen. Among those with citizenship were the Lebous, descendants of the original inhabitants of the peninsula. On May 10, 1914, after a bitterly contested electoral campaign, they had elected Blaise Diagne as the first African to sit in the French Chamber of Deputies. Africans interpreted the declaration of plague three days later and the house-burnings and evictions that followed as the revenge of the defeated white merchant party. Riots and demonstrations broke out in late May and again in October and November. At one point, an irrevocable split seemed imminent. However, both Governor Ponty and Deputy Diagne were moderates interested in avoiding clashes. In the end, most of the people relocated to Medina were Toucoulors and Bambaras, people from the interior, while the Lebous remained in Dakar.[64]

In 1916, all inhabitants of Dakar were declared French citizens. The minister of colonies wrote the Dakar authorities:

> It seems difficult to establish today, between native Senegalese of Dakar and Europeans living in the town, special distinctions. . . . It must be possible to reach the goal one is striving for by using methods aimed not at persons but at categories of buildings subject to particular regulations.[65]

In a report to the minister of colonies in February 1919, the governor-general of AOF still advocated two separate towns based on the different customs and disease susceptibilities of Europeans and Africans. However, he added,

> The words "European town" must be understood to mean "town whose inhabitants accept to be subject to the sanitary regulations applicable to Europeans" and every native who agrees to submit strictly to these regulations will be able to live in the European town on the same basis as any European.

> One should not see in the separation of the two towns any po-
> litical idea of opposition of the races, any tendency to restrict the
> rights of the native population.[66]

Medina was incorporated into Dakar in 1915 "in order to safeguard
the political rights of its inhabitants."[67] Inspector Revel added:

> The toleration granted the inhabitants of Medina must not degen-
> erate into license nor incite the administration and the municipality
> to neglect a group of people most of whom can—events prove it
> every day—be led progressively and patiently toward a new concep-
> tion of hygiene and well-being.[68]

In Dakar, therefore, segregation came to be justified not on
racial grounds but on types of buildings and hygienic habits. The
administration's share in the responsibility for these customs, through
its very unequal distribution of water and sewerage, was ignored.
Dakar had a more liberal political system than existed anywhere in
Africa at the time, but no more egalitarian a society, and no more
enlightened a policy on the distribution of municipal services.

Notes

1. Philip D. Curtin, "Medical Knowledge and Urban Planning in Tropical Africa," *American Historical Review* 90, no. 3 (June 1985): 594–613.

2. Walter Elkan and Roger van Zwanenberg, "How People Came to Live
in Cities," in *Colonialism in Africa, 1870–1960,* ed. Peter Duignan and L. H.
Gann, vol. 4: *The Economics of Colonialism* (Cambridge, 1975), pp. 659–61.

3. S. G. Davis, *Hong Kong in Its Geographic Setting* (London, 1949),
pp. 75–77, 90–95; see also George B. Endacott, *A History of Hong Kong,* rev.
ed. (London and Hong Kong, 1973), pp. 97–98, 116, and David Podmore,
"The Population of Hong Kong," in *Hong Kong: The Industrial Colony. A
Political, Social, and Economic Survey,* ed. Keith Hopkins (Hong Kong, 1971),
pp. 21–24.

4. Endacott, p. 116; Gene Gleason, *Hong Kong* (New York, 1963),
pp. 164–65.

5. Nigel Cameron, *Hong Kong: The Cultured Pearl* (Hong Kong and
New York, 1978), p. 82.

6. Endacott, pp. 180–86.

7. Ibid., pp. 175–88.

8. Thomas L. Tregear and L. Berry, *The Development of Hongkong
and Kowloon as Told in Maps* (Hong Kong, 1959), p. 21.

9. Davis, pp. 72–78; Gleason, p. 165.

10. Tregear and Berry, p. 22; Davis, pp. 72–77; Endacott, pp. 223, 277,

297; T. N. Chiu, *The Port of Hong Kong: A Survey of Its Development* (Hong Kong, 1973), p. 40.

11. These figures include industrial, municipal, and commercial uses of water as well as household consumption. S. W. Goode, *Municipal Calcutta: Its Institutions in Their Origin and Growth* (Edinburgh, 1916), pp. 186, 201, 357–58; Geoffrey Moorhouse, *Calcutta* (London, 1971), p. 283; George Walter Macgeorge, *Ways and Works in India: Being an Account of the Public Works in that Country from the Earliest Times up to the Present Day* (Westminster, 1894), p. 437; W. H. Corfield, *Water and Water Supply*, 2nd ed. (New York, 1890), pp. 27–28; John C. Thresh, *Water and Water Supplies* (London, 1896), pp. 274–83.

12. Davis, p. 71.

13. Ibid., p. 81. Today, much of Hong Kong's water comes from the People's Republic of China; as always, the colony's prosperity depends on the peace and benevolence of its neighbor.

14. Maj. William Wesley Clemensha, *Sewage Disposal in the Tropics* (London, 1910), p. 18.

15. Ibid., pp. 203–4.

16. George Bransby Williams, *Sewage Disposal in India and the East: A Manual of the Latest Practice Applied to Tropical Countries* (Calcutta and Simla, 1924), p. 45. Williams was chief engineer of the Bengal Government's Public Health Department.

17. Ibid., pp. 40–43.

18. George Rosen, *A History of Public Health* (New York, 1958), p. 205.

19. Charles E. Rosenberg, *The Cholera Years: The United States in 1832, 1849, and 1866* (Chicago, 1962), p. 2.

20. Goode, pp. 180–81; Moorhouse, pp. 50, 269.

21. Clemensha, p. 6.

22. Goode, pp. 168–69.

23. Quoted in Rosen, p. 215.

24. Rosen, p. 225.

25. Ibid., pp. 285–89. Joel A. Tarr and Francis C. McMichael, "Water and Wastes: A History," *Water Spectrum* (Fall 1978): 18–25.

26. Goode, pp. 18–19.

27. Ibid., pp. 117–18.

28. Ibid., pp. 188–89.

29. Moorhouse, p. 283.

30. While that may seem logical, it was not the sequence followed by many American cities that increased their water supplies, then reacted in surprise when their drains and open sewers overflowed; see Tarr and McMichael.

31. Edward W. C. Sandes, *The Military Engineer in India*, 2 vols. (Chatham, 1933–35), 2:41–42; Sivaprasad Samaddar, *Calcutta Is* (Calcutta, 1978), p. 81; Goode, pp. 119–36, 170–75; Williams, pp. 97–99.

32. Williams, pp. 12–17, 59.

33. Ibid., pp. 100–113, 183–85.

34. Williams, pp. 7, 102.

35. Clemensha, pp. v–4; Williams, pp. 8, 73, 120–45.

36. Williams, pp. 172–76, 206–15.

37. Goode, pp. 365–66.

38. On the early history of Dakar, see Assane Seck, *Dakar, métropole ouest-africaine* (Dakar, 1970), pp. 111, 273, 290–308; Roger Pasquier, "Villes du Sénégal au XIXe siècle," *Revue française d'histoire d'outre-mer* 47 (1960): 387–426; Albert Boucher, "Le port de Dakar, ses origines, état actuel, extensions futures," *Bulletin du Comité d'études historiques et scientifiques de l'A.O.F.* (1925), pp. 651–75; Robert Delmas, *Des origines de Dakar et ses relations avec l'Europe* (Dakar, 1957); Richard J. Peterec, *Dakar and West African Economic Development* (New York and London, 1967), pp. 39–42; and Claude Faure, *Histoire de la presqu'île du Cap Vert et des origines de Dakar* (Paris, 1914), pp. 125–64.

39. Seck, pp. 139, 210, 294, 310–18; Peterec, pp. 42–43; Pasquier, pp. 421–24; Boucher, pp. 656–57; Afrique occidentale française, Colonie du Sénégal, *Le port de commerce de Dakar* (Dakar, 1918), pp. 12–31. Elikia M'Bokolo asserts that official population statistics are too low, and that the real population reached 30,000 between 1907 and 1914: "Peste et société urbaine à Dakar: l'epidémie de 1914," *Cahiers d'études africaines*, 22, no. 1 (1982): 15.

40. Georges Ribot and Robert Lafon, *Dakar, ses origines, son avenir* (Bordeaux, 1908), pp. 113–17.

41. Accra had less water: from 32 liters per person per day in 1914–20 to 59 liters in 1938; see K. David Patterson, "Health in Urban Ghana: The Case of Accra, 1900–1940," *Social Science and Medicine* 13 B (1979): 253.

42. Archives Nationales Section Outre-Mer (henceforth ANSOM), Travaux Publics 22 (14): Report from Colonial Inspector Revel to minister of colonies, Dakar, July 13, 1918: "Hygiène et assainissement de Dakar. Alimentation de la ville et du port on eau potable."

43. Bruno Salléras, *La politique sanitaire de la France à Dakar de 1900 à 1920* (Mémoire de maîtrise, Université Paris X, 1980), pp. 40–43, 67–73; Gérard Brasseur, *Le problème de l'eau au Sénégal* (Saint-Louis, 1952), pp. 76–81; Afrique Occidentale Française, *Le port de Dakar en 1910* (Dakar, 1910), pp. 50–51 (this is an official pamphlet celebrating the completion of the harbor); Ribot and Lafon, pp. 121–30; F. Heckenroth, *Le problème de la salubriété publique à Dakar* (Dakar, 1921), pp. 86–87, 234–44 (Heckenroth was a municipal public health physician in 1919–21); Afrique Occidentale Française, Agence économique, *Hygiène et assainissement de la ville de Dakar* (Paris, 1930), pp. 3–4.

44. Salléras, pp. 76–81; Ribot and Lafon, pp. 64–65, 107–10.

45. Ribot and Lafon, p. 65.

46. Endacott, p. 216; Cameron, pp. 117–18.

47. M'Bokolo, pp. 18–26. Salléras says 3,686 people died of the plague. The exact number is impossible to ascertain, since many victims were not reported to the authorities but buried privately.

48. Maynard W. Swanson, "The Sanitation Syndrome: Bubonic Plague and Urban Native Policy in the Cape Colony, 1900–1909," *Journal of African History* 18, no. 3 (1977): 408.

49. Salléras, p. 93.

50. Seck, p. 135, n. 1.

51. Patterson, p. 255.

52. On cities in British India, see Anthony D. King, *Colonial Urban Development: Culture, Social Power and Environment* (London, 1976).

53. L. Spitzer, "The Mosquito and Segregation in Sierra Leone," *Canadian Journal of African Studies* 2 (1968): 49–61.

54. J. W. W. Stephens and S. R. Christophers, "The Malarial Infection of Native Children" and "The Segregation of Europeans," *Royal Society, Reports to the Malaria Committee,* 3rd ser. (London, 1900), pp. 4–44; Curtin, pp. 598–99.

55. Albert Wirz, "Malaria-Prophylaxe und kolonialer Städtebau: Fortschritt als Rückschritt?" *Gesnerus* 3, no. 4 (1980): 222.

56. Curtin, pp. 600–602; Patterson, pp. 255–56; Spitzer, pp. 56–60.

57. Ralph A. Austen, "Duala versus Germans in Cameroon: Economic Dimensions of a Political Conflict," *Revue française d'histoire d'outre-mer* 64, no. 237 (1977): 481–82; Wirz, pp. 227–29; L. Rousseau, "Alimentation de Douala en eau potable," *Bulletin de la Société de pathologie exotique* (April 9, 1919): 188–201.

58. Elkan and van Zwanenberg, pp. 663–64.

59. Swanson, pp. 391–400.

60. Ribot and Lafon, p. 160.

61. Salléras, pp. 48–50, 88–90; Seck, pp. 132–33; Raymond F. Betts, "The Establishment of the Medina of Dakar, Senegal, 1914," *Africa* 41 (1971): 143.

62. Betts, p. 145; Heckenroth, pp. 251–56; M'Bokolo, p. 38; Salléras, pp. 94–95; Seck, pp. 134–35.

63. Quoted in Salléras, p. 92.

64. M'Bokolo, pp. 30–44; Betts, "Establishment of the Medina," p. 145–46 and "The Problem of the Medina in the Urban Planning of Dakar," *African Urban Notes* 4, no. 3 (1969): 5–15.

65. Quoted in Seck, p. 137.

66. Ibid., p. 138.

67. Heckenroth, p. 258.

68. ANSOM, Travaux Publics 22 (14): Inspector Revel to minister of colonies, July 13, 1918.

6

Hydraulic Imperialism in India and Egypt

Enlightened despots have always favored public works as monuments to their wealth, public spirit, and administrative prowess. Wherever geography allowed it, great canals and irrigation projects have been triply tempting to such despots as evidence of their power over the forces of nature, as sources of revenue, and as means of turning large numbers of peasants into loyal dependent subjects. Hence, irrigation and kingdoms arose together in the early river valley civilizations. So closely have water control and despotism been united that it is tempting to describe the first as the cause of the second.

European empire-builders of the nineteenth century were imbued with the history of those great hydraulic engineers, the ancient Romans; and none more so than the British public-school boys who grew up to be proconsuls in Queen Victoria's vast domains. It was by chance that Britain acquired, in the course of its conquests, the birthplaces of several early hydraulic civilizations: the South Indian deltas, the Ganges and Indus valleys, the Nile Valley, and, finally, Mesopotamia. (Only China escaped being added to the collection, although Britain at one time did lay claim to the Yangtze Valley.) It was not by chance at all, but by the logic of benevolent tyranny over dry and crowded lands, that British engineers built the largest irrigation projects the world had ever seen. Richard Baird Smith, one of those early engineers, expressed both the practical and the monumental purposes of irrigation works when he called them "more likely, from their relations to the material prosperity of the country, and from their permanent nature, to perpetuate the memory of English dominion in India than any others hitherto executed."[1]

There are many kinds of irrigation systems. Egypt presents the simplest case of all. There, the regular annual flood of a single river provides all the water and most of the soil fertility in an otherwise desert country. India is a more complex land, where many rivers, climates, soils, and peoples interact. It is also the birthplace of modern irrigation engineering, and for that reason we shall begin there.

Referring to the period 1870–1913, the economist W. Arthur Lewis pointed out: "The principal reason why India developed more slowly than almost any other country was simply lack of water."[2] Statistics of the average rainfall on the subcontinent—roughly 1,000 millimeters a year—are almost meaningless because of the enormous variations in time and location. Some areas, like Assam and the Western Ghats, receive 10,000 millimeters or more. At the other end of the spectrum, parts of Sind receive less than 100 millimeters in an average year. Yet regional averages are also misleading; what really counts is variations over time. Almost all the rain falls in the monsoon months from June to October, but some years less than others. One year in five (but unpredictably), the rains are 25 to 40 percent below normal, causing poor crops and hunger. One year in every ten, the deficiency is over 40 percent; then crops wither and die, and people and cattle starve. If one drought succeeds another, famines carry off millions, depopulating whole districts. Nowhere else on earth are there such enormous variations, with such awesome human consequences.

Irrigation, then, serves two purposes: to water lands like Sind and much of the Punjab, where agriculture would otherwise be impossible; and to provide water to vulnerable regions where the average rainfall is adequate (250 to 2,000 millimeters), but the climate is often subject to droughts. These regions include the Gangetic plain and much of the Deccan, in other words two-thirds of the subcontinent.

Unfortunately, the supply of water for irrigation is not equal to the need. The Indus, the Ganges, and their Himalayan tributaries are fed by the melting of the previous year's snow, which brings the rivers to flood between May and October. In the Deccan, however, there is no snow, and the rivers carry only this year's water. The rocky undulating terrain severely limits the extension of irrigable land; only the rich alluvial deltas are fully irrigable. Years of drought, when irrigation is most urgently needed, are precisely when the rivers fail.

From time immemorial, the peoples of the subcontinent have sought to trick nature into watering their crops. In the Indus valley lie the remains of ancient cities and of the inundation canals their inhabitants dug to divert some of the flood waters to distant fields. And almost everywhere one finds wells and tanks built by local communities to preserve last year's water for this year's crops.

The British did not bring irrigation to India. Natives of a rainy land, they approached irrigation hesitantly, deterred by the enormous expenses involved. Periodically stirred to action again by the great famines that hit India, they expanded the irrigation system, laying the foundations for what is today the most heavily irrigated part of the world and the sustenance for half a billion people.

Government irrigation works watered 4.25 million hectares in 1878–79, 7.8 million hectares in 1900, and 11.3 million in 1919–20. The latter figure represented 11 percent of all the croplands of India, while another 11 percent was watered by myriad privately built wells, tanks, and channels. In 1942 the irrigated area had increased to 23.8 million hectares, of which 13.3 million—an area larger than England—received their water from government works.[3] (See Figure 3, Irrigation Canals in India to 1942.)

Precolonial Irrigation and British Restorations to 1837

Long before the British arrived, there was large-scale irrigation in Hindustan. In the seventeenth century, Shah Jehan, builder of the Taj Mahal, had the Hasli Canal dug to water the gardens of Lahore and Amritsar. Another canal, dating back to the fourteenth century, brought water to Delhi from the Jumna (now Yamuna) River. It was rebuilt under Akbar in the sixteenth century, and again under Shah Jehan in the seventeenth. Yet another canal followed the eastern bank of the Jumna River. Though originally built to carry water to royal hunting lodges and gardens, and later to the cities, these works also served the farmers along their banks. In the eighteenth century, as Mogul power disintegrated, irrigation was neglected and whole districts reverted to their natural state of bush and grasslands. In the early nineteenth century, the British found only the Hasli Canal still carrying water, though in sore need of repair.[4]

The first British efforts in irrigation were to bring these derelict

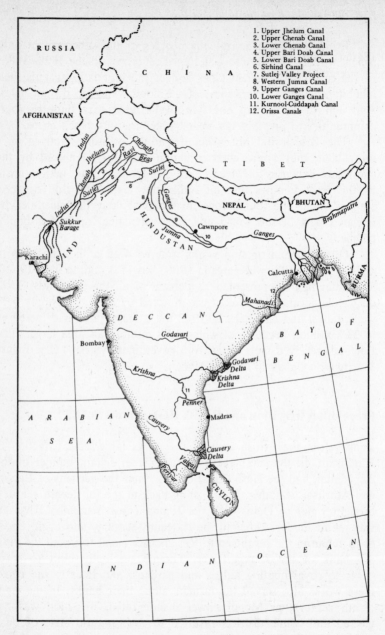

Figure 3 Irrigation Canals in India to 1942

works back into service. The first canals they restored were the Eastern and Western Jumna Canals. Shah Jehan's Delhi Canal, which had been out of service since 1753, was surveyed in 1808–10 and rebuilt under the name Western Jumna Canal between 1815 and 1821. Shortly thereafter, work began on the Eastern Jumna, or Doab, Canal which ran through the interfluvial region (*doab*) between the Jumna and the Ganges. It was opened in 1830.

Very soon these works showed the limitations of existing engineering knowledge, both that of the Mogul Ali Mardan Khan who had originally built them, and of the young officers of the Bengal Artillery, Lts. J. H. Dyas and Proby Cautley and Capt. Richard Baird Smith, who were assigned to rebuild them. Without control over the flow of water or proper drainage of the irrigated land, irrigation caused waterlogging, salt deposits, and a rise in the incidence of malaria. On the Eastern Jumna Canal, the newly appointed engineers learned that too steep a slope made the water flow too fast in places, scouring the canal bed and undermining embankments and bridges; elsewhere the more level stretches silted up. They also had great difficulties with land tenure, distribution rights, water rates, and other administrative questions. Year after year, trudging across the hot plains of Hindustan, they studied the problems, surveyed the land, built and rebuilt. In the process, they laid the foundations of modern hydraulic engineering.

The Eastern Jumna received a permanent masonry headworks in 1836 to regulate the flow of water, and a network of drainage channels to draw away the excess. Cautley, who became supervisor of the Eastern Jumna in 1831, rebuilt the upper stages and installed masonry "falls," lowering the gradient to an average of 28 to 38 centimeters per kilometer. The work was slow because there were only a few engineers and a small staff with little money. That was due to the refusal of the East India Company to borrow money to invest in irrigation, which allowed the work to advance only when the Indian government had money to allocate to its Military Board, which controlled all public works in India. Yet by 1833–34 the fees for milling, timber transport, and cattle-watering rights, as well as for irrigation, had begun to cover the construction and administrative expenses of the canals.

The real test of the canal came in 1837–38, when a drought spread famine throughout upper India but spared the crops on the 366,000 hectares of newly irrigated land. This suddenly made the value of irrigation apparent to the East India Company officials, for whom famine meant both a sudden rise in expenditures on relief and a drastic loss of tax revenues.[5]

Just as the canals of Hindustan were beginning to show results, another project was undertaken on the Cauvery Delta in the Madras Presidency, one of the most fertile regions of India. This delta has two main channels, the Coleroon and the Cauvery, and numerous smaller ones. To irrigate the land, the flow in these two channels must be regulated by weirs or low barrages, called anicuts, which keep the water level constant. The Grand Anicut across the Cauvery dated back to the second century, and was completed by the kings of Tanjore around the year 1000. It was 329 meters long and built of granite blocks packed with clay. Though resting on the sand of the river bed, the Grand Anicut had withstood sixteen centuries of floods.

When the East India Company took over Tanjore in 1803, British surveyors found that the river just above the Grand Anicut had silted up and most of the water was going to sea down the Coleroon. Only two-thirds of the delta was still irrigated, and every year the irrigated area shrank still more. By 1829–30 the situation had become critical, both for the farmers who could no longer grow rice and for the Company, which found its tax receipts shrinking also. In 1834 Maj. Arthur T. Cotton proposed to build a second weir across the Coleroon to divert the waters back into the Cauvery. By 1836 the Upper Anicut, a 781-meter-long masonry weir, was completed. While it did not solve the problem completely—too much water now went down the Cauvery—it reversed the deterioration and proved that great dams could be built successfully on shifting sands. Ten years later Cotton returned to the Cauvery to add a masonry dam on top of the old Grand Anicut. With that, the irrigated area of the delta rose from 270,000 to 410,000 hectares, an increase of 52 percent. As the government's expenditures on the two works totalled £25,000, while its taxes were much higher on irrigated than on dry land, the enterprise proved to be wonderfully remunerative, with a profit of over 23 percent per year.[6]

The Classic Era of Indian Irrigation, 1838–54

The famine of 1837–38 and the great success of the early restorations convinced the East India Company to allow more and greater irrigation schemes. These took place in the same areas as the restorations: the Jumna–Ganges Doab and the deltas of the Madras Presidency. They were associated with the work of two engineers, Proby Cautley in the north and Arthur Cotton in the south, who became folk heroes of the British Raj.

In 1836 Col. John Colvin of the Bengal Artillery first proposed to irrigate the Roorkee plateau between the Jumna and the Ganges, an area without any previous irrigation works. The task of surveying the area was assigned to Captain Cautley, then supervisor of the Eastern Jumna Canal. It was clear from the first that it would be not only the largest project undertaken by the British in India, but also by far the largest irrigation system in the world, three times longer than the Cavour Canal in Italy and with twice the flow of the two Jumna Canals combined.

The Court of Directors of the East India Company tentatively authorized the project in 1841. Cautley still had to overcome major obstacles, both technical and political. No sooner had the work begun than Governor-General Lord Ellenborough decreed that the canal should serve mainly for navigation and only incidentally for irrigation; this would have required a complete redesign, since irrigation canals need to carry much more water than navigation canals, which should flow slowly and include locks.

Cautley's education was minimal; he had received one year's training as an artillery cadet at Addiscombe Military Seminary before being sent to India in 1819. Everything he knew about irrigation and hydraulics he learned on his own.[7] During the hiatus caused by Lord Ellenborough's misguided decree, he took a three-year home leave. Back in Britain, he studied the Caledonian Canal, built by Thomas Telford from 1803 to 1822. On his way back to India in 1848, he stopped off in Italy, where the most advanced irrigation works were then to be found. He noted

> surprise, mixed with a good deal of satisfaction, at the numerous instances in which we, who were entirely separated from all communication with the Italian engineers, had, by the mere process of simple reason, arrived so frequently at precisely the same results, and in so many cases had adopted the same expedients.[8]

Another irrigation engineer, Richard Baird Smith, also made the pilgrimage to Piedmont and Lombardy at the request of the Court of Directors. He too found little to learn:

> As regards the works themselves . . . I do not think the Italians are superior to ourselves; and in regard to the manner in which the

efficiency of the works is maintained, they are, I must frankly say, decidedly inferior: but in the theory of distribution, in points of interior economy connected with the use of water, and in the exactitude and detail of legislation, they are far in advance of us.[9]

The technical aspects of the Ganges Canal divide into two parts: design and construction. In order to irrigate the largest possible area, Cautley proposed to build a single canal for the first 290 kilometers, then two 274-kilometer-long parallel canals, one ending at Hamirpur on the Jumna, the other at Cawnpore on the Ganges. To handle the enormous volume of water—131 cubic meters per second, almost the entire Ganges River at its lowest—the canal had to be 43 meters wide at the top, tapering off to 6 meters at the end of the two branches.

The biggest problem lay in the first 30 kilometers of the canal. In order to carry water to the Roorkee tableland, 25 meters above the plain, the Ganges had to be tapped at its exit from the Himalayan foothills, near the holy town of Hardwar. Between Hardwar and the tableland, however, lay broken terrain and deep ravines. Three seasonal torrents and one river came down from the mountains along that stretch, hurling boulders, sand, and mud. To cross this terrain, Cautley built superpassages, wide aqueducts to carry two of the torrents over the canal. A third torrent was allowed to enter the canal and flow out the other side, with regulators on both banks. The canal was carried over the Solani River on a massive work of masonry 338 meters long with almost 5 kilometers of earthen embankments. To make it strong enough to withstand the yearly floods, the Solani Aqueduct had to be supported on 300 cubes of brick sunk 6 meters below the river bed.

After that the canal flowed unimpeded down the tableland. Here Cautley's problem was to calculate the correct slope in order to avoid the twin evils of canal building, scouring and silting. Using the formulas devised by two French hydraulic engineers, Dubuat and Prouy, he designed an incline of 24 centimeters per kilometer in order to obtain a water velocity of 1.08 to 1.23 meters per second.

While the canal and its upper works dwarfed anything done before, the methods of construction would have been familiar to the Moguls and the ancient Romans. The technique of sinking cubes of masonry into the bottomless sand of a river bed had been known to Indian engineers for centuries. Excavation was done by laborers wielding mattocks and carrying off the dirt in baskets, on the backs of donkeys, or in oxcarts. Cautley did try new methods, but with little success. He imported an English machine to mold bricks, and light rails and side-tilting wagons to haul dirt. He experimented with a tiny locomotive to pull the wagons:

> The engine, however, did not turn out a success; meeting with a bad accident it was discarded at the end of a few months' use. Shortly afterward it was dismounted and utilized for driving machinery in the workshops at Roorkee. Inexperience in management—perhaps a little prejudice against it—and the want of proper facilities for expeditiously carrying out repairs, were no doubt the chief causes of the failure.[10]

The Ganges Canal was inaugurated on April 8, 1854. Right after the ceremonies, Cautley left for England, having lived in India for thirty-two years. Soon the canal revealed a serious defect: the water flowed too fast—1.22 to 1.43 meters per second—whereas its velocity should have been less than 1 meter per second to avoid scouring its bed. Neither Cautley's earlier experience nor Dubuat and Prouy's formulas had been a good guide to building such a huge canal. As Joyce Brown explains, "In reality Cautley, working as he was in the 1850s on large channels carrying high flows through sand, was better placed to add data to the search for empirical design formulae than the existing formulae were in a position to guide the design of so singular a project as the Ganges Canal."[11] Before the problem could be corrected, the Rebellion of 1857 broke out, damaging parts of the canal and turning canal engineers like Richard Baird Smith back into artillery officers. Not until the drought of 1861–62 was the canal given its full flow of water, and its financial and scouring problems were not solved until the 1870s.[12]

Irrigation works are usually judged by the area they water and by their financial returns. On these criteria, the Ganges Canal eventually became an enormous success. By 1919 the canal brought water to over 530,000 hectares and returned 11.7 percent on its initial outlay of £4 million. But there is another way to measure a canal's performance, as Macgeorge explained: "As many lives were saved by it in Bengal during the year 1865–66 as perished during the same terrible year in Orissa. . . . It is estimated that the canal in that year fed little short of 2½ millions of people. In the same year it repaid to the country more than its total cost."[13]

While the Ganges Canal was being built in the North West Provinces, the Madras Presidency was also engaged in canal building. Here again the incentive was famine. The decade 1832–1841 was particularly awful, with four years of famine and three of scarcity. Even as the land lay parched, two great rivers, the Godavari and the Krishna, flowed unimpeded into the sea. Like the Cauvery, they originate in the Western Ghats and flow eastward across the Deccan into the Bay of Bengal, draining catchment areas of 300,000 and 250,000 square kilometers, respectively. Their flow ranges from a trickle during the dry season to over 28,000 and 20,000 cubic meters per second, respectively, at the height of the monsoon. In their deltas were channels to carry their floodwaters to the fields. At their best, these inundation canals watered 41,000 hectares, less than one-fortieth of the potentially irrigable land in the two deltas, and then only for the two months of the year when the monsoon was at its height.

Maj. Arthur Cotton, having just completed the anicuts across the Cauvery River, proposed to do the same for the Godavari. Here, as elsewhere in India, the British authorities were attracted to major projects of this sort and ignored the myriad tanks and channels which peasants used throughout the Deccan. The reasons were both administrative and political. Administratively, Company rule was spread very thin. As David Ludden explains, "Madras concentrated its slim executive ability on very large projects, where administrative costs were relatively low, and where its few engineers could maintain control of project operations." As for the politics, "rulers had traditionally built on a scale embodying their stature, concentrating their patronage on the most productive, river-irrigated lands, where landowners were wealthy and high caste, and where trade and crafts were highly developed."[14]

In 1846 the Company sanctioned the Godavari Delta project, and work began. At the head of the delta four masonry weirs were built, totalling 3,600 meters in length, with another 2,200 meters of embankments. The water thus regulated was distributed through a growing network of canals and distributaries. A land which had once known famines became covered with rice paddies and orchards. The irrigated area grew to 283,000 hectares in 1890–91, and to over 400,000 in 1919–20.

In 1851 the Company sanctioned the Krishna Delta project designed by Major Cotton and Capt. C. A. Orr. A masonry weir over 1,100 meters long and rising 7 meters above the river bed was completed four years later. After that, as in other deltas, irrigation spread through a growing network of channels; by the early 1890s it watered about 150,000 hectares.[15]

The Era of Private Irrigation (1854–69)

To the East India Company and the presidencies of Bengal and Madras, the results of the first irrigation projects were highly satisfying. Not only did they prevent famines with their attendant loss of lives and tax revenues, they paid handsome returns even in rainy years, either through the sale of water in the dry months or through increased tax assessments.

The trouble was that no one really knew how to quantify these highly satisfying returns with any accuracy, nor how to calculate the potential results of future irrigation schemes. Accounting procedures in the Madras Presidency understated the investments required by irrigation works because they assumed that precolonial anicuts and canals were "free." Furthermore, they overestimated the benefits of increased assessments on all delta lands, not just irrigated fields.[16]

The governments of India recognized their inability to calculate the costs and benefits of irrigation, let alone plan future projects with any degree of rationality. In Madras, where 70 percent of public works expenditures went for irrigation, the presidency government established India's first Public Works Department in 1852 to take over these duties from the Military Board. Two years later Governor-General Lord Dalhousie followed this example for all of India. Henceforth policy and administration were in the hands of civilians, though military engineers were to predominate in the execution of new works until the end of the century. These measures did not immediately provide a solution to the major administrative problem of irrigation schemes: their financing. Since such projects were enormously expensive and took decades to bear fruit, they could not simply be paid for out of current budgets but required some form of long-term financing. In the 1850s, however, large amounts of capital could only have been raised in London, for India did not have a money market able to make long-term loans to the government; and the idea that British investors should lend money to the Indian government was still too radical for either party to consider seriously. As a result, before that came to pass, India had to undergo some very strange experiences in canal building and financing.

One source of trouble was Arthur Cotton himself, a man who combined great engineering talent and the enthusiasm of a technocratic visionary with very poor administrative skills. In 1854 he published *Public Works in India: Their Importance with Suggestions for their Extension and Improvement* (London, 1854), in which he argued that India needed navigable rivers and canals rather than railways, and that hydraulic works everywhere would prove as cheap and profitable as in the Madras deltas. Cotton's prestige and the success of his delta projects coincided with a buoyant mood among British speculators, producing a canal mania:

> At a meeting held at Moorgate Street in April 1854, it was proclaimed that the Madras works in aggregate paid a return of 70 per cent., and that they would soon yield upward of 100. In some instances, it was said, the profits had amounted to 140 per cent., and 400 per cent. was not impossible. With such an El Dorado in prospect it was proposed to form a Company of Water Merchants to promote irrigation and navigation canals all over India.[17]

These speculative schemes were greatly aided by the Rebellion of 1857–58. Faced with the depleted finances of the Indian government, the first secretary of state for India, Lord Stanley, found the idea of privately funded canals rather appealing. Meanwhile Cotton had been busy gathering support for further projects. Among them was a grandiose plan to link Karachi, Calcutta, and Madras by 6,400 kilometers of navigable canals and rivers. To carry it out, he supported the formation of the Madras Irrigation Company in 1858. Five years later it was incorporated with a capital of £1 million and a government guarantee of 5 percent. By then it had already completed work on the first section between Kurnool and Cuddapah and had exhausted its capital as well as a government loan of £670,000. The company continued to run deficits until 1882, when the Indian government bought it out for over £2 million.

A similar fate befell the other private venture, the East Indian Irrigation and Canal Company. This firm was founded in 1860 and incorporated a year later with a capital of £2 million to build canals in Orissa. This province was regularly visited by terrible famines, even when food was available in neighboring provinces, because it had no means of transportation other than bullocks, which also starved. The company built several canals, but at a much higher cost than anticipated. Its works were insufficient to prevent the Orissa famine of 1865–66, in which a million people died. In 1869 the government bought it out.[18]

With hindsight—which was exercised by the private canal companies as early as 1866—it is not difficult to see why these ventures failed. They were the result of overoptimism, ignorance, and incompetence. The overoptimism came from false expectations raised by the exaggerated profitability of Cotton's original delta works. A more serious fault was the promoters' ignorance of Indian agriculture. Irrigation worked best in places like the Jumna-Ganges Doab and the Punjab, where the soil was rich but rainfall was never sufficient for food crops, or, as in the Madras deltas, where there was a centuries-long tradition of farming rice and other wet crops. In rain-watered areas, farmers had no incentive to irrigate in years of normal rainfall, and the canal companies could not cover their costs with only the income from years of drought.[19] Another mistake was the companies' ignorance of Indian society:

> In Orissa . . . the land tenure, like that of other parts of Bengal, placed the peasant at the mercy of the Zamindar. . . . Whatever profits might be obtainable by irrigation would go to the absentee Zamindar, while the labour and the loss would fall upon the ryot [peasant] alone. The cultivators of Orissa, though many of them of the Brahman class, were noted even among Hindus for their ignorance, dullness and lack of energy. Their inertness may to a large extent be traced to the land system which the Government had been unwise enough to establish, and the disastrous results of which are now made manifest here and everywhere. To have spent large sums upon an irrigation scheme in a country with a considerable rainfall, and held under a tenure discouraging all improvement, was palpably a blunder—an unpardonable blunder from a financial standpoint reflecting strongly upon the intelligence of those responsible for it.[20]

The canals cost twice their estimate because of engineering errors and overbuilding, or what the Australian irrigation engineer Deakin called "the universal ambition of engineers to preside over massive and handsome masonry headworks."[21] They also suffered from the success of railways, which deprived them of their transportation value. In the end they were only used during severe droughts.

Productive and Protective Works (1866–98)

Though it allowed private enterprise in canals, the Indian government remained conscious that it had abdicated its duty. One of its earliest acts after its transfer to the Crown in 1858 was to study the question of expenditures on public works, which had been handled in such an ad hoc manner by the East India Company. A committee chaired by Maj. Richard Strachey divided these expenditures into "state works" (i.e., barracks, schools, courthouses) and "works of internal improvement"; of these it wrote:

> The obligation of the government in respect of the construction of these works is . . . essentially based on the idea of their being *profitable* in a pecuniary point of view . . . to the entire body politic of the State (both government and community, as partners). If it cannot reasonably be predicted that such a work will be *profitable* in this sense, it should not be undertaken.[22]

This set the tone for the next generation of canals. In 1864 a new secretary of state, Sir Charles Wood, issued a "Minute on Irrigation," declaring that henceforth the government could borrow money to finance "productive" irrigation projects, that is, those that were expected to bring in sufficient income from tax revenues and the sale of water to defray their interest payments. This became policy two years later when the Indian government floated loans in London to remodel the Ganges Canal at a cost of over £500,000.

In 1867 Richard Strachey was appointed inspector general of irrigation for all India, with the responsibility of developing uniform engineering and accounting methods. Soon thereafter numerous plans were issued to irrigate large parts of the North West Provinces and the Punjab at a cost of some £30 million. As in the days of the Company, irrigation schemes were so grandiose and the peasants so lowly that decisions all flowed downhill, like the waters they brought. Deakin, who came from a land of vociferous democracy, was rather surprised:

> The Indian ryot . . . is never consulted in any way or at any stage in the construction. Government initiates designs and executes the work, offering him the water if he likes to take it, and relying only upon his self-interest to induce him to become a purchaser. . . . Upon all "major" schemes the Government acts of its own notion, at its own responsibility, and acknowledges no title in those who use the water to criticize its proposals. In an equally peremptory way it ignores riparian rights, or makes but small compensation for actual injury done or land taken.[23]

In the 1860s there emerged a new approach toward the irrigation-agriculture nexus, one that was in some ways more enlightened, but in others even more despotic. It involved not only bringing water to the farmers, but putting both water and farmers on previously barren lands. The Punjab was especially suited to colonization. It received only 250 to 400 millimeters of rain a year, on the average, and some years as little as 150. Yet it was crossed by five major rivers, the Jhelum, Chenab, Ravi, Beas, and Sutlej, which ran roughly parallel across a land of easy gradients and good soils. Throughout the growing season, these rivers ran full, from melting snows between March and May and from the monsoon between June and October. Here were the ingredients for the largest irrigation system on earth.

The model for the colonization schemes was the Upper Bari Doab Canal, a reconstruction and enlargement of Shah Jehan's Hasli Canal, which supplied water from the Ravi River to the environs of Lahore and Amritsar. The work was undertaken right after the conquest of the Punjab in 1849 to give work to Sikh soldiers after their armies were disbanded. According to Sir Henry Lawrence, lieutenant governor of the Punjab, "the surest means of correcting the roving military spirit which consumed the people was to attach them to the soil."[24] Many errors were made in the design of this canal. Like the Ganges Canal, it was given too steep a slope, and it ended up costing three times its estimate, irrigating far less land than had been anticipated, and running at a deficit for many years. Yet in its social purpose—to settle thousands of people on previously barren land—it was a success.[25]

The Upper Bari Doab Canal was followed by a more ambitious work, the Sirhind Canal. It was first proposed in 1861 by the Maharajah of Patiala, whose lands were to receive some of the water. After difficult surveys and negotiations with the government of Punjab, a treaty was signed in 1869 and work began. At the head of the canal where the Sutlej leaves the Himalayas, the engineers faced the same difficulties that Cautley had encountered on the Ganges: broken terrain, spongy soils, violent floods, and isolation in a remote and sparsely populated district. Deakin describes some of the challenges:

> The officers were required to seek for their raw materials, or buy and test them, dig and burn their own lime, quarry and carry their own stone, find and grind their own soorkee [burnt brick], make their own bricks and kilns, repair and partly design their own machinery, build and work their own railways, make all the surveys, supervise the construction down to the minutest details, train their own carpenters, masons and engineers, catching them first and watching them afterwards. . . .[26]

Despite the references to railways and machinery, traditional construction methods predominated. Three jails were built near the headworks to house the convict labor force: "Practically the whole 900 million feet of excavation in the main line were removed in baskets on the heads of men and women without the use of mechanical appliances of any sort."[27]

Conscious of the flaws in the Ganges and Upper Bari Doab Canals, the project engineer, Colonel Hume, designed the Sirhind with a gentler slope, aiming at a velocity of 0.88 to 0.98 meters per second. His calculations were mistaken, for the canal silted badly and needed to be given new headworks with a still pond to remove the silt. Yet the constant and minute observations recorded along the canal eventually led R. C. Kennedy, chief engineer of the Punjab, to devise the formula for a stable-regime canal, one that would neither silt nor scour. And the Sirhind Canal itself was a success, irrigating 324,000 hectares by 1892 and 648,000 hectares by 1921, with a return of 12 percent on the investment.[28]

Not all the projects of the late nineteenth century were new. Some were extensions or reconstructions of existing systems. Two such works were the Lower Ganges and Western Jumna Canals. The Lower Ganges Canal, which got some of its water from the Ganges River at Narora, joined the Upper Ganges Canal 90 kilometers later, extending the latter by some 800 kilometers. Built between 1871 and 1878, it commanded 480,000 hectares, bringing the total in the Ganges Canal command area to over a million hectares, half of which received water each year.[29]

The full benefit of the Lower Ganges Canal was delayed for a decade by one of India's natural disasters. In 1878 engineers had built an aqueduct to carry the canal over the Kali Nadi River, allowing for a possible flood of over 500 cubic meters per second with a crest 4 meters high, an estimate based on interviews with older farmers. In 1884 the river rose almost 7 meters, carrying 1,100 cubic meters per second, and tore away part of the aqueduct. The following year, just as the aqueduct was being repaired, a cyclonic storm dumped half a meter of water on the area in twenty-four hours, and the Kali Nadi swept down at a rate of 4,000 cubic meters per second, eight times more than the engineers were prepared for. Not only the aqueduct but every bridge over the next 240 kilometers was washed away. After the disaster, the engineers knew to set their sights higher. To build a new aqueduct took 4,000 workers five years.

Finished in 1886, it rested on piers over 6 meters in diameter sunk 16 meters below the river bed. It was quite simply the biggest acqueduct ever built, as befit its opponent, the forces of nature in India. It also cost the enormous sum of £445,700, enough to make the Ganges Canal system run at a low rate of return, despite the benefits it brought.[30]

The Western Jumna Canal, one of the first works of British irrigation engineering in India, had revealed two serious flaws since its opening in 1821: saline efflorescence, which poisoned the soil, and waterlogging, which spread malaria. By the 1870s canal engineers knew how to solve these problems by flushing and draining the land. In 1873 the government authorized them to proceed with the remodeling of the Western Jumna system including the canal, distributaries, and drainage channels. The cost was high—£432,-764—but paid off in the drought of 1876–78 when the canal kept 160,000 hectares watered. It was extended in later years, and by 1919–20 it was irrigating 344,000 hectares and bringing in a profit of 11.25 percent on the total investment of £1,750,000.[31]

Encouraged by the success of these projects, the Indian government was ready by the 1880s to undertake a truly massive irrigation scheme, comparable in scope to the Ganges Canal. This was the Lower Chenab Canal Project to irrigate the desert between the Chenab and Ravi rivers in northern Punjab. The project was sanctioned in 1889 and completed by 1892. From an engineering point of view the location was almost ideal. With few natural obstacles in their way, Maj. S. L. Jacob and his engineers designed a shuttered weir 1,250 meters long across the Chenab River, diverting the equivalent of six times the Thames River. Without silting or scouring, the canal then carried this water to an extremely dry but fertile land, the Rechna Doab.

Revenue officials divided the land into tracts, each with one village and one distributary channel. Within each tract, the land was parceled out into farms of 16 to 20 hectares. Whole peasant communities from the overpopulated parts of the Punjab were brought in to settle these farms. Revenue officials screened the settlers carefully, allowing no young or old people and no debtors or "loafers." True to their heritage, the British officials sought to create a structured society in the new lands: "Grants larger than the ordinary peasant grant are made to hereditary landowners of more substance and of better social status than the ordinary cultivator, while still larger allotments are sometimes conferred on men of means willing to experiment in improved methods of cultivation and irrigation."[32] By all accounts, the scheme was a huge success. The canal irrigated 445,000 hectares, almost as much as the Ganges Canal. In less than ten years the population of the district grew from 8,000 to 800,000 inhabitants. The crops they raised were worth £16 million a year, five times the cost of the canal, and the return on the government's investment reached the awesome figure of 45 percent per year by 1919–20. It was, in its day, the largest and most successful irrigation system in India, and probably in the world. The Lower Chenab Canal and similar projects nearby more than doubled the government-irrigated area of India, from 2.4 million hectares in 1880–81 to 5.43 million in 1895–96. By 1914 the Punjab alone had more irrigated land than Egypt, and India was a net exporter of foodstuffs.[33]

Yet the irrigation boom of the 1880s left most of India untouched and as vulnerable as ever to the vagaries of the weather. The policy of investing in "productive" works was economically wise, but socially dangerous, as became clear in the drought of 1876–78. The famine it caused was one of the most ghastly in Indian history. Despite relief efforts which cost around £10 million, millions of people starved, along with their cows and bullocks, the engines of Indian agriculture.[34] As in the past, such a calamity led to a reassessment of government policy toward public works. In a report issued in 1881, the Indian Famine Commission recommended two sorts of "protective" works, designed not to bring in a profit but to alleviate famines in time of drought. One sort was "famine railways" to link famine-prone areas to the trunk lines or harbors and allow relief shipments when the bullocks failed. The other was "protective" irrigation works, of the sort that farmers had no use for in years of normal rainfall, but desperately needed during droughts. The Orissa Canals, built by the ill-fated East Indian Irrigation and Canal Company, were just that sort of works, though this had not been their builders' intention. Deakin commented on them:

> Thus, after all, a permanent protection has been afforded at no greater cost than would be spent in two years in the spasmodic effort to save life and mitigate disaster. If the works had been undertaken with this end alone they would be pronounced successful. In such an aspect they would have fully justified their existence. The failure has been in theory and prophecy, and the success in fact.[35]

The Indian government accepted the principle of protective works in 1881. In some areas, like the North West Provinces, canals were physically possible but economically unprofitable. In the hinterlands of Bombay and Madras, irrigation in dry years required the storage of water from one year to the next. Small tanks built by villagers already dotted the Deccan landscape, but larger reservoirs required government funding. This led to the building of dams and tanks on a small scale. Unfortunately neither the Indian economy nor the government's budget had much surplus to invest in protective but money-losing projects. These were to be left to the twentieth century.[36]

Despite its tight budget, the Public Works Department built one work that was particularly interesting from an engineering standpoint: the Periyar Project. In southern India, the summer monsoon drops most of its rain on the Western Ghats, a mountain chain paralleling the Malabar coast, leaving little for the lands further east. The district of Travancore, west of the Ghats, receives over 3 meters of rain a year, while Madurai, just to the east of it, averages 800 millimeters, with one dry year out of every two. Throughout the nineteenth century, technical visionaries had dreamt of capturing some of the excess rain that fell on Travancore and diverting it to Madurai.

In 1884 the government approved a plan drawn up by Major Ryves twenty years earlier and revised by Maj. John Pennycuick. It involved a dam on the Periyar River in Travancore and a tunnel through the Ghats to the Vaigai River in Madurai. The work was carried out in an inaccessible gorge 1,000 meters up the mountain and covered with dense rain forest. The dam itself was built of stone and concrete, the first such structure in India, and held 443 million cubic meters of water. To carry water through the mountain required 1.6 kilometers of cutting and a 2-kilometer-long tunnel through solid rock. All this, as usual, was done without benefit of machinery. It cost a million pounds and took eight years to complete (1887–95), yet it ended up providing a modest profit of 5.5 percent to the government and a secure water supply to 50,000 hectares of land.[37]

The Periyar Project attracted attention for another reason: its hydroelectric potential in a region devoid of coal. The multiple-use concept was first proposed by Professor Alfred Chatterton of the Madras Engineering College and reviewed by a committee of engineers including James George Forbes, who was then designing the generators for Niagara Falls. Problems of long-distance electrical transmission and political disputes between Madras and Travancore State prevented this plan from being implemented, but it provided lessons for later hydroelectric projects.[38]

The Indian Irrigation Commission and After (1897–1940)

As the nineteenth century came to a close, large irrigation works seemed to have accomplished a great deal of good. The major rivers of the Punjab and the United (formerly North West) Provinces had been tapped, a region as large as some European nations was being irrigated for the first time, and India had become a food exporter. Other parts of India either had adequate rainfall or railways to bring in food in years of drought.

Then just as complacency was setting in, disaster struck again. The years 1896–97 and 1897–98 were dry. After a year's respite, the drought returned in 1899 and 1900. Famine followed upon famine from 1899 to 1903. Despite £4 million in famine relief, millions of people starved. In 1901, Viceroy Lord Curzon appointed a commission chaired by Sir Colin Scott-Moncrieff "to Report on the Irrigation of India as a Protection against Famine." In its report, issued two years later, the commission gave a balance sheet of irrigation. Of the 91.5 million hectares of cropped land in British India, 5 percent were irrigated by major canals, another 3.5 percent by minor government works, and 11.5 percent by private tanks, springs, and wells. The commission preferred protective railways to protective irrigation schemes and advocated that irrigation funding be concentrated in the areas of greatest potential profit, namely the Punjab, and the Indus Valley. Unfortunately, however, the easiest projects had already been completed, and further progress required storage dams and barrages over the major rivers, both extremely costly. As always, the official mind was attracted to projects of gargantuan dimensions. Great as were the works of the nineteenth century, they were dwarfed by those of the twentieth.

The first of the great irrigation schemes to follow upon the report of the commission was the Triple Canals Project. (See Figure 4, Irrigation Canals in the Indus Watershed.) Its goal was to water the Bari Doab between the Sutlej and Ravi rivers. Unfortunately the Ravi did not carry enough water, and Sutlej water was needed in the southern Punjab and in the State of Bahawalpur. James Wilson, set-

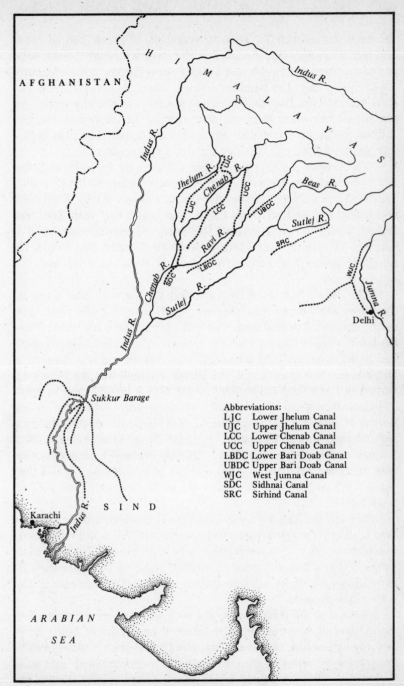

Figure 4 Irrigation Canals in the Indus Watershed

tlement commissioner for Punjab, and Colonel Jacob, retired chief engineer of Punjab and designer of the Lower Chenab Canal, suggested tapping the Chenab and Jhelum rivers further north. According to their plan, water from the Chenab would be carried across the Ravi and into the Bari Doab, leaving the Ravi practically intact. At the same time, water from the northernmost of the five rivers, the Jhelum, would be diverted to replenish the Chenab. The plan therefore involved three major canals and one river crossing.

The project was approved, and work began in 1905 with the first two canals, the Upper Jhelum (from the Jhelum to the Chenab) and the Upper Chenab (from the Chenab to the Ravi). The third one, called Lower Bari Doab Canal, was begun two years later. At Billoki, where the canal crossed the river, the engineers designed a barrage with thirty-five 12-meter-wide steel gates that could be raised or lowered to regulate the flow of both the river and the canal.

The Triple Canals Project marked the start of a new era in irrigation. Unlike previous schemes which involved one river, one canal, and one irrigated area, this one encompassed the entire Punjab basin with its many rivers and regions. All the interests involved had to be consulted and balanced. The engineering was also on a larger scale than ever before; the Billoki Barrage was the largest of its sort in India, and the Upper Chenab Canal carried more water— 331 cubic meters per second—than any in the world. The main canals were completed by 1915, and the 12,000 kilometers of distributaries and channels shortly thereafter. By 1919–20 the system commanded 1.6 million hectares and irrigated 692,000 hectares each year, more than the Ganges Canals. The value of the additional crops grown that year—£9.3 million—was almost equal to the project's total cost of £10.5 million.[39]

Though the Triple Canals Project was completed during World War I, all other major projects were postponed. When the war ended, the government took up irrigation with enthusiasm. The irrigation works of the interwar period resembled those of the prewar, with two differences: technical innovations, and a further increase in the scale of the projects.

Technical innovations affected both the type of irrigation and the method of construction. The tubewell had some of both, for it was dug by machine, and water was pumped up from 8 meters below the ground by diesel or electric motors. Experiments with tubewells began in the Punjab before World War I. They spread slowly, because of the high cost of motors and the lack of electric power, and only became a prominent feature of Indian agriculture after Independence.[40] Though masonry storage dams dated back to the Periyar Project, several more were built in the early twentieth century. Not until the late twenties, however, was a reinforced concrete dam, the Mettur on the Cauvery River, built in India. And the first reinforced concrete barrage, at Trimmu on the Jhelum, dates from 1937–39. Similarly, machines long known in the West such as bulldozers, dump trucks, and graders were not used in India until the 1920s. In a land of cheap labor and little industry, such techniques were simply not competitive.

Though the techniques were slow to evolve, the scale of projects continued to grow. Two in particular broke all previous records: the Sutlej Valley Project and the Sukkur Barrage.

The designers of the Triple Canals Project assumed that the Sutlej waters were reserved for the lands to the south of it, where there were already a few inundation canals. The Sutlej River could provide up to 1,358 cubic meters per second depending on the season, enough to command 1.4 million hectares and irrigate 800,000 hectares each year. Handling all this water, however, required four barrages over the Sutlej and eleven major canals. The project was begun in 1921; three barrages were completed by 1927 and the fourth in 1933.

At that point the Indian government was deeply engaged in an irrigation scheme on the Indus which created conflicts of interest, first over funding, later over water. The mighty Indus had kept the engineers at bay for almost a century, precisely because it was so large and violent and changed its bed with every flood. In Sind, along its lower course, several inundation canals had been restored by the British since 1852; but they only carried water for a few weeks of the year and needed to be cleared of silt after every flood. There was one place, however, where the river could be tamed: at Sukkur, where it crosses a limestone ridge. Every decade since the 1840s engineers had drawn up proposals to build a barrage there, to feed a network of perennial canals. The Indian Irrigation Commission recommended further study of the question. The project was finally approved in 1923.

The Sukkur (or Lloyd) Barrage was 1.6 kilometers long, with sixty-six steel shutters, each over 5 meters high, to regulate the flow of the Indus. When it was finished in 1932, it headed the largest irrigation scheme on earth, capable of distributing up to 1,288 cubic meters of water per second over 2.2 million hectares of land, of which 1.2 million hectares were formerly waste.

Both the Sukkur and the Sutlej Valley projects converted useless scrub, or at best one-crop lands, into fields capable of yielding two crops per year, usually rice or cotton in the summer and wheat or pulses in the winter. The cotton fields of Sind, in particular, needed maximum water from March until June, more than the Indus alone could provide. The Indus basin, a hydrological whole, was split politically into the provinces of Punjab and Sind and the states of Bikaner and Bahawalpur. Water distributed to the farmers in one area was denied those in another. Regional jealousies were exacerbated by religious differences between Hindus, Muslims, and Sikhs. In 1935 the Indian government convened a Committee on the Distribution of Waters of the Indus. As a consolation to the Punjab, this committee recommended yet another plan, the Haveli Project at the confluence of the Jhelum and the Chenab. Nonetheless, the political problems remained. What was, under British rule, a masterpiece of engineering and regional planning was to become a source of disputes after Independence, adding to the tensions between India and Pakistan.[42]

An irrigation system is neither dramatic nor romantic. It takes much longer to build than any other public work and provides no return until it is almost completed. A layman finds it uninteresting to look at unless he remembers the waste lands it replaced. And when it is finally in working order, it only provides what everyone expects to have and then takes for granted: water and food. Hence irrigation has had few poets and publicists to sing its praises compared to those other monuments of the Raj, the railways and the cities.

Yet irrigation has had its critics, and their major complaint was the insufficiency of the British efforts. The economist Romesh Chunder Dutt pointed out that under British rule up until March 1880, £125 million had been invested in railways, but only £12 million in irrigation. This he attributed to the fact that "Englishmen understand railways, and do not understand the importance of irrigation for India."[43] The same argument appears in other works, both Western and Indian. Thus Lidman and Domrese state: "England has rain throughout the year, so the average Englishman recruited and sent to India could have no idea what irrigation means to a country like India."[44] And S. R. Sharma writes: "The British gave priority to the construction of railways over the construction of canals, because the former facilitated British trade in India, whereas the latter benefited agriculture, for which the British trader had little or no concern."[45]

These accusations are certainly true for the nineteenth century. After 1860, irrigation works were built cautiously and only after searching investigations into their profit potential. Railways, in contrast, were created by bold entrepreneurs who felt they could do no wrong since they had a government guarantee of profits to fall back on; or else, if built by the government, they were often designated "strategic" or "famine" railways, hence beyond considerations of profit and loss. By the 1930s, expenditures on irrigation began catching up with railways. In the long run, that is to say by Independence, irrigation works proved to have been a better investment of government funds than railways.

The cultural explanation for the contrast between railways and irrigation is interesting. Yet one need not believe that the minds of Englishmen were so soaked with drizzle that they could not understand the need for irrigation in India. Rather, self-interest played a part in the tilt toward railways. Railways did not just serve traders to transport goods. They also carried every British official in India who, without them, would have had to spend half his tour of duty traveling to and from his post on horseback or in a country boat as in the days of the East India Company. Railways also carried troops and made British power and every British person in India feel safer; if there had been railways in 1857, there might have been no Rebellion. And railways had their advocates back home, for they were built with British steel and British machinery, unlike canals which used little imported equipment.

Irrigation systems, in contrast, did little but feed millions of Indians. Though British administrators were very concerned about drought and famine, none ever actually starved. Both moral values and the profit motive played their part, but their personal involvement with irrigation was simply weaker than with railways. It is not surprising that the British preferred railways, nor that India and Pakistan built more irrigation works in the twenty years after Independence than the Raj had in a hundred.

The irrigation systems which the British built in India were of global importance. When the East India Company sent Capt. Richard Smith to Italy in the early 1850s, he found little to learn.

The army officers who came from Britain in the early years brought with them little knowledge other than the rudiments of mathematics and design and a willingness to experiment. The fundamentals of hydraulic science and the practices of irrigation engineering came out of the great irrigation works of India itself. For India was a laboratory of hydraulic engineering and a school from which this knowledge spread to other lands. Deakin noted that many of Australia's irrigation engineers had begun their careers in India. The reason was that "India has run the whole gamut of irrigation works, and is therefore incomparably superior, as an engineering school in connection with irrigation, to any country in the world."[46] Sir William Willcocks, the man who designed Egypt's Aswan Dam, trained at the Thomason Civil Engineering College at Roorkee on the banks of the Ganges Canal, later worked for the Indian and Egyptian public works departments, and finished his career in Mesopotamia.[47]

Irrigation is unusual among imperial technologies in that it developed in a colonial setting and was transferred from colony to colony by the colonizers. The geographer Aloys Michel summed up this situation: "By the time the Union Jack came down in Lahore, British engineers had not only given the Indus Basin the most extensive irrigation system in the world; they had developed most of the formulas now used everywhere in canal construction and operation."[48]

Egypt, the Nile, and the British

In Egypt, inundation irrigation predates civilization. Every year peasants laboriously built dikes enclosing basins covering from a few hundred to 20,000 hectares. When the Nile rose in July the basins were flooded. Then, as the river subsided, water was retained from six to eight weeks to allow the fertilizing silt to settle and the ground to become thoroughly soaked. By late October the fields were ready for planting. Crops were harvested in the spring and the fields left fallow until the next flood.

There were also canals to bring water to fields far from the river. As the water level fell in the fall, so did that of the canals, and by spring only a few had water in them. Farmers fortunate enough to own lands next to the Nile or the deeper canals used a *saqiya* (a chain of pots turned by an ox or a donkey), a *shaduf* (a bucket on a pole with a counterweight), or an Archimedean screw to lift water onto their fields. In places where the water table lay close to the surface, farmers could draw water from wells. All of these methods required energy, and hence summer crops were uncommon and costly. Thus it had been for fifty centuries or more.[49]

The initiator of modern irrigation in Egypt was Mohammed Ali, who ruled from 1805 to 1849. His goal was to turn Egypt into a military power independent of the Ottoman Empire, and his method was forced industrialization and Westernization. To pay for machinery, fuel, weapons, and other imports, Egypt needed to increase its exports of cotton, in which it had a comparative advantage. But cotton plants need water in the spring; hence Mohammed Ali's interest in perennial irrigation.

Two techniques were used to bring water to the fields when the Nile was low. One was summer canals, which Mohammed Ali began to build as early as 1816. These canals were dug deep enough so that water flowed in them even when the river was at its lowest. To dig them out, the government drafted hundreds of thousands of peasants for two or three months of unpaid labor each year, the same system as used by the pharaohs. Their work was never finished, for each year the Nile flood filled the canals with silt, and each year the peasants had to dig them out again.

To overcome this problem, Mohammed Ali planned to build two barrages on the Rosetta and Damietta branches of the Nile, at the head of the delta just below Cairo. Like the anicuts of southern India, their purpose was to keep the river level constant all year, and, through sluice gates, to distribute its waters as needed between the branches and canals of the delta.

To carry out his plan, Mohammed Ali turned, in this as in so many other fields, to French engineers. Work began in 1833 under Linant de Bellefonds but was interrupted by an outbreak of plague in 1835 and abandoned two years later when Linant became minister of public works. It was resumed in 1843 under another engineer, E. Mougel. But work went slowly, for the laborers, all unpaid conscripted peasants, were unskilled and poorly motivated. In 1847 Mohammed Ali ordered that 1,000 cubic meters of concrete be poured every day, regardless of circumstances. He died the following year, but the foundation of the Rosetta Barrage, poured into running water, was ruined. Nonetheless the work continued under Mougel and later under the Egyptian engineer Mazhar Bey. In 1861 the work was finished, but for two years the engineers did not have the courage to test it. When the gates were closed and the water rose almost 2 meters, ominous cracks appeared in the masonry, so the water was quickly let out again. After that the river was never raised more than half a meter. Nonetheless a section cracked and buckled. The barrage was useless for irrigation, and only served to regulate the flow between the two branches.

It is not easy to pin the blame. The project was perhaps too ambitious for the engineering knowledge of the time, and the art of laying so large a foundation on shifting sand was not yet understood. Also Mohammed Ali was impatient and insisted the foundations be poured faster than they should have been. It was a costly error, paid for by the long-suffering Egyptian peasant, while the benefits were delayed twenty-five years or more.[50]

By midcentury, in any case, the dream of transforming Egypt quickly from a feudal agrarian economy into a complex modern one had failed, destroyed by Egypt's defeat at the hands of France and Britain in 1841, which forced Egypt to open its doors to foreign trade and ruined its budding industries. Mohammed Ali's death in 1848 ended the era of Egyptian national self-assertion. Under his successors Abbas (1848–54), Said (1854–63), and Ismail (1863–79), Egypt was increasingly opened up to foreign businesses and investments, mostly in cotton lands and public works. Cotton output rose sharply from almost none in 1820 to 22,500 tons in 1860, then, thanks to the cotton shortage caused by the American Civil War, to 96,300 tons in 1860 and to 140,600 tons in 1879. This was made possible by increases in perennial irrigation and other investments. As Charles Issawi explained,

> Cotton was the main beneficiary of the government's investment on public works and the magnet drawing private foreign capital to Egypt. . . . And the greater part of imports paid for by cotton exports consisted of consumer goods demanded by the beneficiaries of this rise in incomes. In other words, the large increase in production and exports achieved during this period was absorbed partly by the population growth . . . and partly by a sharp rise in the level of living of the upper and middle classes and a small rise in that of the mass of the population; little of it was reinvested.[51]

Thus a profligate elite and friendly European bankers had made Egypt a slave of cotton long before it was swallowed by the British Empire. Yet when the British invaded Egypt in 1882, they found the task only half finished, because the perennial irrigation of the Nile Valley was still far from its potential.

In 1885, Colin Scott-Moncrieff, under secretary of state for public works of Egypt, wrote an article entitled "Irrigation in Egypt" in *The Nineteenth Century,* in which he asked: "Surely that subject [irrigation] at least is one that the Egyptian understands. . . . Why should England begin teaching Egypt irrigation?" His answer was:

> Irrigation is an art which there is no occasion to practice in England. But there are few forms of agriculture which are not practiced in one or another of Her Majesty's many possessions, and so it happened that from Northern India Lord Dufferin was able to obtain officers possessing the experience required in Egypt.[52]

The officers Scott-Moncrieff refers to were himself and his assistants. Scott-Moncrieff had worked on the Western Jumna and Ganges canals and had taught at the Thomason Civil Engineering College. While traveling through Egypt in 1883 he was offered the post of inspector general of irrigation, and soon thereafter he became under secretary of state for public works, a post he kept until 1890. Edward Sandes, historian of the Royal Engineers, noted:

> Scott-Moncrieff soon realized that he would require able assistants in Egypt, men whom he knew and could trust. The innumerable errors of detail in the irrigation system of the country pointed unmistakably to the necessity of constant supervision by expert subordinates. Naturally, he looked to India for help, and consequently entered forthwith into negotiations for the transfer of four Irrigation engineers to Egypt.[53]

To assist him, Scott-Moncrieff recruited four men with Indian experience. Maj. Justin Ross, who had worked on the Ganges Canal, was put in charge of Upper Egypt. Maj. James Western and Capt. Robert Hanbury Brown, who had worked on various canals in northern India, were given the task of restoring the Nile Barrages and irrigating the delta. Sir William Willcocks, a civilian, came in a subordinate capacity, but later rose to design the Aswan Dam. What these men brought from India was a method of transforming Egyptian irrigation from the basin to the perennial system.

In the course of a year, the Nile varies from a low of 225 cubic meters per second to a high of 14,000 or more. Under the basin system, most of the water ran off to the sea, and what was used for agriculture produced only one crop a year. Yet where water is available, the climate of Egypt allows up to five crops every two years. To put Egypt's farmland under perennial irrigation, however, requires a constant flow of 900 cubic meters per second. In effect, it means storing the flood and letting it out gradually through a series of dams and barrages all along the river from the mountains of East Africa to the delta.

One of the first tasks the British engineers tackled was to investigate the delta barrages. They closed the gates, allowed the water to rise, and measured the width of the cracks. Over the next seven years, they thickened and widened the foundations and installed new gates. In 1890 the water was allowed to rise 4 meters behind the barrages, and the delta canals carried 350 cubic meters of water per second, five times more than they previously had in the low season. As the water in the canals now rose above the level of the fields, even the smaller farmers who could not afford steam pumps had access to water year-round.[54]

By 1890 the British no longer felt they were just temporarily in Egypt to straighten out its tangled finances and put down obstreperous nationalists. Nor were their engineers content to improve a system that had existed since the dawn of history, or fix barrages some Frenchmen had tried to build but botched. Instead, the British began to believe, as they did elsewhere in their empire, that they were the rightful guardians of the Egyptian people, there for a long time to come. And the engineers—especially Colin Scott-Moncrieff and William Willcocks—began to draw up plans for a project of pharaonic dimensions, commensurate with the lengthening timespan of future British rule: the Aswan Dam.

The purpose of the Aswan Dam was to provide Upper Egypt with a perennial supply of water, as the Nile Barrages had done in the delta. The idea dated back to 1860. In 1890, soon after British forces were ensconced in the Sudan and Uganda, William Willcocks was sent to survey the upper reaches of the Nile. Four years later, in a report entitled "Perennial Irrigation and Flood Protection," he recommended a 25-meter-high dam be built at Aswan to store 2.5 billion cubic meters of water. This plan was approved by an international engineering commission, but ran into opposition from irate archaeologists, upset that the reservoir would drown the ancient Temples of Philae. Willcocks suggested to Scott-Moncrieff that the cost of the dam could be met by selling the temples to the Americans and shipping them off to New York, but the idea was rejected. Instead, the consul general of Egypt, Lord Cromer, approved a more modest design of 20 meters, which would only store 1 billion cubic meters of water but would leave the temples above the surface. To this Winston Churchill commented: "The State must struggle and the people starve in order that professors may exult and tourists find some place on which to scratch their names."[55]

The first Aswan Dam was built between 1898 and 1902 by a force of 8,550 workers, of whom 900 were Europeans, mostly Italian miners and stone cutters. It created a reservoir 160 kilometers long and irrigated 270,000 hectares of land. As the Nile in flood carries enormous amounts of sediment, the engineers devised a novel system to prevent the reservoir from silting up. They installed 180 steel gates which were opened during the height of the flood, allowing the silt-laden waters of the Blue Nile from Ethiopia to rush through, then shut them after the flood had passed and the Nile was mainly fed by the clearer waters of the White Nile from Uganda. This system, and the small size of the dam itself, meant that it only handled a third of the water that could theoretically have been used for irrigation.

The early years of the century were prosperous ones for Egyptian cotton. The new dam was so successful that proposals were soon put forth to increase its capacity. In 1908–12 the objections of art lovers were overruled and the dam was strengthened and raised, doubling the capacity of the reservoir and drowning the Temples of Philae. It was raised yet again in 1929–34 to a height of 38 meters, creating a 354-kilometer-long lake with a capacity of 5.7 billion cubic meters.[56]

Aswan was not the only large work built by the British before World War I. They also built the Zifta Barrage on the Damietta Branch in 1901; the Asyut Dam, halfway between Cairo and Aswan, completed in 1910; and the Isna Barrage, 256 kilometers upstream from Asyut, finished in 1912. (See Figure 5.) After the war, they continued their program of irrigation works, but with broader horizons. Since the reconquest of the Sudan in 1898 hydrologists had studied and measured the flow of the Nile and all its tributaries in the Sudan, Ethiopia, Uganda, and Kenya. They installed gauges along the Atbara, the Blue Nile, and the White Nile, and the telegraphs gave the Egyptian Irrigation Service information with which to plan the agricultural seasons. The Sudan Irrigation Service was set up in 1904 as a branch of the Egyptian Service, and did not become independent until 1925. On the principle that the Nile drainage basin is hydraulically one, the colonial rulers favored Egypt at the expense of the Sudan. Even though major dams were built in the Sudan—one at Sennar on the Blue Nile in 1925 and another at Jebel Aulia on the White Nile in 1937—their purpose was primarily to

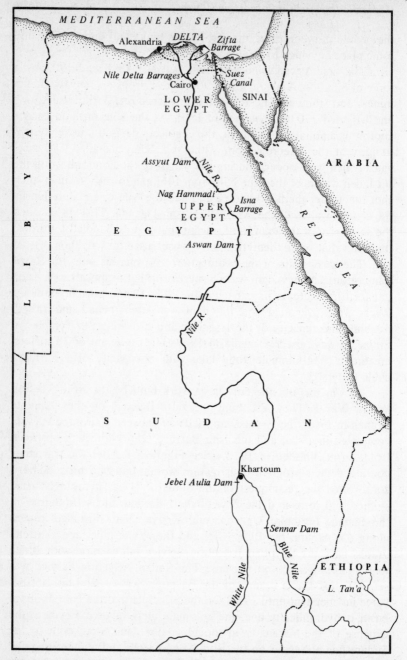

Figure 5 Irrigation Works in Egypt and the Sudan

Table 6.1 Irrigated Land in Egypt, 1821–1937

Year	Cultivated Area (1,000 hectares)	Cropped Area (1,000 hectares)
1821	1,282	1,282
1846	1,581	—
1882	1,998	2,417
1897	2,076	2,825
1907	2,257	3,190
1917	2,230	3,246
1927	2,328	3,579
1937	2,231	3,487

regulate the flow of the Nile into Egypt, and secondarily to irrigate the Gezira region south of Khartoum. This was confirmed by the Nile Waters Agreement signed in 1929 by Egypt and Britain as co-owners of what was then called the "Anglo-Egyptian Condominium of the Sudan."[57]

The result of all these works was to extend the area of perennial irrigation and multiple-crop agriculture in Egypt. John Waterbury gives the figures which are presented in Table 6.1. The cultivated area of Egypt increased substantially until 1907, after which it leveled off. The cropped area, which counts fields twice if they produced two crops a year, expanded until the 1920s, then it too leveled off. At first glance, these figures seem quite remarkable: how many lands of ancient settlement have thus doubled their croplands in a century? On a per capita basis, however, the picture is far less rosy. As the population grew from 4.23 million in 1921 to 15.92 million in 1937, the amount of cultivated land per person fell from 0.3 hectare to 0.14 hectare, while the cropped area per person dropped by one-third, from 0.31 to 0.22 hectare. Given the agrarian nature of the Egyptian economy, only great improvements in the yields per hectare could have compensated for the growth in the population.

Cotton became the dominant crop of Egypt because its yields were so much higher than those of other countries: roughly three times those of the American South, or seven times those of India. But these phenomenal yields peaked in 1897 at 621 kilograms per hectare. After that, intensive cultivation without the addition of fresh silt began exhausting the soil, while over-watering without proper drainage caused waterlogging and salinity. By 1908 the yield had dropped to 335 kilograms per hectare. Both problems were the direct consequence of perennial irrigation.

For decades, the irrigation engineers in Egypt, as in India, had concentrated on providing water on the assumption that the farmers would know how to use it. The drop in yields, however, were a sign that this assumption was flawed. In 1910, therefore, the Egyptian government established a Department of Agriculture to seek solutions to the problems caused by the new irrigation system. A network of drainage channels and pumping stations alleviated the waterlogging and salinity. Chemical fertilizers replaced the natural silting of the basin irrigation system, and by the 1930s Egypt consumed more fertilizer per hectare than any other country.

Yet this was not enough. Cotton is a very delicate and demanding crop which required profound changes in agricultural practices. Research stations and extension services spread knowledge of seed selection, crop rotation, pest control, watering and drainage techniques, and agricultural machines. The larger farms adopted these and other improvements, but at the expense of the poorer farmers who could not compete.

By the 1930s the yields returned to where they had been in the 1890s. Yet this did not solve Egypt's long-range economic problem. Cotton now constituted 80 percent of Egypt's exports. Its value, like that of most other tropical products, was volatile and vulnerable to competition and substitution. The buying power of Egypt's cotton exports, which had multiplied twelvefold between 1848–52 and 1908–12, thereafter fluctuated below its prewar peak. As the population was growing fast, per capita incomes declined. According to Charles Issawi, "Clearly, as far as agriculture is concerned, Egypt may be said to have ended its period of rapid growth in 1914. All subsequent efforts have carried it little beyond the level of that period and future prospects are not bright."[59]

Conclusion

In the drier parts of the tropics, the rewards of irrigation have been known since the dawn of history, but the economies of scale put them beyond the reach of most communities. Hence the symbiotic relationship between powerful governments and large-scale irrigation works. In India and Egypt the British combined the motives of ancient empires with the technologies of the new imperialism. First in India and later in Egypt they took existing irrigation systems as their starting point and, using engineering and scientific practices from the Western industrial world, they developed barrages, perennial canals, storage dams, river basin hydrological policies, and new agricultural practices to go with the new perennial irrigation.

Because of the timing of their conquests, the techniques were first developed in India and later transferred to Egypt. The demand for perennial irrigation was not a result of colonialism, for the nationalist Mohammed Ali felt it more strongly than his contemporaries, the British governors of India. It was the result of free-trade imperialism, especially the industrial revolution in Europe which offered enticing industrial goods in exchange for cotton and other tropical crops. What had to await the new imperialism of the late nineteenth century was the implementation of these projects, which required both benevolent despotism and the latest in engineering techniques.

Yet, if the technology was as similar as local conditions allowed, the results differed somewhat. In India, irrigation systems were designed to protect against famine by growing food even in dry years. As a result they lessened India's dependence not only on the weather but also on the world market. By the interwar period India was on its way toward a balanced (if still very poor) economy. In Egypt, on the other hand, perennial irrigation was the technique by which the country was transformed into a plantation for the cotton mills of Europe. In both cases, the benefits of increased production were largely spent keeping a growing population at a steady or declining standard of living, and to make colonial rule profitable for one more generation.

Notes

1. Richard Baird Smith, *Italian Irrigation, Being a Report on the Agricultural Canals of Piedmont and Lombardy Addressed to the Honourable the Court of Directors of the East India Company,* 2 vols., 2nd ed. (Edinburgh and London, 1855), 1: 305.

2. W. Arthur Lewis, *Growth and Fluctuations, 1870–1913* (London, 1978), p. 205.

3. D. G. Harris, *Irrigation in India* (London, 1923), pp. 8–9; Elizabeth Whitcombe, "Irrigation," in *The Cambridge Economic History of India,* vol. 2: *c. 1757–c. 1970,* ed. Dharma Kumar (Cambridge, 1983), p. 677.

4. Edward W. C. Sandes, *The Military Engineer in India,* 2 vols. (Chatham, 1933–35), 2: 1–2; Alfred Deakin, *Irrigated India: an Australian*

View of India and Ceylon, Their Irrigation and Agriculture (London and Calcutta, 1893), pp. 198–99, 291; S. R. Sharma, "Irrigation," in *The Economic History of India,* ed. V. B. Singh (New Delhi, 1965), p. 164; Harris, pp. 15–16.

5. George Walter Macgeorge, *Ways and Works in India: Being an Account of the Public Works in That Country from the Earliest Times up to the Present Day* (Westminster, 1894), pp. 136–37; Frederick Newhouse, M. G. Ionides, and Gerald Lacey, *Irrigation in Egypt and the Sudan, the Tigris and Euphrates Basin, India and Pakistan* (London, 1950), pp. 40–42; Patrick Fagan, "District Administration in the United Provinces, Central Provinces, and the Panjab, 1818–1857," in *The Cambridge History of the British Empire,* vol. 5: *The Indian Empire 1858–1918,* ed. H. H. Dodwell (Cambridge, 1932), pp. 84–85; Sandes, *Military Engineer,* 2: 3–4; Deakin, pp. 295–302; Whitcombe, pp. 680–89.

6. Robert Burton Buckley, *Irrigation Works in India and Egypt* (London, 1893), pp. 14–15; Henry Cowles Hart, *New India's Rivers* (Calcutta, 1956), p. 17; Richard Baird Smith, *The Cauvery, Kistnah, and Godavery. Being a Report on the Works Constructed on these Rivers for the Irrigation of the Provinces of Tanjore, Guntoor, Masulipatam, and Rajahmundry, in the Presidency of Madras. Published by Order of the Most Noble the Governor-General of India* (London, 1856), pp. 1–17; Sandes, *Military Engineer,* 2: 20–22; Whitcombe, pp. 678–82.

7. On Cautley, see "Sir Thomas Proby Cautley," in *Dictionary of National Biography* 3: 1242–43, and Joyce Brown, "Sir Proby Cautley (1802–1871), a Pioneer of Indian Irrigation," *History of Technology* 3 (1978): 35–89. On the Ganges Canal, see also Buckley, pp. 16–17; Deakin, pp. 179–84; Harris, pp. 20–24; Macgeorge, pp. 148–61; and Sandes, *Military Engineer,* 2: 5–10.

8. Proby Cautley, *Report on the Ganges Canal Works from their Commencement until the Opening of the Canal in 1854,* 4 vols. (London, 1860), 1: 103.

9. Smith, *Italian Irrigation,* 1: 69.

10. Macgeorge, p. 158.

11. Brown, p. 69.

12. Deakin, pp. 179, 188–89.

13. Macgeorge, p. 158.

14. David Ludden, "Patronage and Irrigation in Tamil Nadu: A Long-Term View," *Indian Economic and Social History Review* 16, no. 3 (September 1979): 358–60.

15. On the Godavari and Krishna projects, see Smith, *Cauvery,* pp. 48–138; Sandes, *Military Engineer,* 2: 22–26; Harris, pp. 5–6, 25–31; Newhouse et al., pp. 47–50; and Whitcombe, p. 684.

16. Whitcombe, p. 683.

17. Lionel Jacob, "Irrigation in India," *Contemporary Review* 105 (1914): 803.

18. On the private canal companies, see Deakin, pp. 264–68, 277–78, 283–85; Harris, pp. 32–37; Macgeorge, pp. 172–80, 195–96; and Whitcombe, p. 694.

19. Deakin, p. 268.

20. Ibid., p. 286.

21. Ibid., p. 285; Macgeorge, pp. 170–71.

22. *Report of a Committee . . . on the Classification of Public Works Expenditures* (Calcutta, 1858), p. 12, quoted in Whitcombe, p. 692.

23. Deakin, pp. 233–34.

24. Ibid., p. 199.

25. Buckley, pp. 15–16; Macgeorge, p. 139; Aloys Arthur Michel, *The Indus River: A Study of the Effects of Partition* (New Haven and London, 1967), pp. 58–65.

26. Deakin, p. 213.

27. India, Public Works Department, *Triennial Review of Irrigation in India, 1918–21* (Calcutta, 1922), p. 61.

28. Deakin, pp. 208–19; Michel, pp. 61–72; Sandes, *Military Engineer,* 2: 13–14; J. Allen, "Hydraulic Engineering" in *A History of Technology,* ed. Charles Singer, E. J. Holmyard, A. R. Hall, and Trevor I. Williams, vol. 5: *The Late Nineteenth Century, c. 1850–c. 1900* (London, 1958), p. 542.

29. Harris, pp. 40–43; Brown, p. 82; Macgeorge, pp. 161–62.

30. Deakin, pp. 179–90; Buckley, pp. 16–17.

31. Whitcombe, p. 708; Harris, pp. 16–17.

32. Harris, pp. 50–51.

33. On the Lower Chenab Canal, see Harris, pp. 49–57; Macgeorge, pp. 142–43; Michel, pp. 77–79; Whitcombe, pp. 712–13; and Jacob, pp. 805–6.

34. Deakin, p. 143, says 5,250,000 people starved despite £11,000,000 in relief; Whitcombe, pp. 704–705, puts the numbers at 1,300,000 dead and £9,750,000 in relief.

35. Deakin, pp. 286–87.

36. Buckley, p. 279; Harris, pp. 59–61; Michel, pp. 73–76.

37. Archibald Thomas Mackenzie, *History of the Periyar Project* (Madras, 1899, reprint 1968); Deakin, pp. 257–60; Harris, pp. 63–69.

38. Loren Michael, "The Advent of Hydroelectric Power in India," paper presented to the conference of the Society for the History of Technology, Milwaukee, October 1981; Sandes, *Military Engineer,* 2: 28–29, 45–47.

39. Harris, pp. 71–75; Michel, pp. 84–93.

40. Michel, pp. 451–54; Whitcombe, pp. 722–23.

41. Harris, pp. 90–92; Michel, pp. 445–47.

42. Michel, pp. 104–26; Newhouse, et al., pp. 64–66; Whitcombe, pp. 726–27.

43. Romesh Chunder Dutt, *The Economic History of India in the Victorian Age,* 3rd ed. (London, 1908), pp. 362, 545.

44. Russell Lidman and Robert I. Domrese, "India" in *Tropical Development, 1880–1913,* ed. W. Arthur Lewis (London, 1970), p. 317.

45. Sharma, p. 164. See also Bipan Chandra, *The Rise and Growth of Economic Nationalism in India: Economic Policies of Indian National Leadership, 1880–1905* (New Delhi, 1966), pp. 208–11.

46. Deakin, pp. 148, 321–22.

47. Brown, p. 82.

48. Michel, p. 51.

49. On traditional Egyptian irrigation, see Sir William Willcocks and James Ireland Craig, *Egyptian Irrigation,* 2 vols. (London and New York, 1913), 1: 299–311; Harold E. Hurst, *The Nile: A General Account of the*

River and the Utilization of its Waters, rev. ed. (London, 1957), pp. 38–42; John Waterbury, *Hydropolitics of the Nile Valley* (Syracuse, N.Y., 1979), pp. 26–32; and Newhouse et al., p. 11.

50. On Mohammed Ali and the Nile Barrages, see Julien Barois, *Les irrigations en Egypte,* 2nd ed. (Paris, 1911), pp. 144, 291–95; Edward W. C. Sandes, *The Royal Engineers in Egypt and the Sudan* (Chatham, 1937), pp. 363–68; Willcocks and Craig, 2: 630–37; Waterbury, pp. 32–33; and Hurst, pp. 46–52.

51. Charles Issawi, "Egypt since 1800: A Study in Lop-sided Development," *Journal of Economic History* 21 (March 1961): 11.

52. Colin Scott-Moncrieff, "Irrigation in Egypt," *The Nineteenth Century* 17, no. 96 (February 1885): 342–44.

53. Sandes, *Royal Engineers,* pp. 371–72.

54. Ibid., pp. 373–80; Willcocks and Craig, 2: 637–55.

55. Sandes, *Royal Engineers,* pp. 382–83.

56. Robert L. Tignor, *Modernization and British Colonial Rule in Egypt, 1882–1914* (Princeton, 1966), pp. 220–24; Willcocks and Craig, 2: 718–58; Barois, pp. 221–26, 313–22; Hurst, pp. 28–54; Waterbury, p. 33; Sandes, *Royal Engineers,* pp. 378–84.

57. J. I. Craig, "The Water Supply of Egypt and the Sudan," *Contemporary Review* 127 (1925): 163–70; Tignor, p. 225; Newhouse et al., pp. 18–20.

58. Figures from Waterbury, p. 36.

59. Issawi, p. 16. See also Tignor, pp. 226–34; Newhouse et al., pp. 15–19; and Hurst, pp. 58–65.

7

Economic Botany and Tropical Plantations

Tropical Crops from Plunder to Science

Plants are the wealth of the tropical world and the livelihood of most of its people. It was the seductive and costly spices of the Indies that enticed Europeans to risk their lives upon the oceans. Over the centuries, as the supply of each desired crop increased to meet the demands of Western consumers, new crops stimulated new desires: sugar, coffee, and tea, then cotton, quinine, and rubber. Today's cocaine and heroin are but the latest in a long series of such cravings.

Since the Crusades, Europeans have devised many ways to obtain the wealth of the tropics while minimizing the costs. The first Europeans to sail into tropical waters were pirates toward their competitors at sea, but being weak on land, they had to trade to obtain spices. As soon as they could, however, they substituted coercion for trade. They enslaved Africans to produce sugar, indigo, rice, and tobacco in America, and they forced the natives of Mexico and Peru to provide the precious metals needed to buy Asian products. With more sophistication (if not morality), the East India Company taxed the peasants of Bengal to provide the opium which was exchanged for the tea of China. Thus the old imperialism used the human resources of the tropics to collect the vegetable and mineral ones.

Before alternative means of production could be developed, industrialization hastened the plunder by stimulating the demand for tropical products and by providing more efficient methods of coercion. Industrialized plunder, however, depletes natural resources faster than they can reproduce. The trade in tropical products was too important to the West to allow the threat of depletion to cast a shadow on its long-term prosperity. In the late nineteenth century, plunder and terror in pursuit of rubber from the Congo and the Amazon aroused the same moral outrage that had brought an end to slavery in the mid-nineteenth century. Ways had to be found to produce the same goods more cheaply and with a better conscience. Hence the great interest in scientific research that could improve tropical agriculture.

In the colonial era, two types of tropical agriculture offered the possibility of increasing the supply of crops for the world market: peasant agriculture and plantations. Peasants who produced food for themselves or for local markets could be convinced by threats or promises to grow commercial crops for export. In several European colonies—Senegal, Nigeria, Gold Coast, most of India—colonial administrators were hostile to plantations and leaned toward the "yeoman" ideal of agriculture. To produce both food and export crops, however, required a whole coordinated series of changes: irrigation systems, better seeds, cooperatives and marketing schemes, agricultural education and extension, and legal and social changes. In Burma, the Punjab, and the Gold Coast, the results were spectacular, for awhile. Elsewhere, the results were long in coming. Yet the attempts to reconcile Western agronomy with non-Western peasant societies still go on, if only because there is no real alternative.

In the rest of the tropical colonies, the European administrators looked to planters and agronomists to develop the export crops. In 1909, at the height of the plantation boom, the director of the Peradeniya Botanic Gardens in Ceylon wrote:

> The great development of European planting enterprise in the more civilised and opened-up countries has of course quite revolutionised the primitive agriculture or rather has built up a modern agriculture beside it. . . . Whether planting in the tropics will always continue to be under European management is another question, but the northern powers will not permit that the rich and as yet comparatively undeveloped countries of the tropics should be entirely wasted by being devoted merely to the supply of the food and clothing wants of their own people, when they can also supply the wants of the colder zones in so many indispensable products.[1]

Yet only a few parts of the tropics were propitious for plantation agriculture. In the Caribbean, plantations had long been the norm. In populated regions like Egypt or Java, new irrigation systems permitted a second annual crop. But most modern plantations were established on fertile, well-watered, but vacant lands, such as existed in Ceylon, Malaya, Sumatra, Assam, and Indochina. Here the agricultural entrepreneur could bring together technology and capital from the West and workers from the East and create an outdoor factory for tropical crops.

Our purpose here is to study the application of science to the production of tropical export crops, leaving aside the more complex sociological problem of modernizing traditional agriculture. We shall focus on plantation agriculture and its relationship to economic botany from two different angles: institutions and crops. The institutional side includes the role of botanic gardens in transferring both plants and experts, and the development of multidisciplinary experiment stations in the tropical colonies. On the production side, three examples of tropical export crops—cinchona, sugar cane, and natural rubber—will illustrate the general evolution of the science-crop relationship, and the variations due to biology, economics, and politics.

Plant Transfers and the British Botanical Empire

Until the eighteenth century, plant transfers were as unsystematic, anonymous, and undocumented as the spread of diseases. The deliberate transfer of plants accompanied the rise of systematic botany. In the eighteenth century, European powers with tropical colonies recognized the wealth that had resulted from the unsystematic plant transfers of the past, especially of sugar cane to the West Indies. By the second half of the nineteenth century, almost every domesticated plant had been spread to every other part of the globe where it could grow. But there remained the myriad wild plants which were both valuable and suitable for domestication. In the hopes of discovering new green El Dorados, governments willingly funded botanizing expeditions. Scientist-adventurers fanned out through the tropics searching for cinchona, rubber, sisal, tea, coffee, and sugar cane plants.

In the age of plant transfers, the central institutions were the botanical gardens. The earliest were apothecary gardens to grow medicinal plants, or pleasant places for the wealthy to stroll about admiring the beauties of a nature tamed for man's delight, a favorite eighteenth-century pastime. By the nineteenth century, the larger botanical gardens had become museums of living plants, organized according to the taxonomic systems of Linnaeus, Buffon, and others. They also had herbaria of dried samples and museums of economic botany in which they displayed useful products of vegetable origin.

To complete their collections, botanical gardens exchanged plants and seeds. Their activities were aimed not only at acquiring, but also at diffusing plants and knowledge for the benefit of science, planters, and governments.

The institution of the botanical garden was soon carried by Europeans to their tropical colonies. The earliest were set up by the French in Mauritius in 1735 and Bourbon (now Reunion) in 1769. Britain followed France very closely with Saint Vincent and Saint Thomas (1764), Calcutta (1768), Jamaica (1793), Ceylon (1810), and Trinidad (1818). The Dutch opened one in Java in 1817. After that came, as with European expansion in general, a lull in the mid-century, followed by a rush toward the end. The opening of botanical gardens followed the flag rather closely in Hamma, Algeria (1832); Saigon (1864); and Hanoi (1889); or Kisantu (1895) and Eala (1900) in the Congo. Of the many botanical gardens in the world in the colonial era, a few will suffice to show the part that botany played in tropical plantation agriculture: Kew for the British Empire; Buitenzorg for the Dutch East Indies; and the Muséum national d'histoire naturelle and the Jardin d'essai colonial for France.

Kew, the garden of a royal residence near London, became a center for botanical research in 1772. Its first director, Joseph Banks, was an avid collector of tropical plants who had traveled with Capt. James Cook to the Pacific, as well as to Newfoundland and Iceland. He was also a master teacher who trained the first generation of Kew gardeners. Kew's resources improved even more when Banks inherited the herbarium and manuscripts of Gerhard Koenig, a Danish physician who had spent the years 1768 to 1785 in southern India.[2]

Despite these auspicious beginnings, Kew was overshadowed by the Muséum and other Continental gardens. At that time European botanical gardens collected and classified information about the plant world but had a very limited impact on tropical agriculture, because living plants perished on the long journeys across the tropical seas. Later the Wardian case, a large terrarium invented in 1829 by Dr. Nathaniel Ward, kept delicate plants from drying out and allowed them to be transported safely over long distances. Kew Gardens was the first institution to make full use of the new technique of plant transfer.

In 1841 Parliament made Kew a national institution with an

annual appropriation, in order to create, as Joseph Banks had hoped, "a great exchange house of the empire, where possibilities of acclimatizing plants might be tested."[3] From that year until 1905 it was the fief of a botanical dynasty, the Hookers, around whom revolved the scientific elite of Britain.

Sir William Hooker, director of Kew from 1841 to 1865, was a gifted institution-builder. He turned Kew into a lovely park for well-to-do visitors from London who were fond of plants. With funds from a grateful Parliament, Hooker built up a first-rate library and herbarium, including his own private collection. He also directed the attention of Kew toward tropical botany and the transfer of economically valuable plants.[4]

Sir William's son and successor Sir Joseph Dalton Hooker, a friend of Charles Darwin, was the foremost botanist of his day. An expert on tropical plants, he spent the years 1847 and 1851 exploring Nepal and Sikkim and compiled the authoritative seven-volume *Flora of British India* between 1872 and 1897. He also botanized in the South Seas, Syria, Palestine, Morocco, and the Rocky Mountains. In 1879, Kew acquired the vegetable collections of the India Museum and established a museum of economic botany with samples and illustrations of every possible use for vegetable products. The combination of science and practicality was the strength of British botany. Under Sir Joseph's direction, Kew continued to work on plant transfers, in particular Liberian coffee from West Africa in 1872 and the rubber tree from Brazil to Ceylon in 1876.[5]

In 1885 Sir Joseph was succeeded by his son-in-law, William T. Thiselton-Dyer, an administrator rather than a botanist. As director of Kew, he oversaw the botanical activities of the whole British Empire. Thanks to collectors who had long been sending back seeds and plants from around the world, Kew had over a million species of plants in its gardens and herbarium by 1896 and could identify more plant species than any other institution.[6] Collecting, classifying, and identifying were only the beginning of Kew's responsibilities. Another was to provide information to botanists worldwide through its network of 54 corresponding botanical gardens, 33 of which were in the British Empire. Many inquiries concerned problems of identification; others were more practical, ranging from plant diseases to which "West African palm kernels to carve into coat buttons." Kew was also a publisher of botanical scholarship; by 1879 its floras of the British Empire already comprised twenty-two volumes. In 1887 it began publishing the *Bulletin of Miscellaneous Information,* and in 1892 the *Index Kewensis,* the official nomenclature of all the world's plants.[7]

Kew's functions were more than scientific and economic. Under Thiselton-Dyer it became, avowedly, an imperial institution. To justify its imperial mission, he quoted Dr. Lindley's report to the House of Commons in 1841, which set out the goals of Kew:

> A national garden ought to be the centre round which all minor establishments of the same nature should be arranged; they should all be under the control of the chief of that garden, acting in concert with him and through him with one another, reporting constantly their proceedings, explaining their wants, receiving their supplies, and aiding the mother-country in everything that is useful in the vegetable kingdom.[8]

Thiselton-Dyer thought of Kew as "a sort of botanical clearing house or exchange for the empire." It did not exercise its power through regulations but through the education it offered botanists and gardeners, and its influence upon their careers:

> Without having the least desire to see Kew become a general dispenser of patronage, the Director has felt that nothing deserves more careful consideration than the demand for Colonial botanical officers. . . . Almost everyone who leaves Kew to go abroad keeps up some kind of correspondence with it, and we are generally able to form a good idea of the capacity that each is showing. It would, therefore, be very easy to arrange, as has indeed already to some extent been done, an interchange by way of promotion of botanical officers from one Colony to another. I believe that the experience gained in one Colony would often be extremely valuable when transferred elsewhere.[9]

During his twenty years as director of Kew (1885–1905), Thiselton-Dyer succeeded in carrying out his program. Increasingly, the Colonial Office and many colonial governments sought his advice when filling a botanical or agricultural position, and his nomination usually sufficed to secure the job for his protégé.[10] By the turn of the century, alumni of Kew were serving as directors of botanical gardens, forest departments, and agricultural research stations throughout the empire. The flow went both ways, for it was "a general rule which made it compulsory for all Kew officials to have resided in the colonies."[11] In 1902 the imperial vocation of Kew received official recognition when Thiselton-Dyer was appointed botanical advisor to the Colonial Office.[12]

Around Kew revolved satellite institutions scattered throughout the empire. Small colonies had botanical stations, nurseries that provided seedlings to local farmers. More important places with interesting flora like Hong Kong, Tasmania, and Natal had botanical gardens devoted to collecting local plants and exchanging useful species. A few colonies had large gardens with full research staffs. The Royal Garden at Calcutta, founded in 1786, was larger than Kew and celebrated not only for researching the plants of India and the Himalayas, but also for introducing tea, cinnamon, and tobacco to India. Peradeniya brought cinchona, tea, and rubber to Ceylon. And the Botanic Garden of Singapore turned the rubber tree into the mainstay of Malayan agriculture. These and other gardens were all interconnected by a constant stream of information and live plants and by the rotation of personnel, all channeled through Kew.[13]

Agricultural Research in the British Tropics

In the course of the nineteenth century botany had accomplished the enormous task of collecting and classifying most of the world's plants. By the end of the century, botanists, like other life scientists, turned their attention toward experimental work in physiology, genetics, and ecology. As part of this movement, economic botany also changed. The massive plant transfers of the past centuries had reached diminishing returns, and it became less likely that plant hunters would discover valuable new plants growing wild. Instead, economic botanists turned to improving well-known plants and adapting them to new conditions. This was no simple task. Since agriculture began, there have been two ways to breed better plants. In mass culture, the seeds of the best plants in a large field are chosen for reproduction; for most species this method has to be repeated every year. In pedigree culture, a single plant is selected for its progeny, and the experiment is repeated for several generations until a steady race is obtained. Even under near-perfect conditions, such techniques are slow and costly. The twentieth century added a revolutionary new method of plant breeding: hybridization. By breeding together individuals with desired characteristics, botanists are able to create new varieties that meet their specifications and, in many cases, breed true. These new tailor-made varieties have quickly replaced the old in the environments they were designed for.

These changes required new institutions. Next to botanical gar-

dens which applied one science to thousands of species, there arose
experiment stations to study one crop at a time from the point of
view of several disciplines: plant genetics, physiology, pathology,
mycology, entomology, chemistry, soil science, and agronomy. The
development of experiment stations and the involvement of new
scientific disciplines in tropical agriculture were stimulated by the
appearance of diseases that devastated important crops in the late
nineteenth century: phylloxera of the vine, hemileia of coffee, coco-
nut canker, and sereh of sugar, to name the worst of them.

Given the variety of conditions in the tropics, experiment stations would have been useful for every crop in every ecological zone.
Their high cost, however, limited their numbers, but improved global
communications linked them into networks, each of which specialized in a particular plant and exchanged techniques, plant material,
and personnel. These networks were open to international exchanges
of information, but in other respects they were centered on the great
powers and colonial empires of the period. Botanical gardens, no
longer indispensable as plant-transfer agents, became centers for
theoretical research and the training of scientists for colonial careers.[14]

Agricultural experiment stations, like other scientific institutions,
evolved simultaneously in a number of places. In the British Empire,
their forerunner was the Imperial Department of Agriculture. The
British West Indies, once among the most profitable of tropical
colonies, had become impoverished in the last half of the nineteenth
century. Planters blamed the emancipation of slaves and the rise of a
heavily subsidized European beet-sugar industry. Others saw the
stagnation of cane sugar technology and the declining yields of the
plants themselves as contributing factors. In 1897 Colonial Secretary
Joseph Chamberlain sent a royal commission to the West Indies to
investigate the situation. Upon its recommendation, Parliament voted
to create an Imperial Department of Agriculture, headquartered in
Barbados.

Sir Daniel Morris, the first imperial commissioner of agriculture, was the epitome of the colonial botanist: he had served as
assistant director of the Peradeniya Botanic Gardens (1877–79),
director of the Public Gardens of Jamaica (1879–86), and assistant
director of Kew (1886–98). He introduced two concepts to the
tropics. One was a multidisciplinary approach to tropical agriculture; in his words:

The circumstances of our tropical Colonies and Protectorates have rendered it necessary to provide additional scientific assistance, beyond the domain of Kew, in order to meet these requirements. This is in course of being provided by the chemical and other laboratories and equipments, and by the large staff engaged in experimental research, technical trials, and commercial valuations at the Imperial Institute.

The other was agricultural extension:

It was recognized from the first that local experiment stations distributed on the estates and carried on with the co-operation of individual planters and the scientific officers of the Department were the best and most potent means of demonstrating the lines on which science could be of service to agriculture. This enabled the planters to understand the value of scientific methods of research, and, on the other hand, to bring the scientific workers into sympathy with the difficulties and limitations of the practical side.[15]

The principal interest of the department was sugar cane. In the first years of this century, the entomologist Maxwell Lefroy found a way to prevent the ravages of the moth-borer. The agricultural chemist John Harrison and the botanist J. R. Bovell sought to improve the cane itself, a difficult task, for the plant reproduced by cloning and all sugar canes were essentially the same plant. After careful investigation, they succeeded in observing that rare event, the sexual reproduction of the cane by seeds, and went on to develop several new varieties. These innovations were applied to an estate in Antigua, which registered a gain of 40 percent over traditional methods. By 1914 the British West Indies were out of their depression, in part because the work of the Imperial Department.[16]

Looking back, Morris remarked in 1911: "A gratifying proof of the value of the work of the Imperial Department of Agriculture is the formation of a series of departments on somewhat similar lines in other portions of the tropics."[17] Between 1900 and 1914, agriculture departments and research stations sprang up all over, not so much in imitation of the Imperial Department, but because commerce, administration, and the sciences were suddenly linking up in new combinations everywhere. In the tropics the British were among the most active in spreading the new methods. Geoffrey Masefield, an agricultural administrator, offered an intriguing sociological explanation: "This can perhaps be explained because governors were of the landowning class at home who were accustomed to take a paternal interest in the agricultural improvement of their own estates, and felt it the duty of the upper ranks of society to encourage such improvements wherever they were placed."[18]

In India, Viceroy Lord Curzon opened the Agricultural Research Institute of Pusa in 1903 with a gift of £30,000 from the American philanthropist Henry Phipps. It included research departments for botany, chemistry, bacteriology, entomology, and mycology; experiment farms for agriculture and cattle breeding; and an agricultural college. It developed an improved strain of wheat called "Pusa" and helped provincial departments of agriculture set up seed farms.

Ceylon, which was far more dependent on agricultural exports than India, also acquired a set of new institutions: an experiment station in 1901, an agricultural society in 1904, and a school of tropical agriculture in 1916, all closely tied to the Peradeniya Botanic Gardens. Similarly, the Federated Malay States opened a Department of Agriculture in 1905.

Africa trailed behind for two decades. Its earliest research station was the Biologisch-landwirtschaftliches Institut at Amani in Tanganyika, founded in 1902. Next came the Nigerian Department of Agriculture in 1910. Agricultural innovation in Africa was delayed by ignorance of local social and economic conditions and by the fluctuations of tropical commodity prices between the wars.[19]

The success of the first departments of agriculture spawned two other types of institutions. One was the commodity research station, which applied the interdisciplinary team approach to a single crop. The first of these was the Imperial Sugarcane Breeding Institute founded at Coimbatore, India, in 1912. After World War I others appeared: for rubber in Ceylon and Malaya, cotton in India and the Sudan, tea in Ceylon and Kenya, sugar in Barbados and Mauritius, and cacao in the Gold Coast. For the most part these stations were funded by planters' associations, which attempted to keep their findings to themselves; but never for long, for plants cannot be patented and seeds and information travel fast.

To support these local commodity research stations, there also arose empirewide scientific organizations: the Imperial Bureau of Entomology in 1913, followed after World War I by bureaus of mycology, soil science, plant genetics, and others, all attached to the Imperial Institute of South Kensington and funded by the Colonial Office.[20]

Science and Agriculture in the Netherlands East Indies

The Netherlands East Indies possessed three unusual attributes: Java, a land of inexhaustible fertility; the Javanese people, gifted and industrious gardeners; and the Dutch, skilled at extracting wealth out of the least scrap of land. Together, they made the Netherlands East Indies the envy of the other colonial powers.

To make the most of Java, the Dutch applied not only cunning administrative methods, but also considerable science. The center of their scientific approach was the Botanical Garden at Buitenzorg. When the Dutch crown regained the East Indies after the Napoleonic Wars, it was determined to exploit them more efficiently than the old Dutch East India Company had ever done. On board the fleet which left for Java in October 1815 was Dr. Kaspar Reinwardt, professor of natural history at the Athenaeum of Amsterdam, recently named director of agricultural establishments, arts, and science for the Netherlands East Indies. Reinwardt proposed to start a botanical garden at Buitenzorg (now Bogor), 58 kilometers from Batavia. In April 1817 Governor van der Capellen approved his plan, and work began.[21]

Buitenzorg was a wise choice for a garden. Unlike much of Java, it has no dry season; in fact, it rains there almost every day, seven times more than in Holland. Tropical plants, which everywhere else must be kept indoors and carefully watered, found the place to their liking. For five years Reinwardt traveled throughout the East Indies gathering plants and seeds for Buitenzorg, and he published the description of 1,200 new plant species. When he returned to Europe in 1822, the garden had three European gardeners, a botanical artist, and a staff of forty-three Javanese. It had also begun importing economically useful plants, among them tea, cinnamon, cacao, and tobacco.

Reinwardt was not replaced, however, and the garden's budget was cut. If it survived at all, it was thanks to J. E. Teijsmann, a gardener and self-taught botanist. Under his care, Java obtained the cinchona, which later became one of the East Indies' main export crops. During his years at Buitenzorg, Teijsmann did not cease lobbying for two reforms: the separation of the botanical garden from the palace park, and the appointment of a professional botanist. With the help of friends in Holland, he finally achieved his objective in 1868 when Dr. R. H. C. C. Scheffer was appointed director.

Scheffer's appointment coincided with a profound change in

Dutch colonial rule. The Culture or Cultivation System, introduced by Governor van der Capellen in 1830, had consisted in forcing Javanese farmers to produce export crops for the government; this colonial feudalism clashed with the liberal free-trade ethos of late nineteenth-century Holland and with modern agricultural practices, especially mechanized sugar milling. The Agrarian Land Law of 1870 replaced corvée labor with capitalist property relations and a money economy.

At the same time the Netherlands East Indies government, recognizing the economic value of plant transfers, decided to turn Buitenzorg into a full-fledged scientific institution, with three Dutch gardeners, a botanical writer, and an artist. In 1876 the government purchased a farm called Tjikeumeuh outside Buitenzorg, to be used as an agricultural experiment station. Next to it, an agricultural school was established to train Javanese to become agricultural extension agents, and to teach Dutch officials the rudiments of tropical agriculture. Under Scheffer's direction, his successor wrote many years later, "the botanic garden became secondary, while the experiment station and agriculture school became primary."[22] Buitenzorg concentrated on importing and acclimatizing new varieties of rice, jute, tobacco, peanuts, soy, cassava, wheat, eucalyptus, and coffee.[23]

Buitenzorg's renown as a scientific institution reached its peak under Melchior Treub, its director from 1877 to 1911. He was determined to turn the botanical garden, which had fallen into disarray, into a center for research. This, he knew, would require more money than he could expect to get from the government, so he developed an unusual talent for a European scientist: fund-raising. Treub got the cooperation of local planters by offering to do research on the diseases that attacked their crops. He paid special attention to sugar cane, rubber, and cinchona and published his findings in the planters' journal, *Mededeelingen*. In exchange, the planters paid the salaries of his research staff. In the 1890s, when the planters' associations began setting up their own research centers, Buitenzorg worked closely with them, organizing meetings and exchanging researchers.

Yet Treub knew that both Buitenzorg's reputation and the prosperity of East Indian plantation agriculture ultimately rested on a purer sort of research. As he explained:

No one still thinks that the study of North American flora should be undertaken anywhere else but in America itself. What has hap-

pened in America must necessarily also take place in tropical countries where there are large botanical institutes. There also botanical
research must be carried out on the spot and not in Europe.[24]

Though he remained an active researcher all his life, Treub's greatest
achievement was to provide the setting for the research of others. To
this end, he established a series of specialized laboratories for chemistry and pharmacology, for plant physiology and pathology, and for
zoology.[25] By 1900 Buitenzorg had a staff of 15 Europeans and close
to 300 Javanese. Its budget in 1898 was over 216,000 florins (approximately £18,000), two-thirds the budget of Kew, and more
than the budgets of all French colonial botanical gardens combined.[26]
  Yet all these facilities were not enough to produce good science.
The problem was that Java was on the opposite side of the world
from Europe, where most research was being done. So Treub came
up with a method of keeping Buitenzorg in the mainstream of science: the foreigners' laboratory. He announced in 1884 that Buitenzorg was building a special laboratory for visitors, who would each
have living quarters, a work space with appropriate instruments, and
access to the garden, herbarium, and library. In 1887, while on leave
in Europe, Treub persuaded the Dutch and Swiss governments to
fund this laboratory. Visiting researchers were invited to spend four
to six months at Buitenzorg and publish their findings in the *Annales
du Jardin botanique de Buitenzorg*. From 1884 to 1914 a total of
171 scientists stayed at the foreigners' laboratory. Among them were
50 Germans, 32 Dutchmen, 21 Russians, 15 Austro-Hungarians, and
13 Americans. Between 1914 and 1934 another 81 visitors came;
again the Dutch and Germans predominated with 21 and 19, respectively, followed by 10 Americans, 6 Swiss, and 6 Japanese. Altogether only 6 French and 6 British scientists came to Buitenzorg, for
they had their own research institutes.
  There were no Indonesian scientists at Buitenzorg or elsewhere.
Of the gardeners, Treub said rather ingenuously: "The native personnel is composed of about a hundred individuals, among whom
are three employees who have special botanical knowledge, much
deeper than one would expect to find among the Malays." Buitenzorg was not a place to transfer the culture of Western science to the
people of the East Indies. As long as the Dutch ruled the islands, it
remained a Western enclave in a tropical environment.[27]
  At the turn of the century, the multiplying connections between
science and agriculture produced a proliferation of specialized orga-

nizations, both governmental and private. In the Dutch East Indies the relations between the government and the planters' organizations were probably closer than in any other colony, for in many ways Dutch colonialism was a business and the Indies a huge plantation.

In 1904 the government created a Department of Agriculture under Melchior Treub to oversee both the "institutions of pure science"—the botanical garden and its annexes—and the "institutions of applied science"—the Tjikeumeuh Agricultural Experiment Station, the Institute for Plant Diseases, the demonstration estates and the schools of agriculture. To manage its plantations and market their products, the government created a Department of Government Enterprises in 1907. Finally, in 1934, all of these organizations were brought together into one Department of Economic Affairs.[28]

In addition to the government-sponsored research centers, planters' associations also subsidized research centers for their crops. The best known of these was the East Java Experiment Station (Proefstation Oost Java or POJ), to which we shall return. Its success in fighting the diseases of sugar cane and creating better-yielding varieties stimulated a host of imitators. Two of the biggest were the Deli experiment station of the Union of Tobacco Planters and the research station of the General Association of Rubber Planters of East Sumatra, or AVROS, which pioneered the cloning of rubber trees.

Large estate-owning corporations—Goodyear, U. S. Rubber, Rubbercultuur Maatschappij Amsterdam, Anglo-Dutch Plantations of Java, Klattensche Cultuurmaatschappij—also had experiment stations. Multicrop research stations like the West Java Research Institute of Buitenzorg worked on tea, rubber, cinchona, and sugar cane in close cooperation with the government institutions. All in all, the Netherlands East Indies had the most complete and best-organized botanical and agricultural research network in the tropics. And almost all of it was devoted to commercial export agriculture, rather than peasant subsistence crops.[29]

Tropical Botany in France

Unlike the British and Dutch botanical institutions, which reflected the economic geography of their empires, the French institutions reflected the culture and politics of the metropolis. The most prestigious of French botanical institutions was the Jardin des plantes in Paris. It was the nucleus around which grew the Muséum national d'histoire naturelle, an institution responsible for research in all the natural sciences.

French scientists had long been as active as the British in collecting herbarium specimens and exotic plants. In the eighteenth and early nineteenth centuries, such eminent naturalists as Jussieu, Buffon, and Lamarck gave the Jardin a reputation unequaled by any other botanical garden. But the Jardin des plantes never had the same economic vocation that Kew and Buitenzorg later developed. One can advance a number of hypotheses to explain this phenomenon. France was never as thoroughly committed to international trade as were Britain and the Netherlands. The French elites were more urban and less interested in agriculture than their British and Dutch counterparts. Science was honored apart from its economic usefulness. The Jardin and the Muséum therefore concentrated on the purely scientific approach, which meant, in the late eighteenth and early nineteenth centuries, collecting and classifying information about the natural world.[30]

In the second half of the nineteenth century the Muséum began falling behind. A museum rather than an economic or teaching institution, it retained its dedication to descriptive natural science and taxonomy long after other institutions had shifted their focus toward experimental biology. As a result, it did not keep up with the science faculty of the University of Paris or, in botany, with Kew.

This did not pass unnoticed. Already in 1868 Charles Martins, professor of natural history and director of the botanical gardens of Montpellier—the most serious French rival to the Jardin des plantes— sounded the alarm:

> In natural history, we have been stationary for thirty years, while everything progresses around us. The fatal, inevitable result of such a state of things is decadence. . . . We must have no illusions, French science is declining rapidly while foreign science grows every day. We have not hesitated to renew the weapons of our soldiers charged with maintaining our military preponderance; it is time to renew those of the scientific army, eager, like the other one, to maintain the national honor and to contribute, along with the arts and literature, to the radiance of the true glories of France.[31]

At the time, however, France did not have much of a tropical empire to stimulate research in natural history. Only twenty years later, after the Colonial Exhibition of 1889, did colonialism become popular in France and colonialist institutions arise in Marseille, Bordeaux, and Nantes. The director of the Muséum, Alphonse Milne-Edwards, seized the opportunity to revive the prestige of his institution. To legitimize its newfound colonial vocation, the Muséum pointed to a venerable tradition of botanizing in the tropics going back two centuries and more. This link with the past has been the main point of numerous writings on the Muséum's tropical vocation from the 1890s until the present.[32]

Implementing this policy was another matter. The Muséum occasionally sent collectors to the tropics, generally to South America, to search for rare plants, but seldom to the colonies. In 1890 Melchior Treub wrote an article in the *Revue des deux mondes* expressly to invite French botanists to Buitenzorg, but to no avail, because of what one scientist called "the inertia which retains our young botanists and agronomists in France."[33]

The real work of tropical botany and plant transfer was borne by one man, Maxime Cornu. In 1884 he had been appointed professor of culture at the Muséum, in charge of economic plants and the greenhouses in which to grow them.[34] He actively exchanged seeds, plants, and information with the botanical gardens of Kew, Calcutta, Rio de Janeiro, Peradeniya, Buitenzorg, and others. Every year he published a list of living plants he was willing to send to other gardens and agricultural stations. Unfortunately, he noted, botanical gardens in the French colonies seemed indifferent to his efforts.[35] When he died in 1900, Thiselton-Dyer wrote his obituary for the British journal *Nature:*

> At the moment that Cornu entered on his new duties, France had turned its attention anew to the field in which, in the past, it had done so much—colonial enterprise. Cornu's ambition—and it was a legitimate one—was to utilize the somewhat dormant resources of the Jardin des Plantes in the work, much on the lines of Kew. . . . Cornu threw himself into the work with little short of passion. What he accomplished, both for the French colonies and for the enrichment of the gardens of his own country, with resources more limited than we have at our disposal in England is to me surprising.[36]

Despite Cornu's efforts, the Muséum did not succeed in making itself indispensable, or even particularly useful, to colonial agriculture as Kew and Buitenzorg did. The Muséum continued in its traditional role of gathering and organizing information from the tropics, not transferring it to the tropics. Cornu alone could not hope to rival the large organizations in the Dutch and British empires. Furthermore, the Muséum had no students, nor did it train gardeners for colonial work; hence, it lacked the old-boy network and the moral influence of Kew.

Meanwhile, French colonial lobbyists were developing an interest in botany and agriculture. In 1895 A. Milhe-Poutingon, a member of the Union coloniale française, founded the *Revue des cultures coloniales,* the first French journal devoted to tropical agriculture. Two years later Jean Dybowski, director of agriculture for Tunisia, published a pamphlet advocating the establishment of botanical gardens in the colonies and the creation of a new institution "which would link all our possessions by sending from one to the other the different useful plants."[37] Soon thereafter Henri Lecomte, professor of natural science at the Lycée Saint-Louis, suggested the same idea before the Société de géographie commerciale.[38] The idea was in the air, but the institution was still lacking.

When Colonial Minister Georges Trouillot announced in 1898 that "in the first rank of our concerns [is] the agricultural development of our possessions," Milhe-Poutingon applauded it in an editorial in his journal.[39] After visiting the botanical gardens of Germany, Belgium, and Britain on behalf of the Ministry of Colonies, he wrote in glowing terms of Kew and its powerful and beneficial influence upon agriculture in the British Empire, in contrast to the ineffective French efforts in this field.[40] A commission on colonial experiment gardens, convened by the minister of colonies and presided over by Alphonse Milne-Edwards, recommended that a new experiment station be set up to breed and distribute useful plants to the French colonies.[41]

Finally, on January 28, 1899, President Félix Faure signed a decree establishing the Jardin d'essai colonial. Two days later Jean Dybowski was named to direct it. Two hectares belonging to the city of Paris in the nearby Bois de Vincennes, which the Muséum had a claim to but had never used, was "temporarily loaned" to the new garden.[42] Up to this point the professors of the Muséum believed that they would have scientific if not administrative control over the newcomer. It only took them a few days to realize, however, that Dybowski was beholden only to the Ministry of Colonies. From that moment on and for a decade thereafter the Muséum attempted to unseat the upstart by insisting on the return of "their" land, demanding administrative control over the Jardin colonial, threatening to establish a rival experiment garden, and issuing statements condemning the new creature. Edmond Perrier, director of the Muséum in 1899, waxed indignant: "The Muséum could not accept without protest the cleverly camouflaged spoliation which it has suffered at the hands of the Ministry of Colonies," he wrote.[43] By 1901 accusations were flying back and forth between the Ministry of Colonies and the Ministry of Public Instruction, to which the Muséum belonged. They accused one another of "withholding" or "usurping" the plot which the Jardin occupied and, worse, of interfering with each other's right to transfer information and plants to and from the colonies.[44]

In September 1905 the two ministries issued a joint decree confirming the Muséum's ownership of the disputed land and the Jardin's use of it; on a higher level, they recognized the Muséum's scientific authority and "colonial vocation," while leaving the administrative authority over the garden to the Ministry of Colonies. This too remained a dead letter. The squabble over real estate continued until 1932, when the Muséum gave up its claim to the land in exchange for the right to build a zoo nearby. The deeper hostility, a question of honor rather than land, simmered on for awhile longer, and still occasionally surfaces in passing remarks.[45]

In the midst of all this infighting, Dybowski discovered that no one at the Ministry of Colonies even knew which colonies had botanical gardens; to get this information, he had to send questionnaires to all the governors of French Africa.[46] The Jardin colonial first undertook to supply colonial botanical gardens and experiment stations with plants and seeds. It also analyzed agricultural products and soil samples sent in from French Africa, where no laboratories existed. All of this was done by only 7 agronomists and 6 gardeners, for the garden operated on a shoestring budget of 50,000 francs per year (eighteen times less than Kew), most of it contributed by the colonies of French West Africa.[47]

As in the other botanical empires, plant transfers were soon superceded by the transfer of information and expertise. Dybowski and his staff answered some 2,000 letters a year. Starting in 1901 they issued a newsletter of advice for planters entitled *L'agriculture pratique des pays chauds,* supplemented after 1913 by the journal *L'agronomie coloniale.* This was not enough, for most colonies lacked agronomists to implement the advice. Therefore the Ministry of Colonies set up a school of tropical agronomy, the Ecole nationale supérieure d'agriculture coloniale, to train colonial officials and prospective colonizers. The school and the garden were formally joined in 1921 under the name Institut national d'agronomie coloniale. By the 1920s it was predominantly devoted to training, while research and plant transfers once again receded into the background.[48]

Meanwhile the Muséum, unwilling to abandon the field to its rival, set up a Laboratoire colonial to publish information of use to colonial agronomists and administrators. From 1903 to 1905 it offered lectures on such topics as "Insects Harmful to Sugar Cane," "Drugs to Take to the Colonies," and "Anthropology of Indochina: the Annamese." It also published the *Journal d'agriculture tropicale* from 1901 on. In the 1920s the government endowed two new chairs at the Muséum, one in colonial animal products, the other in colonial agronomy.[49] The competition between the two rival organizations had stimulated more activity than the rivalry with any foreign power, although in comparison to the needs of the colonies, it was still not enough.

Botany and Plantations in the French Empire

In 1841 the director of the Muséum complained to the minister of the navy and colonies: "For a long time the colonial gardens have done almost nothing for the metropole. . . . Only one establishment, and it does not belong to France, corresponds regularly with us. It is the one in Calcutta. It sends us more in a year than all the others in ten."[50] Half a century later Milhe-Poutingon complained: "Our colonies themselves, in gradually becoming dominated by a single crop, let the rich botanic gardens they once possessed decay and in certain cases disappear."[51] Professor Maxime Cornu noted with regret: "Then French colonial gardens, with few exceptions, requested nothing or almost nothing and, as a natural consequence, sent in nothing or almost nothing to the metropole."[52] And Henri Lecomte, after visiting many French colonial botanical gardens, concluded:

> We show, in fact, a well-deserved enthusiasm for all expeditions which aim at extending our colonial domain; we register with a jealous care all our conquests; we bring a certain vanity to the act of planting our flag on every shore; but when it is necessary to develop, to put to use the incomparable colonial domain which we possess, this enthusiasm cools off singularly.[53]

The problem with French colonial gardens was not simply one of "enthusiasm" but also of timing and geography. France had lost its first colonial empire between 1763 and 1815, and only began its second wave of conquests (except for Algeria) in the 1860s. The acquisition of this second colonial empire outran the willingness of Frenchmen to settle, or of capitalists to invest, in it. No sooner had France imposed a military-administrative structure sufficient to permit modern economic development than World War I broke out, postponing everything by a decade.

Geography also hampered development efforts. The lands that France obtained, though immense, were poor in population and natural resources compared to the British and Dutch possessions. The delay between investments and returns was consequently much greater. Eventually France developed the same technical and economic system in its empire as the Dutch and British had in theirs; but it was not until after World War I in Indochina, and after World War II in Africa.

In the late nineteenth century, amidst the chorus of complaints about the decay of botanical gardens in the French colonies, two gardens were always cited as exceptions: Libreville and Saigon. Libreville, a trading post on the coast of Gabon, was the first French settlement in equatorial Africa. In 1887 Governor Bellay encouraged the Reverend Theophile Klaine, an amateur naturalist, to start a garden there. Day-to-day operations were entrusted to Emile Pierre, a graduate of the Ecole nationale d'horticulture of Versailles. Pierre and his successor Chalot corresponded assiduously with Professor Cornu, sent him rare plants from the equatorial rain forest, and in exchange received cloves, cinnamon, coffee, vanilla, cacao, and other useful plants; none of them, however, were planted on a large scale until the 1920s.[54]

In 1900 the French government decided to move the capital of its equatorial African possessions inland, along the Congo River. In anticipation, a new garden was started at Brazzaville, the new capital, and the garden at Libreville was expected to send its plants there. Unfortunately the garden of Brazzaville never amounted to anything, while the one at Libreville, having lost its official support, decayed. An inspector's report of 1919 noted: "After having been the best equipped and most complete botanical establishment on the West Coast of Africa, [it] is in the saddest state of abandon. The nursery is reduced to a minimum and seeds and plants are very rarely given out."[55]

In other French colonies in black Africa, the situation was hardly better. Experiment stations and demonstration farms were established in Senegal, Guinée, Ivory Coast, and Madagascar, but funds were lacking for any serious extension work.[56] Even the valuable peanut crop of Senegal depended on African farmers' using the leftovers of one year's harvest as seeds for the next. In 1936 a conference of colonial governors called for "immediate investments of considerable capital . . . to repair the somewhat selfish attitude which existed in the past toward the colonies, in which the principle of '*self-supporting*' [*sic*] was applied in all its rigor."[57] Only after World War II did France begin to invest in something other than infrastructures in its colonies.

The situation in Indochina began in a similar way but ended very differently. Here too, botanical research waxed and waned with the interests of particular individuals. In 1865 Admiral de la Grandière invited botanist Louis Pierre to create a botanical garden and zoo in Saigon. In his twelve years as head of the garden, Pierre collected plants for an herbarium and information and drawings for a forest flora of Cochinchina and Cambodia, and he also started an experimental farm outside Saigon. Unfortunately when he left Saigon in 1877 the garden was abandoned and the herbarium, left unprotected, was eaten by insects.[58] Some ten years later Resident-General Paul Bert invited Benjamin Balansa, a naturalist and plant collector for the Muséum, to transfer useful plants to Indochina. Balansa brought coffee and cinchona seedlings from Java. When he died the cinchonas did also, although the coffee bushes survived. A third transfer took place in 1897 when a few rubber plants, acquired in the Dutch East Indies by naval pharmacist Raoul, were sent to Indochina and became the ancestors of important natural rubber plantations.[59]

In comparison with the well-funded and highly organized system of plant transfers in the British and Dutch empires, the weakness of these efforts is glaring. There was little organized support from France. Everything depended on the interests and energy of particular governors and botanists. And there were a few planters to insist on more competent efforts.

The situation began to change after 1900. Governor Paul Doumer, a modernizer like Curzon, set up a Mission permanente d'exploration scientifique. The success of plantation agriculture in Java and Malaya stimulated similar developments in Indochina. After 1908, when the first rubber plantations proved profitable, a rubber boom swept through Indochina just as it had through Malaya. And it was the planters and the companies that invested in rubber plantations which gave the necessary impetus for a more sustained administrative action.

For a time, government research efforts were poorly funded and unfocused.[60] Doumer's "permanent" mission died out in 1908, having produced, in six years, a few works of ornithology. During World War I another modernizing governor, Albert Sarraut, took up the cause of scientific research. In 1917 he recruited Auguste Chevalier, France's best-known tropical botanist, as inspector general of agriculture and forests and director of the newly created Institut scientifique de Saïgon. In a report to the governor in 1919, Chevalier praised the work of Buitenzorg, Peradeniya, Calcutta, and Pusa: "Compared to these institutions, we have no establishment in Indochina to study the natural resources of the country and we only have rudimentary agricultural stations to experiment with new crops."[61] To remedy this deficiency, Chevalier called for a botanical research station on the scale of Buitenzorg or POJ, with a budget fifteen times that of the Jardin colonial and thirty times that of the Laboratoire colonial of the Muséum. Nothing came of this proposal. Disappointed, Chevalier left agronomy to graduates of the Ecole nationale supérieure d'agronomie coloniale of Nogent-sur-Marne, whom he looked down on, and turned to pursuits more appropriate to a member of the Muséum, such as compiling a great flora of Indochina and writing articles for the Muséum's learned journals.[62]

Despite bold schemes and resounding titles, the government moved slowly in helping agriculture; in the years 1891 to 1921, the agricultural services of Indochina hired, on the average, 1 agronomist per year. The Ecole supérieure d'agriculture et de sylviculture de Hanoï, founded in 1918, only graduated 3 native agricultural extension agents a year. And graduates of the two secondary schools for farmers preferred to work in government offices.

The midtwenties witnessed a shift in priorities. Starting in 1922, the agricultural services began recruiting 10 Europeans a year, among them 4 graduates of Nogent, 2 agricultural engineers, and 2 veterinarians. In 1925 Governor Martial Merlin replaced the Institut scientifique with an institution of applied science, the Institut de recherches agronomiques et forestières de l'Indochine, or IRAFI. Its first director, Yves Henry, was an agronomist with long experience in Africa.[63] His goal was to provide useful advice to planters and farmers, rather than to do pure research.[64]

Henry's first assignment was to investigate plantation agriculture in Java, Sumatra, and Malaya. On his return he wrote a polemical report calling for "a policy of large-scale colonization."[65] This was happening anyway, without a government policy. The years 1925–30 were the peak of the rubber boom in Indochina. Other crops—tea, coffee, rice—also did well, encouraged by high prices and French tariff barriers. Rubber companies and planters' associations conducted their own research. New plantations were carved out of the jungle. Scores of young scientists and agronomists, including the best students from Nogent, came to Indochina, lured by challenging work and high wages. Planters and agronomists regularly traveled to Malaya, Java, Sumatra, India, and the Philippines to learn new techniques and obtain seeds.

Under Henry's direction, IRAFI built laboratories for agricultural chemistry, genetics, mycology, entomology, and plant diseases. Henry's own studies resulted in a pioneering work on tropical soils.[66] Several small agricultural stations were merged into institutions large enough to do serious research. The Office indochinois du riz, established in 1930–31, applied techniques developed in the Far East, Italy, and the United States to Indochina's most important food crop. Even traditional science got its share of attention, as work on the flora of Indochina continued in cooperation with Buitenzorg and the Muséum.[67] A rubber research institute, the Institut de recherches sur le caoutchouc en Indochine, was established in 1940, just before the Japanese invasion.[68]

For a long time, scientific research in Indochina had been disorganized, unfocused, and erratic, and research back in France was of little practical use to Indochinese agriculture. But in the end it did not matter, because neighboring colonies had developed profitable plants and techniques which enterprising planters could transfer to Indochina.

Cinchona

To understand the connections between science and plantation agriculture, we need to look beyond the institutions to the crops themselves. Three of them, cinchona, sugar cane, and natural rubber, will illustrate the interactions between transfers, selective breeding, and agriculture in the three empires.

Quinine, extracted from the bark of the cinchona tree, was the only defense against malaria before World War II. Without it, Europeans would have found the tropics extremely dangerous. Yet Europeans needed bark in quantities that only their colonies could provide. Cinchona was thus the object of a deliberate, scientifically organized nineteenth-century plant transfer, intimately related to European imperialism.[69]

Until the 1850s, all the world's cinchona bark came from wild trees growing in the Andes. As demand grew, bark-hunters decimated the trees faster than they could reproduce. Naturalists like Alexander von Humboldt feared that the world's cinchona supplies would soon run out. South American bark was also expensive and of unpredictable quality. It therefore became a matter of more than scientific interest to obtain living cinchonas and reproduce them in places under European control. Not by coincidence, it was the European countries with tropical possessions—Britain, France, and the Netherlands—that got involved in the transfer, or theft, of the cinchona.

The French and the Dutch succeeded first. In the late 1840s the explorer Hugh Algernon Weddell sent some *Cinchona calisaya* seeds from Bolivia to the Muséum in Paris. In 1851 the botanical garden at Leyden obtained a plant grown from one of Weddell's seeds. A cutting from this plant survived the trip to Java and was planted in the mountain valley of Tjibodas in 1852, where it grew to give seeds in turn. But to propagate the species required more than one plant. Dutch Minister of Colonies C. F. Pahud asked the naturalist F. Junghuhn to obtain seeds from the Andes, and Junghuhn passed the task on to Justus Karl Hasskarl, assistant gardener at Buitenzorg.

The Andean republics, keenly aware of the value of their monopoly, had long forbidden the export of cinchona seeds and plants. Under false pretenses, Hasskarl got permission to explore the Peruvian Andes for a year and a half. In late 1854 he returned to Java in a Dutch warship with several cases of *Cinchona ovata* (or *Pahudiana*). Some seventy plants reached Tjibodas, where Hasskarl started the world's first cinchona plantation.[70] Two years later, Junghuhn arrived from the Netherlands with several more species of cinchona. From then until 1864, Junghuhn and his staff experimented with different methods of planting, cultivation, and bark removal. They were especially interested in seed germination and in the comparative quinine contents of different species. They started a new plantation at Tjinieroean, at an altitude of 1,566 meters. By the early 1860s over a million cinchona trees grew in Java.[71]

In the 1850s, the British were just as eager as the Dutch to transfer cinchonas. Six saplings were shipped to Calcutta in the early

1850s, but they died. The influx of British troops to India in the Rebellion of 1857 underlined the political urgency of the cinchona transfer. In 1858 Dr. John Forbes Royle of the East India Medical Board and Dr. Thomas Anderson, superintendent of the Calcutta Botanic Garden, persuaded Secretary of State for India Lord Stanley to send teams of collectors to the Andes. Clements Robert Markham, a clerk at the India Office who had explored the Inca ruins of Peru a few years earlier, offered to lead the expedition. Sir William Hooker agreed to build a special greenhouse at Kew to receive the seeds and to send gardeners with the resulting seedlings to India. Three expeditions left England in December 1859. Markham and John Weir, a gardener, went to Bolivia and Peru to get seeds of the *Cinchona calisaya,* the "yellow bark" tree. Dr. Richard Spruce and another gardener, Robert Cross, headed for Ecuador to look for *C. officinalis* and *C. succirubra,* the "red bark" trees. And G. J. Pritchett sought the "grey bark" species. *C. nitida, micrantha,* and *peruviana,* in northern Peru. Their journeys through forests and mountains, eluding suspicious natives and bribing officials, made hair-raising reading to titillate Victorian armchair travelers.[72]

Markham's seeds were the first to reach Kew and germinate, and in August 1860 he wrote a friend: "You will be glad to hear that I have returned here, with a large collection of cinchona plants in good condition, on their way to India."[73] His seedlings, however, did not survive the passage to India in the heat of summer. Pritchett's and Spruce's seedlings, sent in cooler weather, fared better, and in May 1861 Markham wrote: "I returned on Saty last; and after much anxiety & several disappointments, have succeeded in introducing the most valuable species (5) of cinchona into India:— I trust now the experiment is in a fair way of doing well."[74] By July, 2,973 cinchonas, mostly *succirubra* and *officinalis,* were growing at the botanical garden of Ootacamund, a hill station in southern India chosen because its climate resembled that of the Andes. Within two years there were over 100,000 of them, and by 1866 cinchonas covered 20 hectares.[75]

India was now well on its way to becoming a major cinchona producer. But for what purpose? Here the statements made at the time vary, as do historians' judgments. The original objective of the India Office had been to produce quinine "for the treatment of the complaint of Europeans." Markham, in charge of the cinchona project, disagreed. In a memorandum to the Revenue Committee of the India Office in 1865 he wrote: "Did the government undertake

Cinchona cultivation in order that the use of quinine, in some form or another, might be extended to the people of India, now entirely debarred from its use; or did they undertake it as a mere speculation?"[76] He went to India to investigate the possibilities of producing a cheap febrifuge for the benefit of Indians. There he found that three other alkaloids in the bark, quinidine, cinchonine, and cinchonidine, were also effective and that a mixture of them known as totaquine could be manufactured cheaply enough for mass distribution. Markham's idea was restated several times, notably by Thiselton-Dyer: "In India . . . the bark is comparatively inexpensive to grow; the object of Government is not to obtain revenue, but the philanthropic one of supplying to the population a cheap and effective medicine."[77]

The philanthropic policy was only partly successful. In Bengal, totaquine was sold in post offices for a nominal sum, but elsewhere the products of the government cinchona plantations were reserved for British military and civil personnel, and little was available at prices which most Indians could afford.[78] In India, cinchona never became profitable enough to attract private planters.[79]

The Dutch, meanwhile, claimed no philanthropic motives but continued to study cinchonas scientifically. Their methods differed from those used in India; in particular, Junghuhn planted his seeds in the shade, whereas W. G. McIvor, at Ootacamund, preferred to grow cinchonas in the open. Though at first Dutch production lagged behind the British, after a dozen years Junghuhn's trees began producing more bark than McIvor's. However, it was inferior bark. Of the many varieties of cinchona, only three were commercially interesting: *succirubra,* the favorite in India because it could withstand changes in the weather and had a high alkaloid content, though it gave little quinine; *officinalis,* which had a higher quinine yield but was prone to diseases; and *calisaya,* a high-yielding but fastidious plant that only grew well in virgin rain forests. The trees in Java were not of these preferred varieties, but *C. Pahudiana* grown from Hasskarl's seeds, which yielded little quinine. Though they had a head start, the Dutch lagged for many years behind the British in cinchona bark production.

Then came a stroke of luck. In 1865 the English trader Charles Ledger smuggled seeds of a new variety, *C. Ledgeriana,* out of Bolivia. Most of them were planted in India but suffered from the variations in climate. Those that were planted in Java thrived. In 1872–73 the first lots of *Ledgeriana* bark from Java were put on the market. They had the highest quinine content of all and fetched the best prices on the Amsterdam market. To overcome their extreme delicacy and genetic instability, H. W. van Gorkom, director of the government's cinchona plantation, devised a new method of reproducing them by grafting *Ledgeriana* cuttings onto freshly cut *succirubra* trunks. Tjinieroean became a single-crop, multidisciplinary experiment station, a role model for many others founded later.

At last Java had a commercially competitive cinchona. With seeds, cuttings, and technical advice freely disseminated by the government, entrepreneurs opened new plantations throughout the Preanger Regency.[80] In subsequent years cinchona experts perfected their methods. In 1895 a new director of cinchona cultivation, van Leersum, ordered all but the *Ledgeriana* trees cut down to prevent the fertilization of their seeds by pollen from less productive varieties. From then on, quinologists could select plants with ever higher yields: from an average of 5 to 6 percent in the 1880s, to 7 or 8 percent in 1920, and 8 or 9 percent by 1940.[81]

Cinchona production increased rapidly after 1872. By 1916 there were 114 cinchona plantations in the Dutch East Indies, covering 15,500 hectares. Almost all the bark and quinine was marketed in Amsterdam where sales rose from 10 tons of quinine in 1884 to 516.6 in 1913.[82] As more Europeans moved to the tropics and took malaria prevention more seriously, the demand for cinchona bark rose as well, but it was insensitive to price changes. Hence, as vast quantities of bark flooded the world market from 1880 on, the wholesale price of quinine tumbled from £24 a kilogram in 1880 to £1 or 2 in 1913, a drop of 90 percent or more.[83] The price drop did not drive the Dutch planters out of the cinchona business, but it ruined their competitors. In Ceylon the planters uprooted their cinchonas and planted tea bushes in their place; the cinchona area, which had reached over 26,000 hectares in 1883, dropped to 300 hectares in 1910. The same happened in India; even on the government plantations which were subsidized as a public service, many trees were uprooted.[84] The planters of Java, however, weathered the crisis, and by the turn of the century they had captured nine-tenths of the world quinine market.[85]

The Dutch success was due to a combination of geography, science, and perseverance; but it was also a result of that classic response to uncomfortable market fluctuations: the cartel. In the 1890s the planters formed a Cinchona Planters' Association and opened a quinine factory at Bandung. This slowed but did not halt the drop in prices at the Amsterdam cinchona auctions. Therefore in 1913 the planters and the quinine manufacturers agreed to stop the auctions, set the minimum price of quinine at 16.5 florins (£1 8s.) a kilogram, and control supplies accordingly. This was to be administered by the Quinine Bureau of Amsterdam, with representatives of planters and manufacturers.[86]

The temptation was great to use this new power to raise prices and profits. During World War I, Germany was cut off from cinchona supplies, but Allied campaigns in malarial areas made up for the loss, and prices more than trebled by 1921.[87] After the war, governments abandoned notions of laissez-faire, in public health as well as in business. Antimalarial campaigns just getting underway required great quantities of quinine at low prices. While consumption rose, world quinine production—nine-tenths of it in Java—remained steady, driving prices up. This aroused much controversy, some experts maintaining that the Java planters were producing all they could, while others spoke of monopoly and extortion. The issue was charged with moral values because quinine was a life-preserving drug, not just another tropical delicacy. Yves Henry, for example, wrote indignantly about "exorbitant prices for products which are the bread of health of entire peoples, and which have been turned into medicines for the well-to-do."[88]

The polemic had two consequences and an unexpected outcome. The Dutch response to world criticism was a two-tier pricing system. M. Kerbosch, director of the Dutch East Indies cinchona plantations, claimed that since demand for quinine was price-inelastic, it made no sense to place more of it on the world market; instead the Quinine Bureau would keep public supplies steady but offer quantity discounts to government health services engaged in anti-malarial campaigns.[89]

The two-tier system did not satisfy other governments. French Colonial Minister Chaumet wrote to the colonial governors in 1925: "The efforts being made in all our colonies to fight malaria involve an ever increasing consumption of quinine which weighs heavily on the public health budgets and which, nonetheless, is still far from the quantity which ought to be given to the populations." As an example, the previous year the ministry had distributed 4,535 kilograms of quinine in French West Africa and Madagascar, enough for 1 percent of the population; as for supplying the other 99 percent with quinine, "no budget could support such a burden." There were other reasons as well. Quinine purchases used up foreign exchange, a concern to many countries in the protectionist 1920s. Some even argued that world quinine supplies were threatened by the rise of communism in the Netherlands East Indies. Chaumet advocated "a true quinine policy allowing France to free herself from foreign tutelage."[90]

The difficulty was translating these proposals into quinine. Over the years there had been many attempts to grow cinchonas in various parts of the French Empire; they had all failed because of administrative or technical incompetence.[91] After World War I, the bacteriologist Alexandre Yersin imported *C. Ledgeriana* seeds from Buitenzorg and proved that they could grow in Indochina; by the end of the twenties Indochina was supplying part of its own needs in quinine, albeit at more than the world market price.[92] The Indian government plantations only satisfied one-fifth of the requirements of the British Empire. In the 1920s the Botanical Survey of India spent £26,000 opening up new plantations in Burma and Madras, but they failed, and India had to continue importing quinine.[93] The Belgians tried to grow cinchonas in the Congo, but the yields they obtained were too low to repay the costs involved. Italy, having no equatorial colony, bought estates in the Netherlands East Indies, as did the Japanese.[94]

These various attempts to grow cinchonas illustrate the complexity of what economists call comparative advantage. The Dutch began with a geographical advantage, the perfect climate for the fastidious *Ledgerianas*. On that foundation they developed a multidisciplinary scientific approach to a single crop, a symbiosis between government and private enterprise, and one of history's most effective global cartels. All rival projects, even the well-funded British schemes, only nibbled at the Dutch monopoly. What finally killed it was the Japanese conquest of Java in 1941 and the discovery of a chemical substitute, atebrine.

Sugar Cane

Botanically and economically, sugar cane could hardly be more different from cinchona. As undemanding a plant as cinchona is fussy, it will grow almost anywhere in the tropics, provided there is water. It makes but two demands on its growers: huge amounts of unskilled labor during brief intervals, and machinery to process it as soon as it is cut. Hence its historical connection with phenomena seldom found in other branches of agriculture: forced labor and the rural factory. Though it has dominated the economies of whole nations for centuries, the sugar business has always been intensely competitive and has never been controlled by a cartel or monopoly.

In the history of sugar cane there have been two waves of technological innovations spurred on by economic crises: the industrialization of cane processing in the nineteenth century, and the biotechnical revolution of the plant itself in the twentieth. Let us consider these changes as they affected the West Indies and Java.

In the early nineteenth century most West Indian mills replaced their old vertical wooden or stone rollers with horizontal cast-iron rollers which produced more and cleaner juice. The rollers were turned by oxen or, in a few places, by windmills or waterwheels. The juice was then mixed with lime to clarify it and boiled two to four times in open kettles to remove the water. The resulting syrup was left in inverted clay cones for several weeks to let the molasses drip out, leaving crystallized sugar behind. Such methods required large amounts of labor, fuel, and animal power, and the product often needed to be refined again. Yet West Indian sugar dominated the European market because slavery and cheap shipping kept its costs down.[95]

Sugar-cane processing was industrialized when techniques developed in Europe for other purposes were imported to the West Indies. The steam engine, which spread throughout the islands in the early nineteenth century, allowed the use of faster and more efficient sets of shredders and rollers through which cane stalks passed several times. By the end of the century heavy and complex steel machines, most of them British, extracted up to nine-tenths of the juice from the cane.[96]

Another major advance in cane processing was the vacuum evaporator. Unlike the open-pan method, which burned the sugar while boiling off the water at 120° C, the vacuum pan method removed the water at 60° C, leaving white crystals behind. Patented in 1813, it spread to the larger Caribbean estates in the 1830s and 1840s. After passing through the vacuum pans the syrup, or *masse-cuite,* was put into a third machine, the centrifuge. This device, introduced in the 1840s, drained the molasses from the sugar crystals in a matter of hours instead of weeks.[97]

The high cost of these machines put them out of reach of all but the largest estates. Since cane loses its sugar soon after it is cut, the economy of scale of modern processing machines was balanced by the time required to transport the cane from field to mill. Railways extended the area which could feed a mill but consumed fuel and, unlike oxen, produced no fertilizer. The emancipation of slaves and the need for skilled mechanics also benefited the larger estates at the expense of the smaller ones. The central factory provided a partial compromise. Since a large modern factory could process the cane from thousands of hectares, a central factory replaced several mills, thereby gaining control over many surrounding estates. The result was a great transformation in the geography of sugar. The regions which had been most productive in the eighteenth century declined. Small and hilly estates never recovered, except for those on Barbados which survived by producing rum. Meanwhile new estates were opened up in the flat areas of Cuba and British Guiana.[98]

In eighteenth-century Java, most sugar estates were owned by Chinese who used buffalo-powered wooden or stone rollers which wasted most of the juice. A drop in the price of sugar, an increase in that of rice, and the depletion of firewood supplies brought about their demise. They were replaced by European entrepreneurs eager to introduce the newest machines and processes. Wrote the most ambitious of the new manufacturers, Trail and Company, in 1826:

> In embarking on the enterprises we now have on hand, we were sensible of the deficiency of the rude and imperfect machinery by which the manufacture of sugar was carried on here, and therefore determined to import European machinery, with skillful men to conduct the same. . . . We now have three distinct sets of mills, where we employ a European horizontal mill with three cylinders, driven by a six horse-power steam engine; a European eight horse-power mill, with three cylinders, worked by cattle, with six complete sets of iron boilers and iron and copper clarifyers; as also three distilleries, comprising six European copper stills . . . and a suitable complement of fermenting systems for distilling the molasses into Arak and Rum.[99]

Modern machinery was not enough, however. European planters in the then thinly populated region of Batavia had to bring in seasonal workers from the north-coast residencies, and they could never obtain enough to use their equipment to capacity. A different system arose in eastern Java. Rather than use cleared land, planters contracted with villages to grow cane in their irrigated fields, alternating with rice. Thus arose a symbiotic relationship between Javanese rice farmers and European sugar planters, which later spread throughout

Java under the name of Cultivation System.[100] The government encouraged the symbiosis because it both allowed the mill owners to profit from modern machinery and cheap labor and insulated the peasantry from the vicissitudes of the world market. This benign paternalism encouraged ever more intensive farming by a growing population without long-run improvements in labor productivity, a process that Clifford Geertz has called "agricultural involution."[101]

Despite the industrialization of sugar mills in the East and West Indies, the industry remained vulnerable to both biological and economic enemies. The biological enemy was the degeneration of the cane and the depletion of the soil. Cane crops began to falter in Mauritius and Reunion in the 1840s, in Brazil in the 1860s, in Puerto Rico in the 1870s, and in the British West Indies in the 1890s. In Java, the local cane variety, Batavian, was attacked by the sereh disease from 1884 on.

The economic enemy was the sugar beet, which had two advantages. Its production was heavily subsidized in several European countries, and its sugar content had increased threefold in the last of the century. As a result, sugar prices dropped by about half in the last two decades of the century, and beet sugar conquered two-thirds of the world market.[102]

In Java, the outbreak of sereh ruined most cane planters, and their estates passed into the hands of large corporations. Thus the "P & T Lands," owned by a succession of British businessmen since 1813, were sold to the Maatschappij ter Exploitatie der Pamanoekan en Tjiassemlande, a corporation partly owned by the Nederlandsch Indische Handelsbank, and resold in 1910 to a British firm, the Anglo-Dutch Plantations of Java, Ltd. It was the largest estate in Java, with its own railroad and canal, and hundreds of hectares in sugar and other crops.[103]

Corporations were willing to invest in improvements on several fronts. They imported the latest and most expensive machinery. By pouring fertilizers on the cane they doubled both the yield per hectare and the island's sugar production between 1885 and 1900. Most importantly, they invested in research. Between 1885 and 1887 they established three sugar-cane experiment stations on Java, which were merged in 1907 into the Proefstation Oost Java at Pasuruan, the world's foremost sugar-cane research center. By 1924 POJ had a staff of fifty-nine and a budget of over a million florins, more than all the government institutions at Buitenzorg. A similar crisis in the West Indies led the British to set up an experiment station at Barbados in 1887 and the Imperial Department of Agriculture in 1897.[104]

The singularity of the sugar cane, and the reason it fell behind the sugar beet in the late nineteenth century, is that it was thought to reproduce only by vegetative propagation. In effect, all cane plants of a given variety were clones of a common ancestor; outside of Java most of the world's sugar cane was of the Bourbon or Otaheite variety native to Tahiti. All plants of the same variety were vulnerable to the same diseases and could not be improved by breeding.[105] During the nineteenth century, botanists tried to resolve this dilemma by seeking other varieties in nature. Buitenzorg encouraged planters in Java to adopt the Batavian cane. The gardens of Pamplemousse (Mauritius) and Port-of-Spain (Trinidad) collected hundreds of varieties of wild sugarcane.[106] When the Otaheite and Batavian varieties were struck by diseases, more resistant wild varieties such as Tanna, Badila, and Uba were brought in to replace them, but at the price of a lower sugar content.

The breakthrough came simultaneously and independently in the East and West Indies. During the 1886–87 season, three botanists, Soltwedel of POJ and John Harrison and J. R. Bovell of Barbados, discovered the fertility of the cane and the sexual reproduction of seedlings. Techniques of temperature and light control developed at these two stations led to the production of seedlings of a new variety, the "noble" cane *Saccharum officinarum*.[107] The example was quickly imitated at new experiment stations in British Guiana, Hawaii, Mauritius, Reunion, and India. The varieties they created were as disease-prone as previous canes, however, and needed to be replaced periodically by yet newer ones.

During the first decades of this century scientists at POJ led by Professor Jeswiet worked on a more complex technique of cane breeding known as "nobilization." It involved identifying the parentage of seedlings and cross-breeding the wild cane species *Saccharum spontaneum,* which was resistant to sereh but low in sugar content, with the noble *S. officinarum*. The result was 2878-POJ, a hybrid which was both high-yielding and disease-resistant. Meanwhile, the Coimbatore Experiment Station in India was developing a trihybrid of *S. officinarum, S. spontaneum,* and *S. barberi,* which was particularly suited to Indian conditions. After 1940 began a new and more complex phase of development in which canes were custom-tailored to specific environments at experiment stations around the world.[108]

The consequence of these scientific and technical advances was a resurgence of cane sugar, especially from Cuba and Java. Geoffrey Masefield estimated that, thanks to new agricultural and milling techniques, the yield of sugar per acre increased tenfold in the century 1850–1950, an improvement unmatched by any other crop.[109] Figures for shorter periods also show impressive results: as a result of the diffusion of nobilized canes, mean international sugar cane yields rose 26 percent from 1923–24 to 1938–42; and almost all of the 10-million-ton increase in world sugar production between 1909–10 and 1928–29 was due to sugar cane.[110]

Java's sugar production grew from 47,040 tons in 1840 to 2,961,269 in 1930, an annual increase of 4.71 percent on the average. Some of the increase was due to an expansion of the area devoted to sugar, but most of it can be attributed to increasing yields due to technological advances. Improvements in processing raised the proportion of sugar extracted, while new varieties of cane yielded more sugar per stalk. The 2878-POJ cane, which gave over 20 percent more sugar per hectare than its predecessors, spread until it had displaced all other varieties by 1930.

These improvements were accompanied by a concentration of the industry. From 1894 to 1930 the number of factories fluctuated around 180 while the acreage of cane doubled. By 1930 the typical sugar factory cost a million dollars; employed 20 European technicians and 300 full-time and 4,000 to 5,000 part-time Javanese workers; and consumed the cane from 1,000 hectares of land. Since the cane had to be rotated with rice and other crops, this meant that two to three thousand hectares of arable land were bound by long-term contracts to a single factory.[111] This lop-sided symbiosis became even more skewed with the introduction of 2878-POJ. The social costs of this innovation struck sociologist G. H. van der Kolff:

> The sugar cultivation of the estates and the rice and other cultivations of the population were in effect co-ordinated in one large-scale agricultural enterprise, the management of which was practically in the hands of the sugar factory. The demands of the sugar industry therefore received first place in the crop rotation system. For example, farmers were persuaded to plant an early ripening variety of rice for which they were indemnified in money, though they often would have preferred a late-ripening one, so that the field would be clear for cane growing at the time set by the concern. The result in these regions was definitely retrogressive—in place of peasant ingeniousness came a new coolie submissiveness.[112]

At its peak in the late twenties, the Java sugar industry provided over half of the gross export earnings of the colony and attracted investments from all the Western nations. But its Achilles' heel was its dependence on exports to countries which could cut their sugar imports at will. And so, when the Depression struck, the industry collapsed. The number of operating factories fell to thirty-five in 1936; sugar acreage was reduced by half; production fell by 81 percent; and, worst of all, payments to Javanese dropped by 92 percent.[113]

The sugar symbiosis, based on the most scientific form of agriculture and the most efficient colonial administration the world had ever seen, crumbled, leaving behind a large population of poor farmers crowded on a small island. Unlike T. S. Ashton's Chinese and Indians who "increased their numbers without passing through an industrial revolution," the people of Java had indeed passed through an industrial revolution. But it was not one of their own making, and it did them little good.

Rubber

Rubber followed much the same pattern as cinchona and sugar. Here too, Western demand brought Western capital and science together with non-Western lands and labor. Here again was an industry buffeted by the fluctuations of the world market and by rapid technological changes. Yet in the evolution of this crop we find some unexpected variations.

Throughout most of the nineteenth century, rubber was a tropical curiosity. Not until John Dunlop's pneumatic tire (1888) did demand for rubber begin to grow. Britain, which imported less than a thousand tons of rubber in 1850, consumed ten times as much in 1890, and twice again as much ten years later. The rubber boom after 1905 was a product of the automobile industry, particularly in America, where auto sales jumped from 65,000 in 1908 to 187,000 in 1910. In response, rubber prices rose from 60 pence per kilogram in 1901 to over 300 at their peak in 1910.[114]

At the time, almost all the world's rubber came from wild plants: *Hevea brasiliensis, Manihot glaziovii,* and *Castilloa elastica* of South and Central America, *Ficus elastica* of Southeast Asia, and the *Landolphia* vines of equatorial Africa. Of all these, the hevea tree of the Amazon rain forest was preferred because it survived repeated tappings and gave the most latex and the best quality rubber, known as Para. The surge in prices unleashed a wholesale plundering of these wild plants. To gather the precious latex, entrepreneurs enslaved thousands of defenseless inhabitants of the Amazon and Congo forests, leaving both plants and people dead in the wake of their mad rush for rubber.[115]

The sudden surge triggered more than just pillage and bloodshed; it also brought out, once again, the ecnomic value of botany. Long before businessmen thought of it, botanists had anticipated a growing demand for rubber and transferred plants to the European colonies.[116] The first attempt took place in 1873 when Sir Clements Markham and Secretary of State for India Salisbury persuaded Sir Joseph Hooker to organize an expedition to Brazil to get hevea seeds. Of several thousand seeds sent to Kew, only a dozen germinated, and the six seedlings that reached Calcutta died. More expeditions were therefore organized. In early 1876, the British planter Henry Wickham managed to smuggle 70,000 hevea seeds out of Brazil; 2,700 of them germinated at Kew and almost 2,000 seedlings reached Ceylon in September 1876. Other rubber plants were sent to Burma, Mauritius, Australia, and the West Indies. For three years, much of the British colonial botanical establishment was mobilized for the rubber transfers.

The one transfer that mattered was the shipment of twenty-two seedlings from Ceylon to Singapore in 1877. Nine were planted in the garden of Sir Hugh Low, the British resident at Kuala Kangsar in Perak, while the rest remained in the Singapore Botanic Garden. These twenty-two seedlings are the ancestors of almost all the rubber trees in Southeast Asia today. Yet for the first twenty years, they were all but forgotten, for wild rubber was reaching the market in sufficient quantities, and prices were falling. In Malaya, planters interested in quick returns shied away from a crop that took six or seven years to start producing.[117] For a time, only Henry Ridley, superintendent of the Singapore Botanic Garden from 1888 to 1912, cared about the rubber trees. Soon after arriving, he cleared the area around the heveas, planted more seeds, and began experimenting with new ways of tapping. He proved that carefully nurtured and tapped plantation trees could be far more productive than the wild trees of the Amazon. He founded the *Agricultural Bulletin of the Malay Peninsula* to publicize his findings and gave away hevea seeds and plants to anyone who would plant them, until people began calling him "mad Ridley" and "rubber Ridley."[118]

Neighboring countries also experienced long delays before they began producing rubber. Ceylon, having obtained heveas first, began producing rubber in the late 1890s. Buitenzorg had received seedlings from Kew, Ceylon, Singapore, and Malaya, and distributed them to planters in Java. None of them had as high a yield as the trees in Malaya, and this discouraged planters.[119] Indochina got its first heveas in 1897. In the next few years other Europeans began planting seedlings, but on a small scale. At the time, the French government was more interested in getting wild *Landolphia* rubber or starting *Ficus* or *Manihot glaziovii* plantations in West Africa and New Caledonia, and botanical institutions in France had little advice to give the planters in Indochina.[120]

Then came the boom, and everyone who thought he could squeeze latex out of plants rushed into rubber. Unhampered by the long delays that planters faced, the gunmen of the equatorial forests forced their victims to collect as much wild rubber as they could as fast as possible. Fine Hard Para, the product of the wild heveas of the Amazon, continued to be favored despite the scandals surrounding its collection. In the Congo, collectors killed off plants faster than they could grow back. Wild rubber production peaked in 1910–12, declining after that until in 1922 it only accounted for 6.7 percent of exports, nine-tenths of it from South America.[121] What killed off the wild rubber pillage was neither an indignant press nor the outcry of outraged humanitarians, but the sudden flood of Malayan plantation rubber: 6,604 tons in 1910, 44,752 tons in 1914, 131,064 tons in 1918—enough to drive the prices down to a level that no longer made wild rubber collection profitable.

The first hevea plantations in Malaya were started in 1895 by Tan Chay Yan in Malacca and by the Kindersley brothers in Selangor.[122] They were soon joined by others, Chinese with enough money to buy a few acres, or Europeans who had failed in Ceylon and were trying again in Malaya. Rubber was always grown on freshly cut forest land and required more labor than the Malays could provide. The government, wanting to encourage Europeans, granted land in large blocks and encouraged the immigration of South Indians and Chinese, whom the British referred to by the oxymoron "foreign natives." It also extended the roads and railways serving the tin-mining regions of the east coast. All this turned Malaya into a frontier area, a radically different society from such ancient and densely populated lands as India and Java.[123]

A plantation was a costly enterprise. Clearing the land and

planting seedlings cost £ 50 to £ 65 per hectare, followed by a five-
or six-year wait before the first latex was tapped. Few individuals
could afford it. But corporations could, and after 1905 many did.
Some were small companies controlled by the Singapore merchant
houses of Harrison and Crossfield Ltd. and Guthrie and Company.
Others were huge: the Malay Peninsula (Johore) Rubber Concessions Ltd. got 20,000 hectares; Rubber Estates of Johore Ltd.,
10,000; Dunlop Plantations Ltd., 6,500; and the Franco-Belgian
Société financière des caoutchoucs, 5,000. The area occupied by
heveas spread fast: from 2,400 hectares in 1900 to 219,000 in 1910,
and over 900,000 after 1920. By 1910 heveas had displaced all other
export crops and occupied 62 percent of the cultivated land of the
Federated Malay States.[124] The peninsula was fast becoming a land
of monoculture plantations owned by European firms and worked by
Chinese and Indians to provide tires for American automobiles.

The rubber boom triggered a surge of interest in research. The
Singapore Botanic Garden, with its large collection of heveas of all
ages, worked on seed selection, tapping methods, and tree spacing.
It was soon joined by other organizations. In 1900 the Federated
Malay States hired Stanley Arden to direct their experimental plantations, where tapping methods were being investigated. Five years
later the FMS government set up a Department of Agriculture to
advise hevea planters; by 1914 it had twenty-eight European staff
members. Yet to the planters and estate managers, the government's
efforts were insufficient. The Rubber Growers' Association of Ceylon
and Malaya therefore hired its own scientists who developed new
methods of seed selection, tree spacing, ground cover, tapping, and
latex preparation. Their findings were supposed to be reserved for
their members but could not be kept secret.[125]

As with cinchona and sugar cane, the Dutch lagged for a time
in production but were far ahead in research. While scientists in
Malaya and Ceylon were working on planting and tapping methods,
the Dutch were studying the plants themselves. In 1916 scientists at
Buitenzorg succeeded in grafting buds from high-yielding trees onto
ordinary seedlings. This increased the yield from 300–400 kilograms
of dry rubber per hectare to 650 or more but retained the wide variations found in the parent trees, some clones giving many times more
latex than others. At AVROS, the research station of the Sumatra
rubber planters, the geneticist Heusser set out to create a better
hevea by growing pairs of trees from selected seeds deep in the
forests, far from any other hevea that might fertilize them. Artificial

pollination and rigorous selection produced offspring capable of yields up to twice that of good clones. This technique was taken up by the new Rubber Research Institute of Malaya after 1926. AVROS and RRIM did for natural rubber what POJ had done for sugar cane: they rescued a tropical industry from the competition of Western substitutes.[126]

In the interwar period, rubber prices fluctuated wildly because of erratic demand. Prices fell in 1922, recovered in 1925, then began falling again and collapsed in the Depression. The natural rubber industry responded by improving its productivity through technological advances and by controlling supplies through cartels. The first method defeated the second. Britain tried to organize a cartel in 1922, but the Dutch, whose rubber production was still expanding, refused to participate. Malaya's share of world production fell from 50 to 38 percent between 1921 and 1927, while that of the Netherlands East Indies rose from 24 to 38 percent.[127]

The Depression almost ruined the natural rubber business. American tire production, which accounted for over half the world's rubber consumption, fell by almost half between 1929 and 1932. Some estates were abandoned, and others pulled out their heveas and planted other crops. To protect themselves against the worldwide drop in demand, the rubber-producing countries of Asia agreed to the International Rubber Regulation Scheme of 1934, which restricted their exports and forbade new plantings, except in Indochina.

There, the French showed little interest in rubber until 1906. Heveas spread slowly at first, from 1,200 hectares in 1909 to little over 33,000 in 1924. Then came a great spurt until 1930, after which the area planted in rubber trees leveled off at around 127,000 hectares. Rubber production followed with a few years' delay: 3,000 tons in 1920, 8,000 in 1925, 10,500 in 1930, 30,000 in 1935, and 60,000 by the end of the decade. The reason Indochina lagged behind Malaya was partly ecological and partly political. The accessible lands around Saigon and near the coast, where the first plantations were opened, had soils of low fertility and poor drainage. The much superior red soils of northern Cochinchina, Cambodia, and southern Annam were far from towns and covered with jungle. Their exploitation required roads, large amounts of labor, and massive investments. World War I and the subsequent reconstruction of northern France postponed all such efforts for a decade.

Because of this delay, the configuration of the Indochinese rubber industry turned out very differently from that of its neighbors

Malaya and the Dutch East Indies. There was little experimentation and no competition from Asian smallholders. Instead, the industry was dominated by twenty-seven corporations which owned over two-thirds of all the rubber area.[128] From a technical and business point of view, they were astonishingly successful. Auguste Chevalier remarked: "It is, to our understanding, the first case of a modern crop, undertaken on a large scale in the French colonies and which, despite the innumerable difficulties which attend every beginning, has succeeded beyond the most optimistic hopes."[129] The reasons for this achievement are many: vacant lands with good soil, a favorable climate, disciplined labor, and low taxes. But most important of all was the existence of the technology developed in Malaya and the Dutch East Indies. Because the planters of Indochina delayed their expansion until the late twenties, they started with the latest clones and the most productive methods, while their rivals, who had created this technology, were saddled with obsolete trees. As the technical director of one of the large rubber estate companies pointed out in 1937:

> Indochina, a latecomer, profited from the experience of its neighbors to introduce, as early as 1929, grafted plants at an ultra-rapid rate. The year 1931 alone gave our beautiful Colony nearly 170,000 hectares of plants grafted with the best clones from Sumatra and Java. In 1936 it totals 127,147 planted hectares, of which 45,063 in grafts, which is 35.4 percent, whereas our neighbors the Dutch have only 11.1 percent and the English 6.5 percent.[130]

The new plantations were helped through the dangerous thirties by strong protectionist measures. The Indochina government made loans to planters who took trees out of production. Others were granted subsidies by the French government out of duties on imported rubber. The agreement of 1934, which made Malaya and the Dutch East Indies cut back their exports, gave Indochina a quota based on French imports, which exceeded Indochinese exports until 1938. The survival of Indochina's rubber industry was paid for by French consumers, Indochinese taxpayers, and Malayan and Indonesian smallholders.[131]

Conclusion

In this chapter we have considered the relationship between science and the agriculture of three crops in the colonies of three colonial powers. These are not representative cases, but particularly success-ful examples of their kind. Yet in their similarities and differences they offer insights into the process of technology transfer. In all three cases, preindustrial agriculture (in the case of sugar cane) or the collection of wild crops (cinchona and rubber) could not meet the growing Western demand for tropical goods. When slavery and plunder failed, science was brought to bear on this economic prob-lem.

The first response was the simple transfer of plants by collectors and botanical gardens. Then came a phase of research and develop-ment at experiment stations which created better varieties through breeding and cloning, and later through hybridization. Technical advances also improved cultivation and processing. The result was to reduce traditional methods of agriculture or collection to a mar-ginal role. The profits of a science-based agriculture could not long be reserved for its innovators, however. Techniques, whether cul-tural or embodied in seeds, spread swiftly across borders. Price fluc-tuations, brought on by the producers' overenthusiastic response to early profits and an erratic demand, led to schemes to restrict sup-plies. Western science, having improved tropical agriculture, was busy working to undermine what it had created. It almost suc-ceeded with beet sugar in the nineteenth century and with synthetic rubber after 1940, and it has fully succeeded with chemical substi-tutes for quinine.

Yet the three examples differ. Sugar cane became an increas-ingly concentrated industry, dominated by central factories to which the cane growers were tied as day laborers. Cinchona was controlled by a cartel so efficient that other nations could not break its hold. And hevea rubber was taken over by smallholders in some areas and by huge industrialized estates in others. How can we account for such differences?

To a certain extent, they can be attributed to government poli-cies. The Netherlands East Indies government supported research and extension work, while the French colonial governments delayed in helping tropical agriculture. Yet, as it turned out, governments did not have to support research if their neighbors did, and if their nationals saw a profit in importing the new technologies on their own.

Biological differences mattered more. The delicate cinchona was hard to grow outside of Java, and research only accentuated the advantages of this particular environment. The combination of an ecologically sensitive supply and a price-inelastic demand made it a prime candidate for a monopoly, which the Dutch were quick to take advantage of. In contrast, sugar cane was easy to grow throughout the tropics, but its processing was subject to great economies of scale. The result was the rise of modern factories surrounded by a dense, poor population. Though the industry was located in the tropics, it remained culturally alien to the inhabitants of the sugar lands. Research only accentuated this unequal relationship.

Natural rubber is an aberration among tropical crops. Like sugar cane, it was easy to grow. Land, labor, and capital could be substituted for one another in myriad ways, and there seemed to be few economies of scale in growing the trees or in processing the latex. As a result, smallholders and great corporate estates coexisted, their rivalry being played out in the political arena rather than in the market place. The sudden immigration of people of different nationalities into the previously empty lands of Malaya, Sumatra, and Borneo created frontier societies that added diversity to the process. Only in the more tightly controlled Indochina was this smallholders' capitalism repressed, with consequences that only appeared after decolonization.

Notes

1. John Christopher Willis, *Agriculture in the Tropics: An Elementary Treatise* (Cambridge, 1909), pp. 38–39.

2. Isaac Henry Burkill, *Chapters in the History of Botany in India* (Calcutta, 1965), pp. 122–23.

3. Geoffrey B. Masefield, *A History of the Colonial Agricultural Service* (Oxford, 1972), pp. 20–21.

4. Mea Allen, "Hooker, William Jackson," in *Dictionary of Scientific Biography* (New York, 1976), 6: 492–95.

5. William Turner Thiselton-Dyer, "The Botanical Enterprise of the Empire," *Proceedings of the Royal Colonial Institute* 11 (1879–80): 284–86; R. Desmond, "Hooker, Joseph Dalton," in *Dictionary of Scientific Biography* 6: 488–92.

6. Gerald L. Geison, "Thiselton-Dyer, William Turner," in *Dictionary of Scientific Biography* 13: 341; A. Milhe-Poutingon, "Rapport presenté au ministre des colonies sur une mission aux jardins royaux de Kew," supplement to *Revue des cultures coloniales* 3, no. 18 (November 1898): 3.

7. Thiselton-Dyer, pp. 279–80, 287.

8. Ibid., pp. 276–78.

9. Ibid., pp. 291–93.

10. Lucile H. Brockway, *Science and Colonial Expansion: The Role of the British Royal Botanic Gardens* (New York, 1979), p. 85; Louis Gentil, "Les jardins royaux de Kew et leur influence à travers l'empire colonial anglais," *Revue des cultures coloniales* 1, no. 6 (November 1897): 208–9.

11. Milhe-Poutingon, p. 9.

12. Sir Charles Bruce, *The Broad Stone of Empire: Problems of Crown Colony Administration, with Records of Personal Experience*, 2 vols. (London, 1910), 2: 114–15.

13. Kalipada Biswas, ed., *Calcutta Royal Botanic Garden, 150th Anniversary Volume* (Alipore, 1942), pp. 2–12; G. S. Randwaha, K. L. Chadha, and Daljit Singh, eds., *The Famous Gardens of India* (New Delhi, 1971), pp. 13–15; G. Stuart Gager, "Botanic Gardens of the World: Materials for a History," *Brooklyn Botanic Garden Record* 27, no. 3 (July 1938): 181–83, 260–63; D. M. Forrest, *A Hundred Years of Ceylon Tea 1867–1967* (London, 1967), pp. 89–109; Geoffrey B. Masefield, *Colonial Agricultural Service*, pp. 17–28, and *A Short History of Agriculture in the British Colonies* (New York, 1950), pp. 58–61; S. Rajaratnam, "The Growth of Plantation Agriculture in Ceylon, 1886–1931" and "The Ceylon Tea Industry, 1886–1931," in *Ceylon Journal of Historical and Social Studies* (January–June 1961): 1–19 and (July–December 1961): 169–202; Burkill, pp. 66–72, 102–10, 129–34, 167–73, 203–6.

14. John N. Martin, *Botany with Agricultural Applications*, 2nd ed. (New York, 1920), pp. 577–83; Montague Yudelman, "Imperialism and the Transfer of Agricultural Techniques," in *Colonialism in Africa 1870–1960*, ed. Peter Duignan and L. H. Gann, vol. 4: *The Economics of Colonialism* (Cambridge, 1975), pp. 331–34.

15. Daniel Morris, "The Imperial Department of Agriculture in the West Indies," *United Empire* 2 (February 1911): 79–84.

16. Ibid., pp. 76–79; Masefield, *History of Agriculture*, p. 49; Lilian C. A. Knowles, *The Economic Development of the British Overseas Empire, 1763–1914*, vol. 1: *The Empire as a Whole and The British Tropics* (London, 1924), pp. 128–29.

17. Morris, p. 82.

18. Masefield, *Colonial Agricultural Service*, p. 17.

19. Yudelman, pp. 338–50.

20. Masefield, *History of Agriculture*, p. 70; Edgar Barton Worthington, *Science in Africa: A Review of Scientific Research Relating to Tropical and Southern Africa* (London, 1938), p. 308.

21. Melchior Treub, "Kurze Geschichte des botanischen Gartens zu Buitenzorg," in *Der botanische Garten "'s Lands Plantentuin" zu Buitenzorg auf Java. Festschrift zur Feier seines 75 jährigen Bestehens (1817–1892)* (Leipzig, 1893), pp. 23–78, and "Un jardin botanique tropical," *Revue des deux mondes* (January 1890): 162–83.

22. Treub, "Geschichte," p. 71.

23. Treub, "Geschichte," pp. 60–73, and "Jardin botanique," pp. 168–69; Thiselton-Dyer, p. 297; F. A. F. C. Went and F. W. Went, "A Short History of General Botany in the Netherlands Indies," in *Science and Scientists in the Netherlands Indies*, ed. Pieter Honig and Frans Verdoorn (New York, 1945), pp. 392–95; Gottfried Haberlandt, *Eine botanische Tropenreise. Indo-malayische Vegetationsbilder und Reiseskizzen* (Leipzig, 1893), p. 74.

24. Treub, "Die Bedeutung der tropischen botanischen Garten," in *Der botanische Garten*, p. 9.

25. C. G. G. J. van Steenis, "Treub, Melchior," in *Dictionary of Scientific Biography* (New York, 1976), 13: 458–60; Joseph Chailly-Bert, "L'Institut botanique de Buitenzorg," *Revue des cultures coloniales* 3, no. 15 (August 1898): 49–53; Charles J. Bernard, "Le jardin botanique de Buitenzorg et les institutions de botanique appliquée aux Indes néerlandaises," in Honig and Verdoorn, pp. 10–15; F. A. F. C. Went, "In Memoriam," *Annales du Jardin botanique de Buitenzorg*, 2nd ser., no. 9 (1911): i–xxxii; Went and Went, pp. 392–95.

26. Henri Lecomte, "Influence des jardins d'essais sur le développement de l'agriculture aux colonies," *Bulletin de la Société de géographie commerciale de Paris* (1899), p. 31; Milhe-Poutingon, p. 10.

27. Treub, "Jardin botanique," pp. 170–71; K. W. Dammerman, "A History of the Visitors' Laboratory ("Treub Laboratorium") of the Botanic Gardens, Buitenzorg, 1884–1934," in Honig and Verdoorn, pp. 59–75; James Herbert Veitch, *A Traveller's Notes; or, Notes of a Tour through India, Malaysia, Japan, Corea, the Australian Colonies and New Zealand during the Years 1891–1893* (Chelsea, 1896), pp. 83–84.

28. J. S. Furnivall, "The Machinery of Economic Uplift in the Netherlands Indies," *Asiatic Review* 34 (1938): 115–30, and *Colonial Policy and Practice: A Comparative Study of Burma and Netherlands India* (Cambridge, 1948), p. 254; Frans Verdoorn and J. G. Verdoorn, "Scientific Institutions, Societies, and Research Workers in the Netherlands Indies" in Honig and Verdoorn, pp. 425–29.

29. Frans Verdoorn and J. G. Verdoorn, "Scientific Institutions, Societies, and Research Workers in the Netherlands Indies," in Honig and Verdoorn, pp. 425–29; Furnivall, *Colonial Policy*, p. 253; Ch. Coster, "The Work of the West Java Research Institute in Buitenzorg" in Honig and Verdoorn, pp. 55–59; C. J. J. van Hall, "On Agricultural Research and Extension Work in Netherland's Indies," in *Science in the Netherlands East Indies*, ed. L. M. R. Rutten (Amsterdam, n.d. [1929?]), pp. 268–73; Yves Henry, *Rapport d'ensemble sur une mission aux Indes néerlandaises* (Hanoi, 1926), pp. 42–43.

30. A. Davy de Virville, "XIXe siècle. Voyages et explorations botaniques," in *Histoire de la botanique en France*, ed. A. Davy de Virville (Paris, 1954), pp. 349–76.

31. Charles Martins, "Les jardins botaniques de l'Angleterre comparés à ceux de la France," *Revue des deux mondes* (December 1868), pp. 25, 31–32.

32. Camille Limoges, "The Development of the Muséum d'Histoire Naturelle of Paris, c. 1800–1914" in *The Organization of Science and Technology in France, 1808–1914,* ed. Robert Fox and George Weisz (Paris and London, 1980), pp. 235–36; Alphonse Milne-Edwards, "Les relations entre le Jardin des plantes et les colonies françaises," *Revue des cultures coloniales* 4, no. 20 (January 5, 1899): 2–11; Maxime Cornu, *Le Jardin des plantes de Paris (Muséum d'histoire naturelle) et les colonies françaises* (Paris, 1901), pp. 1–12; Paul Lemoine, "Le rôle colonial du Museum," *La terre et la vie* 1 (1936): 3–4; André Guillaumin, "Les jardins botaniques de la métropole et des colonies," *Revue scientifique* 6–7 (June–July 1939): 402–12; Jean Dorst, "Les activités

outre-mer du Muséum national d'histoire naturelle," *Mondes et cultures* 38, no. 4 (1978): 595–602.

33. Treub, "Jardin botanique," p. 183; Lecomte, p. 31.

34. Cornu, p. 5.

35. Ibid., pp. 5–6.

36. William T. Thiselton-Dyer, "Maxime Cornu," in *Nature* 64, no. 1652 (June 27, 1901): 211–12.

37. Jean Dybowski, *Jardins d'essai coloniaux* (Paris, 1897), pp. 38–39.

38. Edmond Perrier, "Préface," in *Conférence de 1917. Nos richesses coloniales,* Muséum national d'histoire naturelle (Paris, 1918), pp. v–vi.

39. *Revue des cultures coloniales* (September 1898), pp. 1–4.

40. Milhe-Poutingon.

41. Archives Nationales Section Outre-Mer (hereafter ANSOM), Généralités 545 (2580): "Procès-verbaux des séances de la Commission des jardins d'essai coloniaux"; Paul Bourde, "Le rapport de M. Bourde sur les travaux de la Commission des jardins d'essai," *Revue des cultures coloniales* 4, no. 22 (February 5, 1899): 70–80.

42. "Le jardin d'essais colonial du bois de Vincennes," *Revue des cultures coloniales* 4, no. 22 (February 5, 1899): 65–70.

43. Perrier, pp. ix–xiv.

44. Papers documenting these arguments are in the archives of the Institut de recherches agronomiques tropicales (Nogent-sur-Marne), dossier "Questions foncières"; Muséum national d'histoire naturelle, Bibliothèque centrale (Paris), carton 2225: "Papiers Edmond Perrier"; and Archives nationales (Paris), carton AJ15, dossier 848: "Jardins et serres, jardins botaniques départementaux et coloniaux." The best brief history of these events is in André Angladette, "La chronique du trimestre. Une vieille et bien curieuse histoire, celle du 'Jardin Colonial' de Nogent-sur-Marne," in *IRAT: Bulletin d'information et le liaison de l'Institut de recherches agronomiques tropicales et des cultures vivrières* 16, no. 1 (January 1982): 253–61.

45. Interview with André Angladette, former director of IRAT, March 13, 1984. See also Dorst, p. 598, and Limoges, p. 238.

46. ANSOM, Généralités 53 (492): letter of June 12, 1899.

47. ANSOM, Généralités 53 (520) and 56 (577); *Revue des cultures coloniales* 8, no. 77 (May 20, 1901).

48. Angladette, "Chronique" and interview. See also Lemoine, pp. 10–11; Georges Schwob, *Exposition universelle et internationale Bruxelles 1910: Les colonies françaises* (Paris, n.d.), pp. 326–35; and Worthington, p. 327.

49. Muséum national d'histoire naturelle, Bibliothèque centrale (Paris), carton 2225: "Papiers Edmond Perrier"; Auguste Chevalier, *L'agronomie coloniale et le Muséum national d'histoire naturelle. Premières conférences du cours sur les productions coloniales végétales et l'agronomie tropicale professé par M. Aug. Chevalier au Muséum national d'histoire naturelle* (Paris, 1930); Limoges, p. 232; Lemoine, pp. 5–16. The *Journal d'agriculture tropicale* was renamed *Revue de botanique appliquée et d'agriculture coloniale* in 1921, then *Revue de botanique appliquée et d'agriculture tropicale* in 1929.

50. ANSOM, Généralités 25 (226).

51. A. Milhe-Poutingon, "La renaissance des cultures coloniales," *Revue des cultures coloniales* 1, no. 1 (June 5, 1897): 1–2.

52. Cornu, pp. 8–9.

53. Lecomte, pp. 17–32.

54. Cornu, pp. 19–20; C. Chalot, "Notice sur le jardin d'essai de Libre-ville," *Revue des cultures coloniales* 2, no. 8 (January 1898): 14–19; and "Rapport sur le jardin d'essai de Libreville," *L'agriculture pratique des pays chauds* 1, no. 2 (September–October 1901): 168–81; Chevalier, pp. 98–99.

55. Luc, "Le directeur du jardin d'essai de Brazzaville à Monsieur l'administrateur de la région de Brazzaville," *L'agriculture pratique des pays chauds* 1, no. 2 (September–October 1901): 182–85; ANSOM, Archives d'inspection: "Mission Bouchat 1900–1901. Service des cultures, no. 16, Libreville" (May 29, 1901) and "Mission Picanon: Service d'agriculture au Gabon, no. 44" (April 8, 1919).

56. Worthington, p. 328; Yudelman, p. 346.

57. ANSOM, Fonds Guernut, Conférence, pp. 1–5.

58. Auguste Chevalier, *L'organisation de l'agriculture coloniale en Indo-chine et dans la Metropole* (Saigon, 1918), pp. 7–9, and "Les améliorations scientifiques et techniques realisées par la France en Indochine," *Revue de botanique appliquée et d'agriculture tropicale* 277 (1945): 1–32; Gager, pp. 265–66.

59. Chevalier, *Organisation*, p. 18; Lecomte, p. 19; *Revue des cultures coloniales* 1, no. 6 (November 1897): 204–5.

60. On scientific institutions and agricultural research in Indochina see André Angladette, "Les recherches agronomiques en Indochine pendant la première moitié du vingtième siècle. Leur impact sur la production rurale. Leur évolution ultérieure," *Mondes et cultures* 41, no. 2 (1981): 189–215; Yves Henry, *Economie agricole de l'Indochine* (Hanoi, 1932), pp. 673–79; P.-B. de la Brosse, "Les institutions scientifiques de l'Indochine française," *Revue scientifique* 69, no. 21 (November 14, 1931): 641–45; Christine Massiou, Sylvie Mouranche, and Manuela Garrido, "Recherches sur les sciences appliquées en Indochine pendant la période coloniale" (Mémoire de maîtrise, Université Paris VII, 1975); and Chevalier, *Organisation*, pp. 20–57.

61. *Courrier d'Haïphong*, April 30, 1919.

62. Angladette, "Chronique," p. 260, and interview; Chevalier, *Organisations*, p. 45 and "Les grands établissements scientifiques du Moyen et de l'Extrême-Orient et la fondation de l'Institut de Saïgon," *La géographie* (April 1920): 306–8.

63. His works include *La question cotonière en Afrique occidentale française* (Melun, 1906); *Eléments d'agriculture coloniale: plantes à huile* (Paris, 1921); and *Eléments d'agriculture coloniale: plantes à fibres* (Paris, 1924).

64. Henry, *Economie*, p. 677.

65. Henry, *Rapport*, p. 26.

66. *Terres rouges et terres noires basaltiques d'Indochine. Leur mise en culture* (Hanoi, 1931).

67. Angladette interview. Angladette studied at Nogent-sur-Marne in 1927–28, went to Indochina in 1929, and headed the Office indochinois du riz from 1934 to 1946.

68. ANSOM, Agence F.O.M. 200 (141); Institut français du caoutchouc,

Le caoutchouc d'hévéa, initiation aux méthodes d'exploitation en Indochine (Paris 1945).

69. On malaria, quinine, and European imperialism in Africa, see Philip D. Curtin, *The Image of Africa: British Ideas and Actions, 1780–1850* (Madison, Wis., 1964), pp. 192–97, 355–62, 483–87, and Daniel R. Headrick, *The Tools of Empire: Technology and European Imperialism in the Nineteenth Century* (New York, 1981), pp. 58–79.

70. The best source in English is K. W. van Gorkom, "The Introduction of Cinchona into Java," in Honig and Verdoorn, pp. 182–90. But see also Pieter Honig, "Chapters in the History of Cinchona. I. A Short Introductory Review" in Honig and Verdoorn, pp. 181–82; Octave Collet, "Notes sur le Quinquina," *Congo* 1, no. 5 (May 1921): 687–88; Emile Perrot, *Quinquina et quinine* (Paris, 1926), pp. 46–49; and Treub, "Geschichte," pp. 48–49.

71. P. van Leersum, "Junghuhn and Cinchona Cultivation," in Honig and Verdoorn, pp. 190–96; van Gorkom, pp. 187–88.

72. See especially Clements Robert Markham, *Travels in Peru and India while Superintending the Collection of Cinchona Plants and Seeds in South America, and their Introduction into India* (London, 1862).

73. India Office Records, Eur. Mss. A882.

74. Ibid.

75. Brockway, pp. 104–17; Burkill, pp. 134–37; Donovan Williams, "Clements Robert Markham and the Introduction of the Cinchona Tree into British India," *Geographical Journal* 128 (1962): 431–42; Sir Frederick Price, *Ootacamund: A History. Compiled for the Government of Madras* (Madras, 1908), pp. 118–28; Randwaha, Chadha, and Singh, pp. 30–31.

76. Williams, pp. 438–39.

77. Letter to the Colonial Office, March 26, 1884, in India Office Records, V/6/312, p. 245.

78. Brockway, pp. 120–33.

79. Historians have also argued the motives of the cinchona transfer. While Donovan Williams fully accepts Markham's humanitarian claims, Brockway rejects them and sees the transfer as being purely imperialistic, that is, designed to enrich Britain at the expense of the Andean republics and reinforce its control over India.

80. The new techniques were publicized by the pharmacologist J. C. B. Moëns in his book *De Kina Cultuur in Azië* (Batavia, 1882), which remained the authoritative work until World War I.

81. Van Leersum, pp. 194–95; Williams, p. 439; Perrot, pp. 50–56; van Gorkom, pp. 187–203; Honig, "Chapters," p. 182; Norman Taylor, "Modern Developments," in Honig and Verdoorn, pp. 203–4; J. Pierpaerts, "Le quinquina," *Congo* 2, no. 5 (December 1922): 671–73; M. Kerbosch, "Some Notes on Cinchona Culture and the World Consumption of Quinine," *Bulletin of the Colonial Institute of Amsterdam* 3, no. 1 (December 1939): 36–51.

82. Collet, p. 695.

83. Perrot, pp. 127, 159.

84. Forrest, pp. 90–92; Bruce, 2:135; Kerbosch, p. 38; India Office Records, V/6/312, pp. 239 ff.

85. Two other uses for cinchona bark hardly affected the world quinine

market: totaquine was sold only in India, and the aromatic wild cinchona
barks of the Andes were used to make tonic water and *vin de quinquina*.

86. Perrot, pp. 120–31; Collet, pp. 697–99; Taylor, pp. 204–6.

87. Perrot, p. 159.

88. Henry, *Rapport,* pp. 40–41. For other criticisms of the Dutch monopoly, see Perrot and Collet. For a defense, see Pierpaerts and Kerbosch.

89. Kerbosch, pp. 45–48; Perrot, pp. 156–57, 169.

90. Chaumet, "Le Ministre des Colonies à MM. les Gouverneurs
généraux, Gouverneurs des colonies et Commissaires de la République au
Cameroun et au Togo," *Comptes rendus de l'Académie des sciences coloniales*
(June 15, 1925), p. 237; Dr. Paul Gouzien, "La question du quinquina et les
colonies françaises," *Comptes rendus de l'Académie des sciences coloniales*
(January 19, 1927), pp. 321–23.

91. ANSOM, Généralités 25 (234); Perrot, pp. 59–65, 80.

92. Perrot, *Quinquina,* pp. 71–79, 173–74, and "Commission du quinquina.
Rapport," *Comptes rendus de l'Académie des sciences coloniales* (May 18,
1927), pp. 507–11; Gouzien, p. 320.

93. Perrot, *Quinquina,* pp. 56, 160–73; Kerbosch, pp. 42–44.

94. Pierpaerts, pp. 667–706; Collet, pp. 685–708; Gouzien, p. 214; Perrot,
Quinquina, pp. 68–69; Furnivall, "Machinery," pp. 119–20.

95. Arthur Chapman Barnes, *The Sugar Cane* (London, 1964), pp. 5–7;
George Thomas Surface, *The Story of Sugar* (New York and London, 1910),
pp. 159–61.

96. Barnes, p. 9; Masefield, *History of Agriculture,* p. 49; Surface, pp.
156–62; Noel Deerr, *The History of Sugar,* 2 vols. (London, 1949–50), 2: 546–
53, 578–83; R. W. Beachey, *The British West Indies Sugar Industry in the Late
19th Century* (Oxford, 1957), pp. 62–64.

97. Barnes, pp. 10–11; Masefield, *History of Agriculture,* pp. 48–49;
Deerr, 2: 559–73; Beachey, pp. 68–72.

98. Beachey, pp. 63–68, 81–86; Morris, p. 77.

99. Letter from Trail & Co. to Nederlandsche Handelmaat-schappij,
quoted in G. R. Knight, "From Plantation to Padi-Field: The Origins of the
Nineteenth Century Transformation of Java's Sugar Industry," *Modern Asian
Studies* 14, no. 2 (1980), p. 182.

100. Knight, pp. 177–204.

101. Clifford Geertz, *Agricultural Involution: The Process of Ecological
Change in Indonesia* (Berkeley, 1963), pp. 82–85.

102. Thomas Lough, "The Brussels Sugar Convention," *Contemporary
Review* 83 (1903), p. 75; Barnes, p. 12.

103. Wilfred Hicks Daukes, *The "P. & T." Lands [Owned by the Anglo-
Dutch Plantations of Java, Ltd.]: An Agricultural Romance of Anglo-Dutch
Enterprise* (London, 1943); Daukes was chairman of this company; Knight,
pp. 183–85.

104. George Cyril Allen and Audrey G. Donnithorne, *Western Enterprise
in Indonesia and Malaya: A Study in Economic Development* (London, 1957),
pp. 26–31, 83–84; J. S. Furnivall, *Netherlands India: A Study of Plural Economy,* 2nd ed. (Cambridge, 1944), p. 200, and "Machinery," pp. 120–21; H. C.
Prinsen Geerligs, *The World's Cane Sugar Industry, Past and Present* (Altrin-

cham, 1912), pp. 119–20; H. D. Rubenkoenig, "Problems of the Java Sugar-Industry," *Asiatic Review* 24 (1928): 370.

105. The best summary of the process of sugar cane breeding is Robert Evenson, "International Diffusion of Agrarian Technology," *Journal of Economic History* 34 (March 1974): 58–64.

106. Lecomte, "Influence," pp. 22–24; Thiselton-Dyer, "Botanical Enterprise," p. 305.

107. Evenson, pp. 60–61; Masefield, *Colonial Agricultural Service*, pp. 82–83; van Hall, pp. 270–71.

108. Evenson, pp. 61–63; van Hall, p. 271; Biswas, p. 229; Y. N. Sukthankar, "The Present Position of the Sugar Industry in India," *Asiatic Review* 34 (1938): 176.

109. Masefield, *History of Agriculture*, p. 51, and *Colonial Agricultural Service*, pp. 82–83.

110. Evenson, p. 64; C. J. Robertson, "Cane-Sugar Production in the British Empire," *Economic Geography* 42 (April 1930): 135.

111. Prinsen Geerligs, pp. 120–21; van Hall, p. 271; Allen and Donnithorne, p. 84; Barnes, p. 73; Rubenkoenig, pp. 370–71.

112. G. H. van der Kolff, "An Economic Case Study: Sugar and Welfare in Java," in *Approaches to Community Development,* ed. Phillips B. Ruopp (The Hague, 1953), p. 195. See also Geertz, pp. 86–88, and Allen and Donnithorne, pp. 30, 84.

113. Julius Herman Boeke, *The Structure of Netherlands Indian Economy* (New York, 1942), p. 73.

114. Lim Chong-Yah, *Economic Development of Modern Malaya* (Kuala Lumpur, 1967), p. 74; J. H. Drabble, *Rubber in Malaya, 1876–1922: The Genesis of an Industry* (Kuala Lumpur, 1973), p. 74; James C. Jackson, *Planters and Speculators: Chinese and European Agricultural Enterprise in Malaya, 1786–1921* (Kuala Lumpur, 1968), p. 216.

115. Ruth Slade, *King Leopold's Congo* (London, 1962); Barbara Weinstein, *The Amazon Rubber Boom, 1850–1920* (Stanford, Calif., 1983).

116. On the hevea transfer, see: Drabble, pp. 1–5; Colin Barlow, *The Natural Rubber Industry: Its Development, Technology, and Economy in Malaysia* (Kuala Lumpur, 1978), pp. 18–20; J. G. Bouychou, "L'origine de l'hévéa d'Extrême-Orient. I. Introduction et distribution," *Revue générale du caoutchouc* 8 (August 1956): 711–18; and H. C. Chai, *The Development of British Malaya, 1896–1909* (Kuala Lumpur, 1964), p. 154.

117. Barlow, pp. 20–21; Drabble, pp. 5–6.

118. Drabble, pp. 6–9; Barlow, pp. 21–22; S. S. Pickles, "Production and Utilization of Rubber," in *A History of Technology,* ed. Charles Singer, E. J. Holmyard, A. R. Hall, and Trevor I. Williams, 5 vols. (Oxford, 1958), 5: 774; J. W. Purseglove, "History and Function of Botanic Gardens with Special Reference to Singapore," *Tropical Agriculture* 34, no. 3 (July 1957): 57; "Introduction of Para Rubber in the Malay States," *United Empire* 2 (November 1911): 798–99.

119. Bouychou, pp. 712–18; T. A. Tengwall, "History of Rubber Cultivation and Research in the Netherlands Indies," in Honig and Verdoorn, p. 344.

120. Henri Lecomte, "Le caoutchouc et la gutta-percha dans les colonies

françaises" in *Conférence de 1917. Nos richesses coloniales,* Muséum national d'histoire naturelle (Paris, 1918), pp. 170–78; Auguste Chevalier, E. Girard, A. Hallet, L. Jacque, and E. Rosé, *L'hévéa en Indochine* (Saigon, 1918), p. 19; Charles Robequain, *L'évolution économique de l'Indochine française* (Paris, 1939), pp. 223–24; H. Hamet, "Etude sur le caoutchouc au Soudan" in Société des ingénieurs civils de France, *Mémoires et comptes rendus des travaux* 53, no. 6 (March 1900): 282–306; J. Lan, *Notes sur l'hévéa brasiliensis en Cochinchine* (Saïgon, 1911), p. 3.

121. John H. Harris, "Native Races and Rubber Prices," *Contemporary Review* 104 (1913): 649–55; Drabble, p. 220.

122. Barlow, p. 25; Jackson, pp. 218, 243–44; Chai, p. 155; Lim, p. 104; Bouychou, pp. 712–13.

123. Jackson, pp. 220–24, 234–35, 263–67; Barlow, pp. 27–29.

124. Drabble, p. 91; Barlow, pp. 29–35; Jackson, pp. 225–45.

125. Drabble, pp. 42–47, 86–91, 119–21, 150–51; Barlow, pp. 74, 115–16, 148–49, 161–67; Jackson, pp. 216–17.

126. Barlow, pp. 74, 114–16; Bouychou, p. 716; Tengwall, pp. 347–48; David G. Fairchild, "Rambles in Sumatra" in Honig and Verdoorn, p. 87; Henry, *Rapport,* pp. 29–30.

127. Lim, pp. 76–82; Barlow, chap. 3; Tengwall, pp. 349–50.

128. Pierre M. A. Michaux, *Caoutchouc. L'hévéaculture en Indochine, son évolution* (Paris, 1937), p. 14; ANSOM, APC 36: Papiers Cibot, including: Henry de Lachevrotière, "Nos plantations. La plus grande plantation d'hévéas de l'Indochine" and "Rapport de Monsieur Van Pelt sur la plantation de Locninh visitée le 14 janvier 1931."

129. Chevalier et al., *Hévéa,* p. 3.

130. Michaux, p. 4. See also André Bourbon, *Le redressement économique de l'Indochine, 1934–1937* (Lyon, 1938), p. 121, and Robequain, p. 232.

131. Robequain, pp. 223–29; Michaux, pp. 3–7; Henry, *Economie,* p. 562.

8

Mining and Metallurgy

One of the charges leveled at colonialists is that they were too eager to exploit the natural resources of the colonies for their own purposes. Another is that they contributed too little to the industrialization of their colonies, or even hindered it. The mining and metallurgical industries provide evidence for both of these conflicting positions.

In the European empires of the nineteenth and twentieth centuries, there arose four major metals industries: South African gold, Malayan tin, Central African copper, and Indian iron and steel. The first of these helped turn South Africa into a settler colony and then into an independent nation. It was not, therefore, a colony in the same sense as the other lands we are studying here. The other three cases are much more comparable, both as industries and as colonial situations.

Copper, tin, and iron are ancient metals, known since prehistoric times. They are also modern metals, smelted by industrial means and for industrial purposes. Beyond the technological similarities are three very different histories. The differences are partly economic, but in the case of colonial industries, they are cultural and political as well.

One crucial difference is that the industrial nations of Europe had ample supplies of coal and iron ore and exported iron and steel products to the rest of the world. In contrast, European supplies of copper and tin were essentially depleted by the 1870s, just when the demand for these metals for electrical equipment and food canning was surging ahead. For a time, the demand for copper was met from North and South American sources, but after World War I the colonial powers were eager to develop their own supplies for political reasons. Once Cornish tin was exhausted, further tin supplies were found in only a few places: Malaya, Sumatra, Bolivia, and Nigeria. The colonial powers made every effort to develop the resources under their control.

The distribution of ores, then, accounts to a certain extent for the interest of European entrepreneurs and investors in the copper and tin deposits of the colonies, and their lack of interest in iron ores. Yet this is only the beginning of an explanation, for all three industries arose in our period. To understand this phenomenon, we must turn to the other protagonists: colonial administrators and non-European entrepreneurs. The three cases are very different.

In the Belgian Congo, the copper industry was a purely European enterprise, an exotic enclave in which Africans participated only as workers. There was simply too great a chasm between the complex and costly copper industry and the Africans' small-scale political and economic organizations. The industry was foisted upon Africa by Europeans.

Malayan tin was a simpler metal with a more complicated history. Tin ores are much easier to mine and smelt than copper. Furthermore, Malaya had attracted immigrants from China who were more enterprising and ingenious (though less organized) than Europeans. Hence, there ensued a tug-of-war between European companies and techniques and Chinese miners and their methods. Not until the 1920s did the Europeans win, briefly, with a combination of new technologies, larger investments, and political manipulation.

The case of iron and steel in India is the opposite of Congolese copper. European investors showed no interest in it. Colonial administrators were mildly interested for fiscal and military reasons but lacked the technical and managerial competence to succeed. So it was Indians who seized the opportunity to build a modern steel industry. In this case, technology was drawn from the West by Indian entrepreneurs with Indian capital. These examples therefore illustrate three very different kinds of technology transfer under colonial circumstances.

Malayan Tin and Chinese Technology in the Nineteenth Century

Tin has two main uses: one, known for thousands of years, is in the manufacture of alloys such as bronze and pewter; the other is the plating of sheet steel for cans ("tins" in British parlance) and oil drums. The latter use had to await the development of the steel, food-canning, and oil industries in the last third of the nineteenth century. Until 1871, Cornwall supplied most of the tin for Britain, then the world's foremost producer of tinplate. As demand grew, the tin mines of Cornwall were rapidly depleted. Malaya, long a supplier to the traditional tinsmiths and alloy-founders of China and India, replaced it as the world's first source of the metal for industrial uses.

From an economic point of view tin is a passive commodity, subject to a demand over which the producers have no control. In the short run the demand is price-inelastic, because the cost of tin is only a small fraction of the price of the items it is used in. At the same time it is income-elastic, because demand for tin fluctuates violently with the business cycles in the industrial countries. Hence, it is a risky business which tempts producers to form cartels in the hopes of keeping production in line with consumption. In the long run, however, the demand for tin is vulnerable to technological changes, both more efficient uses such as electroplating, which spreads it thinner, and substitution by aluminum and plastics. Though tin consumption trebled between 1871 and 1895, it only doubled from then until 1930, and only increased slightly from 1930 to 1960.[1]

The fluctuations in the world's consumption of tin parallels the fluctuations in consumption of many other tropical raw materials demanded by the industrial West. What is surprising is that the industrial West entered the tin-mining business after the demand had leveled off. In the boom period itself, before World War I, tin mining in Malaya was in the hands of Asians.

Cassiterite, the tin ore, occurs as lodes in the granite hills and in the alluvial sand and gravel washed down by the monsoon rains. For centuries, Malay farmers had panned for tin in the streams, using shallow wooden bowls called *dulangs*. They also shoveled tin-bearing soil into ditches where running water carried away the lighter sand and gravel, leaving the heavier ores on the bottom. Their methods were crude and their labor unorganized. Tin mining was essentially a part-time family occupation after the harvest was in. In the words of historian Wong Lin Ken, "They had neither the commercial shrewdness nor the aptitude for hard and sustained work so essential for the success of any business undertaking."[2]

The Chinese did, however. In the nineteenth century Malay chiefs in need of funds had encouraged the immigration of Chinese to the tin-rich regions of Perak and Selangor. These migrants displaced the Malays from tin mining because their technology and organization were particularly suited to the conditions they found. They were not miners from Yunnan, but mostly rice farmers from Kwangtung, adept at handling water and soil. These were just the skills they needed to get rid of the water that collected in open-cast mines and to wash the tin-bearing soil. To bring water to the mines, they installed bamboo pipes and dug channels from streams in the hillsides. And to pump water out of the open-cast mines they built *chia-chias,* chain-pumps powered by water wheels which could lift up to 16 tons of water per hour. These devices worked during half the year, when there was neither too little water to run the pumps, nor so much rain that the mines were flooded; together with a few simple tools—a hoe, two baskets hung from a bamboo pole, steps carved into a log—chia-chias allowed Chinese miners to dig 10 meters below the surface, much deeper than the Malays could.

Concentrating the ore was also done with wooden devices: the dulang and the *palong,* or sluice-box. Until the 1880s the ores were smelted at the mine because of high transport costs. Mine owners built small smelters of clay called Dreda furnaces, with piston pumps to provide the blast; they usually lasted one season. After midcentury, larger, more efficient brick Banka furnaces were built to last five years. In both cases, the main cost was hardwood charcoal. These devices, like those used in mining and washing the ores, were simple and required almost nothing from the outside world.

But they were labor-intensive. By all accounts, working conditions in the mines were appalling. The miners stood knee-deep in water at the bottom of pits, shoveling gravel under the tropical sun. Or they climbed up and down ladders all day, carrying loads of sand and gravel to the surface. As one Westerner commented: "A deep Chinese mine with its hundreds of coolies working far below the surface irresistibly suggests a very badly damaged ant hill."[3]

To prevent their workers from escaping, mine owners locked them up at night. More effective were opium and gambling, which the mine owners provided in order to addict their workers. The owners were also important members of the secret societies which arranged the immigration of Chinese to Malaya and controlled their lives. Chinese mining was based on the disciplined labor of the miners as much as on ingenious mechanical devices.[4]

The situation in the mining districts in the late nineteenth century was anything but static. The endemic warfare between the Malay chiefs soon involved the Chinese as well, as alliances of secret societies fought one another for control over the tin deposits. When the turbulence threatened to cut tin supplies or, worse, to attract French or German intervention, the British who controlled the Straits Settlements felt obliged to step in. Starting in 1874, they imposed residents, unofficial proconsuls, upon the various sultanates of western Malaya.

The first decades of British rule saw a tremendous upsurge in tin production, from 6,000 tons in 1871 to 50,000 in 1895.[5] By 1883, Malaya was the world's foremost tin producer. This was not due to Western enterprise or technology, despite several attempts by Western entrepreneurs to gain a foothold in the industry, but was almost entirely the result of Chinese initiatives. In this process, the rivalry between the two technological systems and the attitude of the government are especially interesting.

Almost no Westerners attempted to mine for tin in Malaya before 1874 because political conditions were too dangerous. That year the Malayan Peninsula (East India) Tin Mining Company was floated just weeks after the British takeover of Selangor. Despite official blessing, it failed a year later. This scared off others for a few years. In the early 1880s there appeared a number of other Western companies: in 1881 the French-owned Société des mines d'étain de Pérak; in 1882 the Hongkong and Shanghai Tin Mining Company and the Rawang Tin Mining Company, both owned by Western merchants in China; in 1883 the Australian-based Sandhurst Tin Mining Company and Melbourne Tin Mining Company; and in 1887 the Pahang Tin Mining Company. By the midnineties they had all failed, except for the Pahang Company.[6]

Part of their problem was difficulty controlling the Chinese miners. But the main reason seems to have been extravagant management. A British administrator, Sir Frank Swettenham, explained why:

> European mining is done by companies, and company's money is almost like government money. It is not of too much account because it seems to belong to no one in particular and is given by Providence for the support of deserving expert and often travelled individuals. Several of these are necessary to start a European mining venture and they are mostly engaged long before they are wanted. There is the manager and the sub-manager, the accountant, the engineer, the smelter. . . . Machinery is bought, houses are built, in fact the capital of the company is spent . . . and then—if things get so far—some Chinese are employed on wages or contract,

the former for choice, to remove the overburden. After possibly a series of great hardships to the staff and disasters to the company, it is found that the tin raised is infinitesimal in value when compared with the rate of expenditure, and the longer the work goes on the greater will be the losses. This is usually discovered when the paid up capital is all but exhausted. The company is wound up and the State gets a bad name with investors, and the only people who really enjoy themselves are the neighboring Chinese miners who buy the mine and the plant for an old song and make several large fortunes out of working on their own ridiculous and primitive methods.[7]

Chinese mine owners responded to the European competition by adopting new equipment. Some was Western, such as the steam pump which Sir Hugh Low, the British resident in Perak, installed in a Chinese mine for demonstration purposes in 1877. Though a steam pump cost many times more than a chia-chia, it could pump water from a greater depth, and hence allowed mine owners to re-open mines abandoned because of flooding. By 1892 three hundred steam pumps were in use in the larger mines, while smaller mines made do with the older device.

The Chinese also innovated on their own. One invention was the *lanchut keechil,* a coffin-shaped wash box some 3 meters long. Unlike the old wash box it replaced, the lanchut keechil did not need running water but only a small pool, and it could be operated by three men. Its invention in 1891 led to a flight of mine workers away from the established mines to marginal areas with little water. Similarly, in 1892 Chinese miners introduced a system of underground mining called *ta lung,* by which parallel shafts were dug into the hillside and then the earth excavated between them until the hillside collapsed. It was a dangerous but cheap way to get ores.[8]

In Malaya, the Chinese were newcomers like the Europeans, energetic, ingenious, and greedy. Though poorer and without an industrialized homeland to supply and support them, they succeeded in countering their Western rivals with innovations that were more appropriate to the geological and labor conditions of Malaya. As late as 1914 they produced three-quarters of the country's tin, while Western firms accounted for only one-quarter.

The government of Malaya helped the Chinese mine owners with a number of measures designed to keep labor cheap and docile: encouragement to immigration, the sale of gambling and opium permits, and the discharge ticket system, which made it illegal to hire a miner before his previous contract was expired. The government also sold low-cost mining concessions and built roads and railways into the mining districts. Until 1896, as Wong points out, "probably because Western enterprise had so dismally failed to work the tin resources, the British administrators did so much to induce the entry of Chinese labour and capital into the mines that they were actually. accused of being pro-Chinese by disappointed and envious Western miners."[9]

The Western Takeover of Malayan Tin

In the Sino-Western rivalry over Malayan tin, the mistakes made by Westerners were only temporary, while their advantages—access to European capital and a fast-changing technology—grew stronger with time. The first area in which Western entrepreneurs gained a foothold was smelting.

Until the 1890s the Chinese had dominated tin smelting and refining as they had mining because tin ores were smelted at the mine. Toward 1880 the older furnace types were displaced by more efficient designs. The *relau semut,* a natural-draft furnace, needed no pump and little labor but required hardwood charcoal; hence it was used in remote areas where labor was scarce and timber abundant. Elsewhere, Chinese mine owners introduced the *relau tongka,* a clay furnace standing on a three-legged iron pot imported from China, which used softwood charcoal.

Yet even in densely forested Malaya, charcoal-based metallurgy was self-defeating, because it consumed the trees on which it depended. By the 1890s charcoal metallurgy was being displaced throughout the world by coal and coke, the fuel of the West. This happened in Malaya in 1887 when the newly founded Straits Trading Company built a large coal-fired reverberatory furnace at Singapore. The new railway network made it more economical for mine owners to sell their ore to this company than to process it themselves. Furthermore, it produced a more refined tin—99.85 percent pure— which captured the European market. So efficient was the new smelter that the company eventually received shipments of ore from as far away as Australia and the Congo.

Yet the Chinese did not give up smelting without a fight. Many installed steam-powered fans on their tongka furnaces to cut labor costs. In 1897 the merchant Lee Chin Ho built a second reverberatory furnace at Penang called the Eastern Smelting Company. By

1910, it was smelting 29.2 percent of all the tin shipped from the Straits, and the following year it was bought by a British firm.[10]

Much the same happened, with some delay, in the mining industry. The Chinese share of Malayan tin production gradually declined from 78 percent in 1910 to 49 percent in 1929 and to 34 percent in 1935. The causes of their displacement by Western firms are a complex tangle of technological changes, business practices, cultural values, and government policies.

Chinese mining methods, for all their ingenuity, could only operate profitably with cheap docile labor and rich ore deposits close to the surface. In contrast, the power of Western mining techniques was their ability to extract metal profitably from ever lower grades of ore in ever less accessible deposits. At the heart of this rise in productivity was the introduction of bigger, more complex, and expensive machines, with teams of experts to run them and business organizations to finance them.

The transition from a Chinese to a Western system of mining was due, in the first place, to a geological factor: the exhaustion of the surface deposits which the Chinese were so efficient at mining, and the existence of ores which only Western equipment could reach. In 1892 the engineer F. D. Osborne, working for the Gopeng Consolidated Tin Mining Company, introduced hydraulic mining, a system first used in the gold fields of California. At a cost of 50,000 Malayan dollars (about £6,000), he had water piped a distance of 10 kilometers downhill. A monitor, or huge fire hose, ejected a stream of water at the hillside mine face, washing away some 300 cubic meters of ore-bearing soil a day. Not only was this method faster and cheaper than the Chinese ta lung system, it was also easier to use, permitting the companies that introduced it to hire Malay or Indian miners. By 1900 nine such monitors were in operation, and after that it was adopted by Chinese mine owners as well.[11]

In 1906 the engineering firm Osborne and Chappell introduced the gravel pump, which sucked not only water but also the soil from the bottom of flooded mines. As it could work to a depth of 20 meters, it allowed the reopening of flooded and abandoned open-cast mines. Being fairly inexpensive, it too quickly spread to the more prosperous Chinese mines.

Yet there remained tin-bearing soils in low-lying areas like the Kinta valley which were covered with swamps and inaccessible to all the techniques so far described. To mine them required bucket dredges, devices first used in California, New Zealand, and Australia before they appeared in Malaya. The dredge imported by the Malayan Tin Dredging Company in 1912 was a barge 46 meters long by 11 wide, with a chain of buckets that could scoop up the bottom of swamps down to a depth of 15 meters, at a rate of over 2,000 cubic meters a day, and wash out the ores on the spot. Not only did bucket dredges open up new deposits, they also made it profitable to mine lower-yielding soils than ever before. And unlike the monitors, which washed away whole hillsides and ruined the land downstream, bucket dredges were fairly gentle on the environment. However, they had to be imported from Britain at a cost of millions of Malayan dollars and were only worth using on the largest concessions. For that reason they caught on slowly. Only after World War I did firms invest in dredges and their support systems. By 1925 the forty-four dredges in operation produced 20 percent of Malaya's tin. By the late 1930s even bigger dredges, which could dig down 40 or 50 meters below the water level, accounted for half of Malayan tin production. It is these machines which eclipsed the Chinese methods.[12]

The displacement of Chinese by Western techniques accounts only partly for the displacement of Chinese by Western firms. Other causes include the business culture in the two communities and the policies of the government. Business organization was crucial because the new machines, especially the dredges, were so costly that only joint-stock companies could afford them. Here Chinese business methods were a drawback, as Wong explains:

> The Chinese were reluctant to reorganize their mining companies into joint-stock companies, without which it would be difficult to raise the large capital required to start and operate mines with the new mining techniques. In Perak the Mines Department tried in 1905 to show to the Chinese how much they stood to lose by refusing to follow the times, but it failed to break down the Chinese ignorance of the practice of joint-stock companies, as well as Chinese conservatism, individualism, and clannishness, which had all combined to make them reluctant to change their organization. In 1914 there was not a single Chinese mining company operating on the limited liability principle.[13]

The role of the British administration in all this is subject to different interpretations. Wong sees the government as favoring Western methods rather than Western people:

> Though developments in the period after the 1890's undoubtedly favoured the entry of Western capital into the tin industry, the evi-

dence does not point to the conclusion that the changes in policy were initiated with the object of discriminating or weakening Chinese mining enterprise. The outcome of these changes was not the result of discrimination but was rather the consequence of the failure of the Chinese miners to adapt themselves to the new situation. Indeed, the administration took the trouble of demonstrating to the Chinese miners how much they stood to gain by modernizing the organization and working of their property.[14]

Li Dun-jen, on the contrary, blames the British squarely:

How the British capitalists captured the tin enterprises from their Chinese subjects is interesting not only because it shows that in free competition the stronger capitalist often swallows the weaker one; it also indicates that the capitalists of the colonial power, supported by their own government, could easily squeeze out of business their colonial subjects, whose voice could not be heard or was ignored in the determination of official policy.[15]

Numerous policies affected the tin industry. In the 1890s legislation curtailed two of the most exploitative Chinese labor practices, the secret societies and the discharge ticket system, with the result that labor costs began to rise. Mining codes and inspectors made it more difficult for mines to dump their tailings on agricultural land downstream or to use dangerous methods like ta lung. The opium and gambling farms were abolished in 1901 and 1912, respectively. After 1906, mining properties which were not being worked could be "resumed" and sold to those possessing "sufficient capital to work with labor-saving devices." Water supplies came under government control. Concessions and mining permits were issued for larger areas to firms with more capital than previously. These policies, enacted for sound humanitarian or environmental reasons, did not hurt tin production as a whole but only the smaller, predominantly Chinese-owned mines. It is no coincidence that they appeared at the same time as Western labor-saving, capital-intensive methods.[16]

The downfall of the small Chinese mining entrepreneur came during the Depression. As long as prices were high (£313 per ton in 1926), producers could sell at a profit. When prices collapsed to £132 per ton in 1931, the Malayan government joined the Netherlands East Indies, Bolivia, and Nigeria in a cartel to restrict production. With consumption down by a third, only big firms with dredges, whose costs per ton were little more than half those of small mines, survived. The cartel accelerated a process already well underway, under the pressures of a changing technology.[17]

Opening the African Copperbelt

In some ways, the story of copper in Central Africa resembles that of Malayan tin. Like tin, it was one of the earliest metals used by humans and a basic material of industrialized societies. Impelled by a growing demand in the industrial countries for boilers and electrical wiring, Western enterprises turned Central Africa into a major supplier of copper for the world market. In other ways, however, the development of Malayan tin and African copper stand in sharp contrast. The production of copper in Africa jumped directly from a small-scale, traditional African technology to one of the most highly mechanized and large-scale industries on earth. There were no intermediate stages and no competition from any third technological system as in Malaya. Three reasons account for this. Geologically, most copper-ore deposits lie too deep and are too complex to be exploited by any but the most mechanized methods. Culturally, there were only two groups in Central Africa: the indigenous Africans and the Europeans; no other immigrants created a competing industry nor were any invited in by the colonial authorities. And finally there is the matter of timing. The effective occupation of Katanga by the Belgians and Northern Rhodesia by the British dates from the turn of the century. By the time the first copper was poured from a Western furnace in the Copperbelt in 1911, mining and smelting technology had matured, and large corporations, well supplied with funds and engineering talent, had replaced the lone prospectors and starry-eyed speculators of a previous generation.

The Copperbelt covers about 36,000 square kilometers, two-thirds of it in Katanga (now Shaba), the rest in Northern Rhodesia (now Zambia). For fifteen centuries or more, Africans mined and smelted copper there. The products they made of it—wire, weapons, utensils, and ornaments—were traded throughout southern Africa and were known to the Arabs and Europeans.

Africans dug open-cast mines 5 to 10 meters deep, occasionally as deep as 20 meters. The ore was sorted by hand and washed in streams. Smelting, a craft surrounded by mystery and ceremony, was done in small clay furnaces which produced up to 12 kilograms of copper per firing. The fuel was charcoal made from hardwood, which was rare in the savannas and had to be carried from afar by human porters. By the late nineteenth century this traditional industry had almost vanished, killed off by the slave trade and the depletion of ore deposits and hardwood trees.[18]

The ore which Africans smelted was malachite, a bright green carbonate with a copper content as high as 57 percent, which could be smelted at low temperatures without fluxes. It was produced near the surface by the weathering of other, more complex ores. Most of the deposits in Katanga were of malachite and other oxides such as azurite and cuprite with average yields of 15 percent. Far below the surface, especially in Northern Rhodesia, were copper sulphides which were harder to reach, had a lower yield, and could only be processed by industrial methods. The relationship between ores and metallurgy explains why deposits that were depleted from the Africans' standpoint looked promising to Europeans, and why Katanga was developed before Northern Rhodesia, even though it was further from the sea.

Though explorers had noted the presence of malachite outcroppings and African mine sites, the extent of Katanga's copper deposits were not readily apparent. In 1892–93 the explorer-geologist Jules Cornet noted the existence of copper ores but thought they were too remote and low-yielding to justify the expense of developing them. In 1898 Capt. Charles Lemaire, leader of another expedition, reported: "The mineral treasures which have been for a long time so liberally ascribed to Ka-Tanga did not reveal themselves to us."

Other prospectors thought differently. In 1899 Robert Williams, an associate of Cecil Rhodes, founded Tanganyika Concessions, Ltd. and obtained a concession from the British South African (Chartered) Company to prospect in Northern Rhodesia. He sent out an expedition under George Grey (brother of Foreign Secretary Sir Edward Grey) which discovered the Kansanchi deposit south of the Congo border. While there, the prospectors did a little clandestine investigation in Katanga itself. On the basis of their reports, Williams approached King Leopold of Belgium and obtained a concession from the Comité spécial du Katanga, an affiliate of the Congo Free State government, to prospect in a 150,000-square-kilometer area of Katanga. In 1901 Grey led another expedition with fifteen Europeans, fifty Africans, and two years' worth of supplies. They discovered the enormous deposits of Kolwezi, Kambove, and what was to be the Star of Congo mine. These discoveries shaped the future of Central Africa for years to come.[19]

To exploit these deposits, Tanganyika Concessions, the Comité spécial du Katanga, and the Société générale (a Belgian bank) founded a new company, the Union minière du Haut-Katanga. This arrangement was the forerunner of a long series of partnerships be-

tween the "portfolio state" (as critics called the Congo) and the Société générale, which alone accounted for 60 percent of private investment in the colony.

In 1906 the Union minière was granted a concession to mine
copper in a 15,000-square-kilometer area. By the end of that year its
prospectors had found over a hundred ore deposits, with yields running as high as 33 percent and an average of 12.5 percent—a rich
find indeed. The problem was getting equipment in and copper out.
At the time, the nearest railhead was at Broken Hill in Northern
Rhodesia. When Prince Albert, heir to the Belgian throne, visited the
Congo in 1908, he took a steamer to Cape Town and a train to
Broken Hill, and then traveled the last 500 kilometers on foot and by
bicycle. For heavier equipment, other means were used: locomobiles,
huge steam tractors that slowly dragged four or five freight cars over
dirt paths and made one round trip a year during the dry season.

These were temporary expedients. Under Robert Williams's direction, the railroad was extended from Broken Hill to the Congo
border in 1909, and to Elisabethville, near the Star of Congo mine,
in 1910. Only then could industrial mining begin in earnest.[20]

Katanga Copper, 1911–1940

Because of its different ores, the mining and metallurgy of copper are
much more complex than those of tin, and they changed radically in
the period 1911–40. Two kinds of metallurgy were transferred to Katanga. The first was the nineteenth-century technology of smelting the
ore in a furnace, upon which the industrial complex of Lumumbashi
near Elisabethville was based. The other system, in which the ores
were processed by chemical and industrial means, was introduced in
the 1920s to Katanga's second industrial complex at Panda-Jadotville
(now Likasi).

As soon as the railway reached Elisabethville, a water-jacketed
smelter was erected at Lumumbashi. In it, high-grade ores were reduced with coke imported from Europe. In June 1911, in front of
Robert Williams who had come from London to witness the event,
the first copper flowed from the furnace. In its first six months the
smelter produced almost 1,000 tons of copper, proof that this was
from the start a large-scale industry.

At first the ores were collected and washed by hand, a major
bottleneck. In 1913 the Union minière planned to expand production

and began to import heavy steam (and later electric) shovels capable of removing up to 1,000 tons a day, as much as 300 men with shovels. Most mines were enormous open pits in which the giant shovels filled whole trains with ore. Only at Kipushi, near the Rhodesian border, was there a mine shaft.

In the years 1913–18, as ore production increased, more water-jacketed furnaces were added to increase smelting capacity. Coal for the coke ovens was imported from Wankie in Southern Rhodesia until 1922 when a coal mine was opened in the Congo.

In 1914 the American metallurgist A. E. Wheeler, who had worked for the Anaconda and Great Falls Copper companies, surveyed an area near Panda, 150 kilometers from Elisabethville. He reported that the deposit, though extensive, contained ores that were too low-grade for the smelting process then in use. Four years later, however, Katanga's high-grade ores were already running out. The Union minière, eager to increase its share of the world copper market at a time when prices were high, decided to introduce new methods of processing medium and low-grade ores.

One such method was the gravity concentrator, which crushed the ores and separated them in shaking machines. In 1921 the company built a gravity concentration plant at Panda near newly opened mines. Another was the reverberatory furnace, which could smelt finer particles of ore than the water-jacketed furnaces and burned powdered coal instead of coke. The resulting copper matte was then passed through a Bessemer-type converter, which refined it to blister copper up to 99.4 percent pure. This degree of purity was still insufficient for electrical wiring, which must be 99.9 percent copper. Until the late twenties, Katangan copper was shipped to the United States for further refining. In 1919 the Union minière spawned a Belgian affiliate, the Société métallurgique de Hoboken, to refine Katangan copper, tin, cobalt, and uranium. By the process of thermal smelting, the Congo produced over 90,000 tons of copper in 1925, putting it in third place after the United States and Chile.[21]

Another method of concentrating the ore, flotation, extracted up to 90 percent of the ores from low-grade deposits. In this process, the ores are first ground to a fine powder, then mixed with oil and water and agitated. The ore particles stick to the oil, while other substances do not; hence the oil lifts the ore particles to the surface of the mixture, where the ore-rich froth can be skimmed off and smelted in a reverberatory furnace. Originally developed at the turn of the century to concentrate the sulphide ores of Chile and Australia, it took several years of experimentation before this method was adapted to the oxide ores of Katanga.

Metallurgists knew yet another method of obtaining copper: leaching and electrolysis. Electrolysis had long been used in Europe and America to refine impure metals. Heavy anodes of blister copper and thin cathodes ("starting sheets") of refined copper were placed in a bath of dilute sulphuric acid and copper sulphate. When a strong electrical current was applied between them, the anodes shrank as copper migrated to the cathodes and impurities fell to the bottom. The result was 99.98 percent pure copper.

With oxide ores it was possible to avoid entirely smelting by leaching—that is, dissolving concentrated ore in sulphuric acid—which then formed the electrolytic bath and deposited almost pure copper on the cathodes. The combination of leaching and electrolysis had first been used industrially in the United States and Chile during the war, and it proved to be a commercially viable way to process low-grade sulphide ores, provided there was cheap electric power. This method appealed to the Union minière because it would reduce the dependence on imported coal and the need to send copper to America for refining. However, an industrial leaching and electrolysis installation could only work in conjunction with plants to produce sulphuric acid and other chemicals. In other words, an entire integrated industrial complex would have to be built in the middle of Africa.[22]

In 1921 the Union minière built a pilot leaching plant at Panda which produced 4 tons a day. Two years later, once the technical and design problems were overcome, the company decided to create an industrial complex at Panda. It included gravity concentrators and a flotation plant to concentrate the core; four reverberatory furnaces able to produce 60,000 tons of copper a year; and a leaching and electrolysis plant at Shituru with a capacity of 30,000 tons a year. In addition, the Union minière created a number of affiliates to supply its needs: the Société générale des forces hydro-électriques du Katanga (Sogéfor), which built a hydroelectric dam on the Lifira River; the Société générale de chimie (Sogéchim) to make fatty acids, sulphuric acid, and other chemicals; the Charbonnages de la Luéna for coal; as well as a construction company, a flour mill, and other enterprises.[23]

All of this took time, and not until 1929–30 was Panda ready to produce at capacity. Between 1926 and 1930 the Union minière's production of copper rose from 80,639 to 138,949 tons a year. By then about 17 percent of its production was electrolytic. Then came the Depression, and the Union minière, a partner in a cartel called Copper Exporters, Inc., reduced its production. Most of the reduction was in thermally produced copper, however, while the more valuable electrolytic copper's portion of the company's production increased to 45 percent by 1945.[24]

Throughout this period and for years thereafter, two problems hindered the Katangan industry more than any other copper mining venture: transportation and labor. These explain why Katangan copper was no more than competitive with the United States, Chile, and Canada, despite much richer ores.

The first railway which reached Elisabethville in 1910 connected Katanga to Salisbury in Southern Rhodesia and Beira in Mozambique, a journey of 2,600 kilometers. As early as 1902 Robert Williams had sought a shorter route. The Portuguese government granted him a concession to build a railway across Angola to Benguela and Lobito Bay on the Atlantic, a distance of 2,100 kilometers. However, endless negotiations and World War I delayed construction, and the Benguela Railway did not link up to the Katangan rail network until 1931. By then the Belgians had built a third line within the Congo, but it required reloading onto river steamers at Port-Francqui and back onto the railroad at Leopoldville. These three competing railroads did little other than transport copper one way and supplies the other, and their costs were therefore high.[25]

The labor problem was, if anything, more severe. It involved four elements which could, within limits, be substituted for one another: Belgians, other whites, Africans, and machines. Their cost was only one consideration among many. Others included a racial policy which reserved the best jobs for whites, political discrimination in favor of Belgians, and a bias toward labor-saving equipment. There was, at the time, no bias toward Africans.

In the first years Belgium could not furnish enough mining and metallurgical engineers, and the technical director of the Union minière, Robert Williams, naturally sought talent where it was most abundant. The first director of the company, P. K. Horner, was an American, as were the technical managers and even the steam-shovel operators; in 1920 Americans constituted only 4.8 percent of the white workers but 42 percent of the highest-paid staff. Other mine workers and technicians came from South Africa, the Rhodesias, or Britain. The Belgians feared that the large number of Anglo-Saxons in Katanga—where, for a time, the linqua franca was English—would lead to another Jameson Raid. Hence they made every effort to Belgianize the area. By 1914, 53 percent of the Europeans were Belgians, but because of the war, the proportion dropped to 22.5 percent in 1917. After some labor troubles with white South African workers in 1918–20, the company began dismissing non-Belgians and recruiting Belgian workers and technical and administrative personnel. It did so by offering free transportation, housing, medical care, and a low-cost suburban lifestyle. In response to these efforts, the number of white employees rose from 900 in 1920 to 7,500 in 1937. Yet the company deliberately avoided attracting settlers as had happened in Rhodesia and South Africa.[26]

Toward Africans, the attitude of the Union minière and its affiliates was decidedly mixed. To begin with, Katanga was very thinly populated, with only two inhabitants per square kilometer. Hence almost all labor had to be imported, fed, and housed. Before and during World War I, when much of the mining was done by pick and shovel, the company recruited workers from Rhodesia, Ruanda-Urundi, and the lower Congo on one-year contracts. It paid them just enough to subsist and pay their taxes, but not enough to feed their families who remained home. Despite the threats of the tax collector, this system did not furnish enough workers, and for a time there was talk of bringing in 5,000 Chinese coolies. Instead, the company decided to mechanize its operations. As it explained, "In order to economize this labor force, the company is committed to develop more and more the use of every mechanical means to replace hand labor. . . . the new mines and plants are equipped according to the latest in technical progress, in order to use the minimum number of natives."[27]

Mechanization reduced the need for unskilled workers only to increase the demand for skilled ones, who were hard to find anywhere in the world. The company rejected suggestions that it train Africans: "The Congolese, too primitive, was not yet prepared to master the least skilled work. It would require a slow, patient and progressive training of 25 years before the best of them could be entrusted with machines formerly driven by Europeans."[28]

In 1928, however, the company inaugurated a new policy of "stabilization." Instead of recruiting unskilled young men on one-year contracts, it offered three-year contracts to men with families. Africans were trained as locomotive drivers, machinists, and laboratory technicians. To induce skilled workers to remain in Katanga, they were given bonuses if they married and settled in the mining towns. Housing, schools, and medical care were provided under the

supervision of Belgian religious orders. As a result of this policy, the
ratio of annual recruits to total African workers dropped from 96
percent in 1921–25 to 7 percent in 1931–35.[29] The treatment of African workers in the big enterprises of Katanga was pervaded with a
smug paternalism, as described by a company historian:

> If the mining industry extracts from the African soil a part of the
> wealth it contains—much of which it discovered anyway—most of the
> profits it makes are directly or indirectly returned to the native pop-
> ulations in the most precious and durable forms: order, peace,
> health, education, and the possibility of progressing toward a better
> existence.[30]

Iron and Steel in India: The Demand Side

The story of iron and steel in the colonial world is more complex
than that of copper and tin because the ferrous metals were not in
demand by the West—on the contrary, the European powers supplied their colonies with ferrous metals—and therefore both the demand and the supply had to come from within the colonial world.
This only happened in India.

In nineteenth-century India, ferrous metals had three uses: to
supply a widespread but low-level demand among Indians for tools
and hardware; to fulfill the army's requirements for weapons; and,
after 1853, to meet a huge demand for iron (and later steel) rails
and railway equipment. India supplied enough iron to meet the first
two needs but did not develop the industry to satisfy the railway demand until World War I.

The connection between the needs of one industry and the
growth of another is known as a backward linkage.[31] The success of
a backward linkage in stimulating a supplier-industry in the same
country as the customer-industry will depend on several factors: the
existence of native entrepreneurs, their access to capital and technology, their costs compared with those of foreign competitors, and the
policies of the government. When a sufficient demand exists but does
not give rise to a domestic industry, there is a leakage of the back-
ward-linkage effect to foreign suppliers, and a loss of what could
have been a stimulus to economic development. All countries beginning to industrialize have been conscious of this effect and have
hastened to protect their infant industries with tariffs, subsidies, or
state enterprises. That India did not was just as much a political

choice. Before we turn to the history of the Indian iron and steel industry, let us therefore consider the major consumer of ferrous metals in India, the railways.

During the railway booms, rails were one of the major products of the iron industry. In 1848, 27 percent of the puddled iron production of England and Wales went into rails. The new steel industries which arose in the sixties and seventies were even more dependent on their railroad customers. Until the 1890s over half the steel produced in America went into rails; in the year 1881 the rail mills used 94 percent of America's steel. If to this we add the railways' other uses of iron and steel for locomotives, rolling stock, bridges, buildings, and the like, the total is even higher; Duncan Burn estimated that in the 1860s the railways consumed two-thirds of Britain's iron production.[32]

How much iron and steel went into the railways of India? This is a difficult question, to which we can only give an approximate answer. Let us divide the Indian railways into two sorts: the standard-gauge lines which used rails weighing 42 to 50 kilograms per meter, that is, 84 to 100 tons per kilometer of single track; and the narrow-gauge lines, which used lighter rails weighing 20 to 30 kilograms per meter or 40 to 60 tons per kilometer.[33]

The number of kilometers added each year to the Indian rail network was, on average, as follows:[34]

1853–57	107	1898–1902	1585
1858–62	612	1903–7	1165
1863–67	573	1908–12	1170
1868–72	542	1913–17	917
1873–77	809	1918–22	414
1878–82	828	1923–27	674
1883–87	1436	1928–32	1047
1888–92	1293	1933–37	44
1893–97	1150		

In the boom years of railway construction between 1883 and 1912, the railways added, on the average, 1300 kilometers of track per year. Since about half the added track was standard-gauge, the railways required between 54,600 and 64,350 tons of new standard-gauge rails, plus 26,000 to 39,000 tons of new narrow-gauge rails: a total of 80,000 to 103,350 tons of new rails each year, in other words, the output of a fair-sized steel mill. Long before it had one,

India was consuming enough steel to keep at least one steel mill in business.

By 1913 India had over 57,000 kilometers of track, excluding sidings. This represents approximately 4 million tons of metal for the original rails alone. To this we must add the sidings, which Morris and Dudley estimate constituted 27.5 percent of all track in 1946–47;[35] replacement rails on existing tracks; iron crossties, much used in India because of the tendency of wooden ones to rot; bridges and other superstructures, which were made of steel rather than masonry as in Europe or wood as in America; and finally, locomotives and rolling stock: all in all, 2 or 3 million more tons of iron and steel.

Given the size and growth of the Indian rail network, the world's fourth or fifth largest between 1880 and 1940, how much of its iron and steel was Indian-made? Table 8.1 gives the production of the Indian industry, in thousands of metric tons.[36]

Though India had roughly 6 percent of the world's railway mileage, it barely reached, in its best years, 2 percent of the world's pig-iron production, and less than 1 percent of the world's steel production. In ferrous metals, it was on a par with Italy and Poland, well below Luxembourg.[37]

Furthermore, India's iron and steel industry only began to meet a significant share of domestic demand in the 1930s, many decades after the railway boom was over. The railways' great demand for ferrous metals—the classic backward linkage—had leaked abroad.[38]

Table 8.1 Indian Iron and Steel Production, 1900–1940

Year	Pig Iron	Year	Pig Iron	Finished Steel	Year	Pig Iron	Finished Steel	Heavy Rails
1900	36	1914	239	68	1928	1069	280	
1901	36	1915	246	77	1929	1414	419	
1902	36	1916	249	94	1930	1194	441	
1903	36	1917	252	116	1931	1073	457	
1904	42	1918	251	132	1932	928	434	
1905	46	1919	322	136	1933	894	449	39
1906	48	1920	316	115	1934	1127	560	36
1907	40	1921	374	127	1935	1364	637	79
1908	39	1922	325	114	1936	1566	688	65
1909	40	1923	498	153	1937	1577	623	85
1910	37	1924	684	252	1938	1670	679	76
1911	50	1925	894	325	1939	1601	738	92
1912	60	1926	935	366	1940	1867	817	103
1913	60	1927	1158	436				

Since India produced little iron and no steel before 1914, 96 percent of railway supplies had to be imported: 70 percent from Britain, much of the rest from Belgium. Britain's exports of iron and steel to India rose from 82,300 to 475,500 tons a year between 1873 and 1889, paralleling the growth of railroads. By then India had become Britain's best customer for iron and steel products, and these products represented almost one-tenth of India's imports in 1913.[39]

To assert that heavy industries could have arisen somewhere, it is not enough to show that there existed a demand for their products; one must also demonstrate that their costs were low enough to withstand foreign competition, within the bounds of government support. What the historical record shows is that India became the world's lowest-cost producer of pig iron once the ferrous-metals industry got started. It was also able to produce steel, but only with the same kind of government help that other countries' heavy industries received. Let us now look at this historical record.

Indian Iron before 1914

India was famous for its iron and steel long before the coming of the Europeans. In the Middle Ages, the swords of Damascus were probably made of Indian *wootz* steel. Yet traditional Indian iron-making techniques were among the most primitive in the world. In much of India iron was made by nomadic people called Agarias who gathered ore by hand from open pits and made charcoal from trees felled nearby. They smelted the ore in small furnaces a meter or two in height, quickly made of mud and cow dung. They forced a draft into the furnace by rocking back and forth on a pair of goatskins. Since furnace temperatures were low, cast iron was unknown in India. As the ironworkers could not transport the ore or charcoal more than a few miles, they soon exhausted the local fuel and ores and had to move their works every few years.

Such primitive smelting methods kept labor productivity very low. Twenty men operating a furnace could make 50 to 100 kilograms of raw iron per day. In the early 1850s the average iron furnace in the Birbhum district of Bengal produced 24 tons of wrought iron per year; elsewhere, the average production per furnace may have been around 5 or 6 tons.[40]

This ancient iron industry sufficed for the needs of eighteenth-century Indian society, but it could not meet the British demand for weapons and other iron goods. In the course of the nineteenth century, the British tried several times to produce iron in India by European methods. Yet the methods they imported were not industrial, but the obsolescent techniques of preindustrial Europe.

The first operating European-style ironworks in India were set up in the Madras Presidency in the 1830s. The founder of the grandly titled Indian Iron, Steel and Chrome Company was a retired East India Company official named Joshua Marshall Heath. His firm's output was of high quality; in fact, some was shipped to Britain and used in the Menai and Britannia iron bridges. Its methods, however, were but minor improvements over indigenous ironworking. Nearby forests were felled for charcoal, and wood had to be imported from Ceylon at great cost. Oxen powered the bellows and other equipment and pulled the carts. Unable to compete with cheap British iron, the firm barely limped along on government grants and loans it could not pay off. Briefly revived by a group of Madras businessmen in 1853, it finally ceased production in the early 1860s and was liquidated in 1874. Such was the fate of an enterprise using seventeenth-century technology in competition with the large-scale coke-fueled ironworks of mid-nineteenth-century Britain.[41]

British interest in iron making was not quenched by the failure in Madras. The native industry of Birbhum aroused considerable attention. Despite a negative report from the Geological Survey of India, a Calcutta firm, Messrs. Mackay and Company, opened the Birbhum Iron Works in 1855 and leased the nearby forests. Its operation produced enough high-grade pig iron to drive the native ironworkers out of business. But the firm had too little capital to purchase a puddling and rolling mill, and thus it could not turn out rails, boiler plates, and other finished iron goods. Depletion of the forests soon drove up the cost of charcoal. Like Heath's operation, it succumbed to British competition.[42]

Between 1855 and 1879 the Indian government, faced with the high cost of building unprofitable strategic railways, tried to relieve the pressure on its budget and on the balance of trade by developing a domestic iron industry. In 1861 it sent Colonel Keatings of the Indian army to Sweden to study charcoal iron making—proof that the techniques had been forgotten in Great Britain. Keating brought back with him a Swedish ironmaster named Mitander. With a subsidy of 50,000 rupees (£5,000), Mitander set up a blast furnace, a charcoal oven, a rolling mill, and calcining kilns for limestone. The next year the government cut off the subsidy, the works closed, and

Mitander went home. The same fate befell a number of other attempts, at Kumaon (United Provinces), Barwai (Indore), and Nahm (Punjab), to set up small ironworks under government auspices. Finance Minister Trevelyan explained the reason for the shift in policy: "It is a misdirection of the resources of India to enter into competition with England in this branch of industry. . . . By setting up Government Iron Works we are competing, at the public expense, against the English iron trade and the English mercantile community."[43]

Up to this point, all attempts to make iron in India had involved the use of charcoal. Charcoal was much preferred to coke because it produced a better iron and India still had vast hardwood forests. Britain had switched to coke in the eighteenth century only because of the depletion of the British forests. Yet by the late nineteenth century, coke-iron was so cheap that it displaced charcoal-iron, even in well-forested countries like Russia and Canada. Looking back in the late 1880s, F. R. Mallet, superintendent of the Geological Survey of India (GSI), wrote: "Numerous attempts have been made to manufacture iron on the English system in India, but nearly all of these have been unsuccessful and have long since been abandoned; one of the chief causes of failure being the difficulty of keeping large furnaces supplied with charcoal."[44]

By the 1870s, three factors converged to make experts consider seriously the use of coke in an Indian iron industry: the long string of failures using charcoal; the phenomenal growth of coal-based iron and steel industries in Europe and America; and the discovery of important coal deposits by the GSI.

The British had known of coal deposits in India as far back as 1774. The huge Raniganj coal field, north of Calcutta, was discovered in 1815, but it served only the marginal demands of river steamers until the East Indian Railway connected it to Calcutta in 1855. In 1836 the East India Company had appointed a Committee for the Investigation of the Coal and Mineral Resources of India. Nine years later, D. H. Williams of the British Geological Survey was sent to India "for the purpose of making a geological survey of those districts in which coal fields are situated." These efforts led to the creation of the GSI in 1851, under the direction of Thomas Oldham, professor of geology at Dublin University, who discovered the Gondwana system, one of the world's largest coal reserves, in 1867.[45]

Indian coal was of poor quality. Whereas British coal averaged

68 percent fixed carbon and produced 7.8 million calories per kilogram, Indian coal had only 52 percent carbon and produced between 6.1 and 7 million calories per kilogram. Indian coal also contained between 10 and 30 percent ash, compared with 2.7 percent for British coal. Yet coal was so abundant in India and labor costs were so low that its price in Bengal fell from 10.5 rupees per ton in the 1840s to 3.4 rupees in the 1890s, while that of imported coal rose from 13.5 to 17 rupees. Indian production rose from 100,000 tons in the late 1850s to 16 million tons in 1914. Though steam engines needed twice as much Indian as British coal to produce the same energy, Indian coal was not only competitive in eastern and southern India, it was also exported to Southeast Asia. Only in western India and the Arabian Sea did transport costs favor European and South African steam coal.[46]

For steam use, different coals could be substituted for one another, depending on the price. For metallurgy, however, different coals posed different technical problems, and Indian coal was especially difficult to deal with. The first attempts to make coke-iron in India took place in 1874–75. One was a government experiment, directed by the German metallurgist Ritter von Schwartz, using iron ore from the Chanda district of the Central Provinces and coal from the Wawora coal fields; it failed because of the poor coking quality of the coal. The other attempt was a private venture. The GSI had long advocated the use of Raniganj coal to make iron, and in 1874 the Calcutta managing agency Rutherford and Company set up the Bengal Iron Works at Barakar in the Raniganj coal fields. For several years it produced 20 tons of iron a day, mostly for the government. The government refused its request for long-term contracts, loans, or a dividend guarantee, however, and when orders ceased in 1879, the works were closed.[47]

These attempts, like the many charcoal-iron projects that preceded them, were doomed by technical errors, undercapitalization, and foreign competition. Behind these business errors, however, lay a more fundamental political question. An industrial ironworks had to be large in order to take advantage of economies of scale in coke ovens, furnaces, rolling mills, and transportation. To ensure a market for its output, such a plant required a commitment of some sort from either the railways or the government. The Indian railway companies had their regular suppliers in Britain; the directors of the Bombay, Baroda and Central Indian Railway, for example, instructed their agents in India to discourage the local purchase of railway materiel.

Railway materiel constituted a major part of India's imports, fluctuating between 3 percent in the 1870s and 6.5 percent in the decade 1900–1909.[48]

The government had long subsidized the smelting of iron in India, but inconsistently. For half a century it had opted for small-scale, low-budget experiments, the purpose of which was not so much to succeed as to be terminated with minimal losses in the event of a change of policy. The government's hesitation to develop an iron industry was a result of the split system of authority between the viceroy on the one hand and the India Office on the other.

The purchase of supplies by the Indian government came under a body of regulations known as the stores policy.[49] In 1858, the secretary of state for India had ruled that Indian government purchases had to be made through the Stores Department of the India Office in London. In 1863, to encourage British manufacturers still more, Indian import duties were reduced from 10 to 1 percent on iron and eliminated on machinery. At the time they were formulated, these policies were of little consequence because there were no manufacturers of railroad supplies in India. Yet in the long run, the policy hampered the emergence of Indian industries which could have competed with the British.

Indian government officials were sensitive to the heavy burden which railroad purchases placed both on the government budget and on the balance of payments. The issue came to a head during the viceroyalty of the Marquis of Ripon (1880–83). In 1881 the government purchased the defunct Bengal Iron Works with the intention of giving it to a private firm with sufficient capital to operate it and supporting it with long-term contracts. This plan was vetoed by Secretary of State for India the Marquis of Hartington, however, so Ripon decided to operate the works as a state enterprise. Ritter von Schwartz was appointed director, and skilled workers were brought from Europe. A new blast furnace, installed in 1884, raised production fourfold to 31,000 tons of pig iron a year. A foundry turned out pipes, sleepers, bridge piles, axle boxes, agricultural implements, and other castings.

Ripon's purpose was to reduce India's trade and budget deficits by stimulating import substitution industries. In 1882 he wrote:

> The Government of India have, for sometime past, had under special consideration the importance of developing the iron industry in India. The advantages which such development would afford both to State and the public—by cheapening the cost of railway construc-

tion and maintenance, and of works for improving the water supply; by substituting metal for more perishable materials in buildings; by reducing the home charges and their concomitant loss by exchange; by creating for the population non-agricultural employment; and by increasing the means for profitable investment of capital, are too well known to require lengthened exposition.

Nevertheless Lord Ripon was well aware of political realities, as his dispatch of January 23, 1883 indicates:

> It is, we presume, certain that the establishment of the iron and steel industry would be viewed with disfavor by the persons interested in the manufacture of these articles in England. In this connection it is by no means improbable that even the most legitimate efforts to develop and encourage local industry will be represented by those interested in the matter as though such efforts involved the adoption of a protective policy on the part of the Government of India.[50]

As often happened in that benevolent dictatorship, the government of India, Ripon's policy lasted as long as he was in office. In 1884 he was replaced by Lord Dufferin, a man more inclined to leave industry to private enterprise. In 1887 the India Office insisted that the Indian government limit to the utmost "the local purchase of building materials not produced in India, such as iron, steel, tools and plant, and especially of machinery." Even engineering firms located in India which used imported machines and materials were not considered to be bona fide Indian manufacturers; hence, they would have to bid on government contracts in London, not in India.[51]

Forbidden by the India Office to invest any more money in the Bengal Iron Works, the government sold it in 1889 to Martin and Company, a Calcutta managing agency. The government agreed to purchase 10,000 tons of iron a year for ten years, but only if it cost 5 percent less than English iron, "to disarm the home manufacturers' opposition." In 1900 the firm added a third blast furnace, raising its capacity to 75,000 tons a year. The firm now had a plant large enough to produce pig and cast iron that could compete with imports. Its production rose to 25,000 tons in 1901 and to 72,000 tons in 1914.[52] A modern iron industry had finally taken root in India, albeit fifty years later than it could have, given the country's raw materials and demand for iron. Despite the firm's new name of Bengal Iron and Steel Company, or BISCO, India still failed to produce steel.

Background of the Indian Steel Industry

During the nineteenth century, an iron industry in India seemed almost inevitable. Steel was another matter entirely. A steel mill requires costly and complex equipment, and it cannot grow from small beginnings but must be built big from the start.

The first attempt to make steel by modern methods was a military project. Since the 1890s the Indian army had imported the steel artillery shells which its ordnance factories were not equipped to make. In 1891 Maj. R. H. Mahon wrote a *Report on Cast Steel in India,* in which he advocated casting steel shells at the Cossipore Ordnance Factory. The director general of ordnance, Major General Walker, approved the idea, as did the India Office. Under Major Mahon's direction, the Cossipore factory installed a steel furnace and bar rolling mill in 1896, the first such plant in India. It had no commercial significance, however, being devoted entirely to weapons manufacture.[53]

The first commercial venture into steel making was a mill built by the Bengal Iron and Steel Company in 1905. Though it had a capacity of 20,000 tons, the firm received orders from the government for 600 tons of steel in seventy different sections, negating any economies of scale that might have existed. After only eight months the pig iron, which had too high a phosphorus content, damaged the furnaces. Having lost 500,000 rupees (£36,666) on the venture, the company shut down the mill. In 1906 the *Report of the Stores Committee* noted: "It seems improbable that any such industry can be profitable, if largely dependent on private demands, especially in view of the very considerable imports from the United Kingdom and the Continent." BISCO never again tried to make steel. In 1919, facing facts, it changed its name to Bengal Iron Company.[54]

The Indian steel industry thus began inauspiciously. It was not for lack of demand, however. In the 1870s and 1880s the railways switched to steel rails, which were safer and more durable than iron, but the Indian railway companies bought their rails in Britain. The government plant was restricted to casting shells, and private British efforts had failed to create a steel industry. The industry that finally arose was not the work of Europeans but of Indians, in particular the industrialist Jamsetji Nusserwanji Tata and the geologist Pramatha Nath Bose.

Tata, a Parsi from Bombay, had made his fortune as a cotton manufacturer after 1860. Not content to be the wealthiest industrialist in an otherwise backward nation, he actively sought to modernize India by developing technical education and the electrical power and steel industries. Tata's vision was industrial, capitalist, and nationalist. Having named one of his mills the Empress in 1877 in honor of Victoria's coronation as Empress of India, he named his next one, nine years later, the Svadeshi, or "Native Self," Mill.[55]

In the early 1880s Tata read Ritter von Schwartz's *Report on the Financial Prospects of Iron Working in the Chanda District*. Not so readily discouraged by difficulties as government officials, he sent samples of coal and iron ore from the Chanda district to be tested in Britain. When the tests proved encouraging, he went to Sir John Henry Morris, chief commissioner of the Central Provinces, to ask for a concession to mine the deposits in the area and to build a 72-kilometer-long railroad from Wawora to the nearest GIP (Great Indian Peninsula) trunk line. The request was denied, and Tata had to postpone his plans until more propitious times.[56]

Until 1899 the iron industry was located in places like Chanda and Birbhum, where surface deposits of ore had been worked for many years. Little was known of India's vast underground iron-ore deposits, largely because of government obfuscation. The GSI, in its narrow focus on coal and precious minerals, had deliberately ignored iron ore. Until midcentury, surveying was restricted to Britons. The GSI recruited its first Indian apprentice in 1873 and appointed its first Indian to a graded post in 1880. Though much scientific work was done in India, it consisted almost entirely of British scientists using India as the object of their field research.[57]

Private prospecting was discouraged until the mid-nineteenth century. By the end of the century, individuals—but not companies—could obtain prospecting licenses. Licenses were limited to a 10-square-kilometer area, and a distance of 12.8 kilometers had to separate any two prospecting areas licensed to the same individual. Then, once an area was explored, the government could auction off the mining rights to it. Based on a misguided concept of fairness, the regulations effectively discouraged even the most sanguine prospector.[58]

All this changed quite suddenly in 1899, when George Curzon became viceroy. He was determined to modernize India in order to strengthen the British Empire and counter the growing flood of manufactured goods imported from Germany and Belgium. To that end, he removed the onerous regulations which had hampered prospecting and mining. That same year, Major Mahon, now superintendent of the Cossipore Ordnance Factory, published a *Report upon the Manufacture of Iron and Steel in India,* in which he advocated a large modern steel mill in India. These two events reawakened Tata's interest in steel. On a trip to London, he visited Secretary of State for India George Hamilton, who told Tata he favored Indian industries developed with Indian capital. He also wrote Curzon: "I want to associate increased investment of British capital there with a simultaneous action on the part of the Government in developing industrial enterprise."[59]

These verbal encouragements meant that the India Office and the government of India would no longer stand in the way of Tata's plans; they did not mean that the government would help. In 1902 Tata returned to Britain and asked Hamilton for a pledge that the government would purchase some of the products of his proposed steel mill. This Hamilton refused.

While in Britain, Tata studied the iron and steel industry. He had a sharp eye for industrial machinery, as he had proved thirty years before in equipping his textile mills. This time he was looking for the best steel-making methods. He did not find them in Britain. From Britain he traveled to Germany to see the Dusseldorf Industrial Exhibition, from where he wrote his son Dorabji on September 5, 1902: "We are all surprised at the superiority and cheapness of all German machines and articles, as compared to English."[60]

From Germany Tata sailed to the United States. He had been there once before, to visit the Columbian Exposition in Chicago in 1892, and had met George Westinghouse and Senator Mark Hanna. This time he traveled to Cleveland, where Hanna showed him several steel plants and introduced him to steel company officials. In Pittsburgh he discussed the Niagara hydroelectric project with Westinghouse and visited the Homestead and Duquesne mills of the Carnegie Steel Corporation. As was evident to an astute observer like Tata, the American steel industry was then at the leading edge of this technology. Already in 1890 American Bessemer converters had an average output double that of their British counterparts. Not only were American furnaces larger, they were also pushed harder, with blast pressures almost double that of British furnaces. Because British engineers were conservative, their machines lasted longer but their products cost more. Britain was falling behind in the most basic of all industries.[61]

Among Tata's many acquaintances, the most useful was Julian Kennedy, of the metallurgical engineering firm of Julian Kennedy,

Sahlins and Company. Tata asked him to design a steel mill, and Kennedy recommended the consulting engineer Charles Page Perin to look into the raw materials situation. Perin agreed to work for Tata, but first he sent his assistant, the geologist C. M. Weld, to India to prospect for ore. Back in Bombay in late 1902 Tata, exhausted from his travels, handed the steel project over to his son Dorabji. When J. N. Tata died in May 1904, the project was well underway.[62]

In the early 1900s, the metalliferous regions of India were overrun by prospectors, most of them looking for manganese. In 1903, Dorabji Tata, C. M. Weld, and J. N. Tata's nephew, Shapurji Saklatvala, began by prospecting for iron ore in the Chanda district, the area which had attracted the elder Tata's attention years before. While they were there, the commissioner of the Central Provinces, Sir Benjamin Robertson, showed them a report entitled *The Iron Industry of the Western Portion of the District of Raipur,* published in 1887 by P. N. Bose of GSI.[63]

Pramatha Nath Bose was the first Indian to occupy a graded post in the Geological Survey of India. In his youth, while a student at the University of London, he had agitated and campaigned for Indian rights. The India Office, wanting to be rid of him but having no teaching position in India, instead offered him a post as assistant superintendent in the GSI. Soon after his return to India in 1880, Bose discovered and described the iron ores of the Raipur district. At the time, this attracted no attention, and Bose turned to other tasks. While working for the GSI, he kept alive his interest in the industrialization of India, perhaps through the influence of his father-in-law, the economic historian and nationalist Romesh Chunder Dutt. In 1886 Bose wrote a pamphlet entitled *Technical and Scientific Education in Bengal.* Five years later he organized the first Industrial Conference in Calcutta and helped found the Indian Industrial Association, an affiliate of the Indian National Congress, to lobby for technical education, industrial information, and government support for new industries. He also started a soap factory and a coal mine, but they both failed, like most other Bengali industrial ventures of the time.[64]

In his 1887 report on the Raipur district, Bose had described the ores as a rich hematite containing up to 72.92 percent iron. When Dorabji Tata, Weld, and Saklatvala read the report, they decided to investigate the area. They found two hills of iron ore so pure it rang under their boots. Hearing of their discovery, Sir

Thomas Holland, head of the GSI, came and reported that the hills contained 2.5 million tons of ore with an average iron content of 67.5 percent, richer than the iron ores of Britain (28–30 percent), Germany (32 percent), the United States (45–50 percent), and even Sweden (64 percent). It was easier to mine, being found in hills rather than underground. It was also low in sulphur and phosphorus, thus easy to smelt.[65]

However, Raipur was far from any source of coal. Bose had written in his report: "A charcoal furnace on a large scale could possibly be maintained here to advantage." Tata, Saklatvala, and Weld, however, were not about to make the same mistake that had doomed so many previous experiments. What was required to make a steel mill succeed was a combination of iron ore, coking coal, flux, and water close enough to each other and to major markets to keep transport costs within reason. Raipur did not possess this combination.

The Tata family spent, altogether, four years and £35,000 looking for that combination. Their careful research, in fact, is what set them apart from all other attempts to make iron and steel in India, which had been undertaken in a cheap and haphazard manner. All previous experimenters had located their plants near promising ore deposits, then looked for fuel nearby. The Tatas reversed this process. After many tests in European and American laboratories, they realized that suitable coking coal came only from the Jharia coal fields, and that their best chance would be to find good ores near the coal fields.

Once again, it was Bose who came to their aid. In 1903 he had resigned from the GSI to protest the appointment of Thomas Holland, his junior, as director general, against the usual rules of seniority. He had then become state geologist for the Maharajah of Mayurbhanj, a small principality in Orissa. There he discovered in 1903–4 the richest hematite deposit in the world, Gurumaishini Hill.[66] In February 1904 he wrote Jamsetji Tata, telling him of the new discovery and pointing out that the Mayurbhanj deposits were closer to the Bengal coal fields than the Raipur hills. Perin, hearing of the discovery, came from New York. Dorabji Tata, Saklatvala, Weld, and Perin visited the area and decided that the Tata steel works would use Mayurbhanj ore.[67]

Having found the right combination of raw materials, the Tatas obtained something equally precious: the aid of the government. In 1905 the Department of Commerce and Industry contracted to buy

20,000 tons of steel rails a year for ten years. Furthermore, the government granted railroad connections to the East India Railway's trunk line, low freights for the steel mill's raw materials and finished products, legal assistance is obtaining land and machinery, and various other favors.

There has been some debate over the causes of the government's positive attitude toward the Tata enterprise. Certainly Lord Curzon, an imperialist more than an administrator, saw the need to strengthen the British Empire, and this required steel. Britain's commercial position in India had been eroding for some time. Already in the 1890s India had begun importing more steel from Belgium than from Britain. German steel was also penetrating the Indian market, and Sir Thomas Holland commented that unless India developed its own industry, "it will soon become as much a market for German as for British goods." The Tatas were the beneficiaries of the policy shift, as they were later to acknowledge: "The very generous concessions made to our enterprise which more than any others have made an enterprise like the Tata Iron and Steel Works possible."[68]

But first a steel mill had to be built, and for that the Tatas needed to raise money. In 1906, therefore, Dorabji Tata and Perin went to London, but they found that British bankers were reluctant to invest in a new enterprise unless they could control it. The chairman of the British Railway Board, Sir Frederick Upcott, even told Perin: "Do you mean to say that the Tatas propose to make steel rails up to British specifications? Why, I will undertake to eat every pound of rail they succeed in making."[69]

Disappointed but not discouraged, Dorabji Tata returned to India. In 1907 he issued a prospectus offering shares worth 23,000,-000 rupees (£1.53 million) in the Tata Iron and Steel Company. The first stock issue sold out in three weeks, almost entirely to Indians. Part of this success was due to the sound reputation of the Tata family and its contacts with the government, which promised high returns at a reasonable risk. Part of it was the *swadeshi* movement, a popular economic nationalism which encouraged wealthy Indians to invest in Indian enterprises rather than the traditional land and jewelry; thus the Maharajah Scindia of Gwalior contributed £400,000 toward Dorabji Tata's working capital. There was also an ethnic factor at work: the Parsis, who constituted only 0.03 percent of the Indian population but were the dominant business class of Bombay, bought 36 percent of the shares of TISCO, the new Tata Iron and Steel Company.[70]

The Tata Iron and Steel Company

For the site of their steel plant, the Tatas chose the village of Sakchi (now Jamshedpur) situated between the iron deposits of Gurumai-shini and the Jharia coal fields, 225 kilometers west of Calcutta. Construction began in 1908, according to plans drawn up by Julian Kennedy and Charles Perin and under the supervision of Kennedy's partner Axel Sahlin. They imported two 200-ton blas furnaces, four 40-ton open-hearth furnaces, 180 coke ovens, a steam-powered blooming mill, a rail and structural mill, and a small bar mill from Germany and the United States.[71] The first furnace was blown in December 1911 and the first steel ingots rolled in February 1912. A year later the plant began producing steel rails. The Railway Board set up a laboratory at Sakchi to test them. By 1916 TISCO was producing 10,000 tons of steel rails and sections a month.

For TISCO, World War I was a godsend. India was cut off from Germany and Belgium, and British supplies became scarce. Steel imports fell by 84 percent from 1,040,000 tons in 1913–14 to 165,000 tons in 1917–18. Imports of railway supplies fell by 93 percent. Meanwhile, the demand for steel soared as the war effort put an increasing strain on the railways and the military called for ever more munitions. Its stores policy in abeyance, the Indian government now purchased all of TISCO's output. As the Indian Industrial Commission of 1916–18 explained: "In consequence of the increased difficulties of obtaining from Europe stores for war and essential purposes, the necessity of stimulating the local manufacture of munitions became a matter of vital importance."[72]

For several years TISCO grew up in a totally protected seller's market. Though the government paid less than the market price, the lost profits turned out to be a wise investment in government goodwill for the future. No infant industry could have asked for a happier childhood. TISCO's managers, Perin, Dorabji Tata, and the economist B. P. Padshah, took advantage of the market to modernize and expand the plant. As the British Ministry of Munitions had forbidden the export of steel-manufacturing equipment during the war, TISCO turned again to American suppliers. TISCO's pro-American bias, due originally to the influence of its American engineers and managers, was reinforced by wartime necessity and the nature of the raw materials. The inspectors for the Industrial Commission who visited the plant in January 1917 wrote: "The steel works, designed and erected by Americans, possess the characteristic features of American practice—a large output and the application of labour-saving machinery to the utmost extent possible."[73]

Among the new machines imported from the United States were larger blast furnaces, Bessemer converters and open-hearth furnaces, and an electric blooming mill. Special coke ovens were designed for TISCO to produce the hard coke required by the larger furnaces. The firm also purchased nearby collieries, dolomite and limestone quarries, and iron and manganese mines. By 1916–17 they had raised the plant's capacity to 200,000 tons, while its actual output rose from 31,000 tons of steel in 1912–13 to 181,000 tons in 1917–18, more than half of India's consumption. TISCO was now the largest industrial establishment in India.[74]

To overcome the problem of poor-quality coking coal, they had to adopt the duplex process used in the mills of Gary, Indiana, and Buffalo, New York. The ore was reduced in a Bessemer converter, and the resulting steel was poured into an open-hearth furnace to remove the phosphorus introduced into it by the coke. This made the steel more expensive than either the Bessemer or the Siemens-Martin open-hearth steel produced in Europe and America.

During the war, Perin and Tata had laid plans for the "greater extensions" of their plant, designed to boost its capacity to 500,000 tons of steel a year. The necessary equipment was to be imported from the United States, and the expansion was to cost $70 million.[75] But the greater extensions took longer to build than expected. Meanwhile, peace ended TISCO's cozy monopoly of the Indian steel market. In 1921 the *Asiatic Review,* an imperialist journal, compared the Indian steel industry favorably with that of Britain, noting that the raw materials needed to make a ton of pig iron cost half as much in India as in Britain, and that Indian labor, though still inefficient, was improving. The competition, however, was no longer from British but from German and Belgian steel made from battlefield scrap at prices neither Britain nor India could compete with. Though Indian pig iron was among the cheapest in the world, Indian steel was more expensive than its competitors because of the duplex process and the overvaluation of the rupee in relation to Continental currencies.[76]

The war had swept away the traditional British beliefs in free trade and laissez-faire. As early as 1915 Viceroy Lord Hardinge admitted:

> It is becoming increasingly clear that a definite and self-conscious policy of improving the industrial capabilities of India will have to be pursued. . . . After the war India will consider herself entitled to demand the utmost help which her Government can afford to enable her to take her place, so far as circumstances permit, as a manufacturing country.[77]

And the Indian Industrial Commission concluded:

> It appears to us that, in the interests of Indian industries, a radical change should be made in the methods of purchasing in India Government and railway stores. The existing system has been handed down from a time when India was almost totally dependent upon Europe for manufactured goods; but it is unsuited to modern conditions and has had a deterrent effect on attempts to develop new industries in India.[78]

The violent swings of the postwar economy led to further policy changes. In 1920 the Stores Purchase Committee recommended that the government buy through a Stores Department in India instead of the India Office in London, and that it encourage infant industries with orders at favorable rates.[79] After the war, TISCO turned to the government for protection against its dangerous new competitors. As a result the Indian Fiscal Commission presided over by Sir Ibrahim Rahimtoola recommended "discriminating protection" in 1921. The Tariff Board, set up in 1924, turned its attention first to the steel industry. On its recommendation the Indian government passed the Steel Industry (Protection) Act of 1924, which raised duties on imported steel from 2.5 to 33.3 percent ad valorem and provided subsidies for Indian rails. The next year the secretary of state for India transferred control over stores purchases to the Indian government, which required that bids on government contracts be submitted in India and in rupees, and instructed purchase officers to prefer Indian to foreign goods. A further drop in the price of Belgian steel led to the Steel Industry (Protection) Act of 1927, which raised tariffs on Continental steel while lowering them on British steel. Meanwhile, further subsidies were granted on steel and rails.[80]

The result was to build a high wall around the Indian steel market, behind which TISCO proceeded with its greater extensions. It raised its capacity to 610,000 tons of pig iron and 580,000 tons of steel annually by the mid-1920s, five times its original capacity. When the Depression reached India in 1930, TISCO was in a good position to resist, technologically, economically, and politically. Government orders for rails dropped from 121,600 tons in 1929–30 to 37,000 tons in 1932–33. Other orders for steel also fell off. Total steel consumption dropped from 1,678,085 tons in 1929 to 823,825 in 1933. Yet it was the foreign firms that lost the most. With the aid of tariffs, TISCO increased its share of the Indian steel market from 14 percent in 1920–21 to 73 percent in 1938–39; by then it produced 99 percent of the rails purchased in India.[81] TISCO had become one of the largest and most modern steel mills in the British Empire. India had at last obtained its backward linkage by political means.

Conclusion

Since the Industrial Revolution, the mining and metallurgical industries have become too complex and large-scale to evolve gradually from the artisan to the industrial stage; Chinese tin mining in Malaya was the last successful attempt. Strictly economic considerations such as the location of raw materials, the demand for the products, and the costs of production and transportation do not suffice to account for the success or failure of metallurgical industries in the colonies. Politics and culture were just as instrumental.

Tin and copper stand in contrast to iron and steel. The world demand for tin and copper gave sufficient economic motivation for foreigners to develop these industries in Malaya and the Congo. The metals industries that arose were intrinsically part of the world economy and only accidentally part of the local ones; in other words, they were enclaves. In both cases, the technology transfer was foreign-driven, with indigenous peoples playing a very incidental role.

Yet there were also political and cultural differences between them. In Malaya, the British administration was fairly passive at first, limiting itself to imposing order, and later to supervising mining operations and preventing gross abuses. It encouraged the most efficient producers, the Chinese in the late nineteenth century, and the Europeans in the twentieth. As a result, not one but two streams of foreign technology were transferred to Malaya in competition with each other. In Katanga, only the Union minière was encouraged, or even tolerated, by the authorities. The actions of the company and those of the government dovetailed so nicely that one can speak of a corporate colonialism or "portfolio state," as opposed to the petty-capitalist colonialism of Malaya.

In India the economic motivation to create an iron and steel industry was lacking. Therefore politics and culture did not simply influence this industry, they created it. And they did so after a fairly long delay. Forty years separated the first Indian railway boom of the 1860s from the opening of a modern coke-iron industry, by which time the railways had switched to steel. From the 1880s on, when steel rails began replacing iron and railroad construction reached its peak, a modern iron and steel industry would have been viable in India. The raw materials were plentiful, as Bose showed; entrepreneurship and capital could have been forthcoming, as the Tatas proved; and the technology could have been imported as it was for the railways. What was lacking was a consistent attitude on the part of the government. Before such an attitude finally appeared, thirty years went by.

What caused such long delays? One explanation is the poverty of India, which placed obstacles in the path of industrial development. Yet poverty was no obstacle to the creation of a great rail network, nor to the rise of the cotton and jute industries. Culture has also been blamed; colonials in particular liked to dwell upon Eastern "otherworldliness" or "tradition" as obstacles in the path of "progress." Yet Indians belong to many cultures, and among them certain groups are as entrepreneurial, in the Western sense, as their European counterparts. And "Indian culture" did not prevent the rise of an indigenous cotton industry.

If culture and poverty played a part in delaying the rise of industries, it was in a distant way. As explanations, the decisions of the elite that ruled India, and the values that led to these decisions, are more specific. We have seen several. One is free trade, which the British erected into a dogma before World War I and which therefore became the bugbear of Indian nationalists. All other countries which built rail networks comparable to India's used import duties to ensure the rise and survival of their heavy industries; and in India too, those industries that finally arose required protection. The stores policy, which affected the government's own purchases, long deprived potential Indian enterprises of the surest customer they might have had. The Indian government's commitments to buy iron from BISCO and steel from TISCO are ample evidence of the impact of government purchases. Prospecting rules also delayed industrialization. The GSI, so dedicated and competent in its scientific studies and in its search for coal, showed no interest in iron, even when great deposits were discovered and published by its one Indian geologist, P. N. Bose.

Thus the delay resulted largely from deliberate government policies. But what were the motives for these policies? It would be simple to blame British industries eager to hold onto their hunting preserve, the captive market of India. Yet the rise of the Bombay cotton and the Calcutta jute industries, in direct competition with powerful British interests, casts doubt on this explanation. British industries were only one of several pressure groups that influenced government policies in India. There were others, including the viceroys and secretaries of state; the Indian Civil Service; the British business community in India; and the Indian nationalists.

India was a benevolent despotism of a peculiar sort: it had not one but two despots, the viceroy in India and the secretary of state in London, neither of whom held office for very long. Hence the policies of the Indian government moved by fits and starts, from dynamic action to near-paralysis and back. Viceroys with powerful personalities like Dalhousie, Lawrence, and Curzon could start impressive programs and accomplish much in a short time. One is tempted to agree with W. Arthur Lewis when he says: "It seems almost an accident whether the government should be helpful or adverse to development. This is true even of colonial governments. . . . Much depended on the personality of the colonial governor."[82]

At other times, a stalemate arose between the viceroy and the secretary of state. The iron industry in particular, which normally needs years to develop, fell victim to these periodic stalemates. The result was a lack of direction of which the Industrial Commission complained in these terms:

> This account of the efforts made by Government for the improvement of Indian industries shows how little has been achieved, owing to the lack of a definite and accepted policy, and to the absence of an appropriate organization of specialized experts. . . . Much valuable time has been lost, during which substantial advances might have been registered.[83]

Yet neither constitutional rigidities nor personal conflicts can account fully for the hesitancy with which the Indian government approached industrialization. The vacillations, rather, reflect a real contradiction in the British position. On the one hand the British Empire was founded on the relations between Britain and India, in which Britain supplied manufactures, transportation, administration, and defense, and in exchange India supplied tropical products and manpower. In this system other colonies formed a defensive perimeter around India and an extension of the Indo-British economic system. On the other hand, outside the British Empire there was, as Curzon and his successors realized, a world of rival nations with growing industries, powerful navies, and gangster ethics.

What India needed to participate in the defense of the empire was the very industries that would help reduce India's dependence on Britain and Britain's hegemony within the empire. It is not that the rulers of British India were confused, but that the realities of world politics presented them with an intractable conflict between the interests of Britain and those of the British Empire, both of which they were committed to defending. They responded with hesitant procrastination.

Behind the transient political appointees who ruled India stood a powerful bureaucracy, the Indian Civil Service. Drawn from the gentry of Britain, its members were educated in both the humanities and the natural sciences, and they combined the qualities of an aristocracy with those of a cultured intelligentsia. This group of men, heirs of a social class that had lost its preeminence in Britain itself to the "Manchester men" and London merchants, went to India as enlightened despots, experts in the administration of fairness, ruling over a race of simple peasants and wealthy landlords. To do so they had to withstand the pressures of commerce as well as the winds of revolution.

Neither as gentlemen nor as intellectuals did the members of the ICS have much respect for technological change per se, except insofar as it was useful. Certain technologies contributed directly to their authority and efficiency and to the perpetuation of British rule. Others brought security, comfort, and status to those who used them. Railways, telegraphs, harbors and steamships, urban amenities and, lastly, automobiles and aircraft were all of this nature. The heavy industries, in contrast, were remote, not a part of the education or experience of civil servants. Thus the poor showing of the government's efforts at founding an iron industry.

What was true of government officials was also true, in a somewhat different way, of the British business community. They were of course subject to the profit motive, but their forte was international trade; hence the Scottish domination of the Bengal jute industry—an export business—as contrasted with the Bombay cotton mills—an import-substitution activity. The metallurgical and engineering industries offered, from the point of view of British businessmen in India, less profit and more risk than investing in trade or manufacture for export. Their conservative attitude was summed up by John Keenan: "Most of them felt, like the diehard fellows with me on the boat, that Indians were all right in their place, but steel making was a cut above them. A big cut. A steel industry in India would not only compete with the English mills but it wasn't practicable. Heath had proved that."[84]

Only Indians themselves, and then only a tiny minority of Indians, found their profit motives sufficiently reinforced by nationalism to warrant taking risks in heavy industries. A steel industry, like the cotton industry before it, was more than a manufacturing process and a business: it was an import-substitution activity, in other words, a swadeshi enterprise. P. N. Bose recognized this when he wrote:

> The aggressive imperialism of modern Europe is based upon industrialism. It is chiefly in the interest of their industries, that the greater powers of the West are anxious to dominate the peoples of the East. If these peoples made a vigorous well-concerted effort to develop their resources on Western methods, and supply their own wants, their markets would cease to be exploited in the way they now are by Western manufactures, and their lands would cease to be the happy hunting ground of Western enterprise. Western imperialism would then die a natural and peaceful death at least in its present highly objectionable militant form.[85]

Here, then, is the cause of the delay. Heavy industries in India did not respond to the economic forces engendered by the railways because they were distorted by political and cultural forces. They had to await two shifts in values: on the part of the British political system, the realization that such industries would be more useful to the empire against outsiders than detrimental to Britain within it; and from the point of view of creating such an industry, the appearance of Indian entrepreneurs driven by patriotism as well as by capitalism.

Notes

1. Lim Chong-Yah, *Economic Development of Modern Malaya* (Kuala Lumpur, 1967), pp. 37–43 and appendix 2.1; Ernest S. Hedges, *Tin in Social and Economic History* (London, 1964), pp. 41–45.

2. Wong Lin Ken, "Western Enterprise and the Development of the Malayan Tin Industry to 1914," in *The Economic Development of South-East Asia: Studies in Economic History and Political Economy,* ed. C. D. Cowan (London, 1964), p. 131.

3. Lennox A. Mills, *British Rule in Eastern Asia: A Study of Contemporary Government and Economic Development in British Malaya and Hong Kong* (Minneapolis and London, 1942), p. 177.

4. On Chinese mining technology before the 1880s, see H. C. Chai, *The Development of British Malaya, 1896–1909* (Kuala Lumpur, 1964), pp. 166–74; Ooi Jin Bee, "Mining Landscapes of Kinta," in *Readings in Malayan Economics,* ed. T. H. Silcock (Singapore, 1961), pp. 351–56; Wong Lin Ken, *The Malayan Tin Industry to 1914* (Tucson, Ariz., 1965), pp. 47–60; and Lim, pp. 44–47, 120–22.

5. Lim, appendix 2.1.

6. Wong, "Western Enterprise," pp. 133–40.

7. Sir Frank Swettenham, *About Perak* (Singapore, 1893), p. 34, quoted in Lim, p. 165, and Chai, p. 50.

8. Chai, p. 167; Wong, *Malayan Tin Industry,* pp. 56–59, 196–98.

9. Wong, *Malayan Tin Industry,* p. 238.

10. Ibid., pp. 154–67, 227–29; Chai, pp. 170–71; George Cyril Allen and Audrey G. Donnithorne, *Western Enterprise in Indonesia and Malaya: A Study in Economic Development* (London, 1957), pp. 158–60.

11. Chai, p. 167; Ooi, pp. 354–56; Wong, *Malayan Tin Industry,* p. 201.

12. Wong, *Malayan Tin Industry,* pp. 200–219; Lim, pp. 49–63; Harry F. Bain, *Ores and Industry in the Far East* (New York, 1933), p. 197; B. T. K. Barry and C. J. Thwaites, *Tin and Its Alloys and Compounds* (Chichester, 1983), pp. 22–27.

13. Wong, *Malayan Tin Industry,* p. 218.

14. Ibid., pp. 237–38.

15. Li Dun-jen, *British Malaya: An Economic Analysis* (New York, 1956), pp. 50–51.

16. Ibid., pp. 51–52; Wong, "Western Enterprise," pp. 147–49.

17. On the cartel of the 1930s, see William Y. Elliott et al., *International Control in the Non-Ferrous Metals* (New York, 1937), pp. 89–106; Mills, pp. 180–83; and Li, pp. 52–56.

18. Walter Cline, *Mining and Metallurgy in Negro Africa* (Menasha, Wis., 1937), pp. 56–75; Kenneth Bradley, *Cooper Venture: The Discovery and Development of Roan Antelope and Mufulira* (London, 1952), pp. 32–36; J. Austen Bancroft, *Mining in Northern Rhodesia: A Chronicle of Mineral Exploration and Mining Development* (n.p., 1961), pp. 27–39; Charles Perrings, *Black Mineworkers in Central Africa: Industrial Strategies and the Evolution of an African Proletariat in the Copperbelt, 1911–1941* (New York, 1979), pp. 5–6; Francis L. Coleman, *The Northern Rhodesia Copperbelt, 1899–1962: Technological Development up to the End of the Central African Federation* (Manchester, 1971), pp. 170–73.

19. Simon E. Katzenellenbogen, *Railways and the Copper Mines of Katanga* (Oxford, 1973), pp. 19–24, and "The Miner's Frontier: Transport and General Economic Development," in *Colonialism in Africa,* ed. Peter Duignan and L. H. Gann, vol. 4: *The Economics of Colonialism* (Cambridge, 1975),

p. 374; Simon Cunningham, *The Copper Industry in Zambia: Foreign Mining Companies in a Developing Country* (New York, 1981), p. 31; Bancroft, pp. 48–50; Coleman, pp. 6–12.

20. A. Marthoz, *L'industrie minière et métallurgique au Congo belge* (Brussels, 1955), p. 19; R. E. Birchard, "Copper in the Katanga Region of the Belgian Congo," *Economic Geography* 16 (October 1940): 429; Union minière du Haut-Katanga, *Le Katanga, pays du cuivre* (Liege, 1930), pp. 3–5; *Union minière du Haut-Katanga,* 2nd ed. (Brussels, 1956), pp. 66–92; and *Union minière du Haut-Katanga, 1906–1956: Evolution des techniques et des activités sociales* (Brussels, 1957), pp. 11–22; Jean-Luc Vellut, "Mining in the Belgian Congo" in *History of Central Africa,* ed. David Birmingham and Phyllis M. Martin, 2 vols. (London and New York, 1983), 2: 126–62; Katzenellenbogen, "Miner's Frontier," pp. 361–75; Bancroft, pp. 46–63; Bradley, pp. 60–73.

21. R. Chadwick, "New Extraction Processes for Metals" in *History of Technology,* ed. Charles Singer et al., vol. 5: *The Late Nineteenth Century, c. 1850 to c. 1900,* pp. 72–101; V. van Lint, *Le Congo et les carrières coloniales d'ingénieur* (Brussels, 1925), p. 60; Eugène Prost, *La métallurgie en Belgique et au Congo belge, Historique—situation actuelle* (Paris, 1936), pp. 281–97; J. Quets, "Métallurgie" in *Livre blanc: Apport scientifique de la Belgique au développement de l'Afrique centrale,* Académie royale des Sciences d'Outre-mer, 3 vols. (Brussels, 1962–63), 3: 1085–1104; Perrings, pp. 31–33, 74–75, 245–48; Vellut, pp. 135–40; Union minière, *Evolution,* pp. 102–30, and *Union minière,* pp. 91, 103, 107–9, 123, 146–47, 178–79.

22. Copper Development Association, *Copper: Its Ores, Mining and Extraction* (Kendals Hall, England, 1951), pp. 15–39; Chadwick, pp. 74–84; Coleman, pp. xv–xx; Perrings, pp. 44–45; Prost, pp. 283–90; Quets, pp. 1085–91; Union minière, *Evolution,* pp. 78, 106–12.

23. Union minière, *Katanga,* pp. 5–25; *Evolution,* pp. 106, 130, 162–63; and *Union minière,* pp. 148–52, 174; Prost, pp. 286–92; Katzenellenbogen, "Miner's Frontier," p. 405.

24. Perrings, pp. 245–48; Elliott et al. pp. 412–36.

25. See Katzenellenbogen, *Railways,* passim.

26. L. H. Gann and Peter Duignan, *The Rulers of Belgian Africa, 1884–1914* (Princeton, 1979), pp. 199–200; Cunningham, pp. 32, 45–46; Union minière, *Union minière,* pp. 99–105, 112, 124–26.

27. Union minière, *Katanga,* p. 45. See also Perrings, pp. 24–31, 61–62, 90–94.

28. Union minière, *Evolution,* p. 300.

29. Ibid., pp. 220–38.

30. Marthoz, pp. 58–59.

31. See Albert O. Hirschman, *The Strategy of Economic Development* (New Haven, 1958), p. 100.

32. On this linkage effect in various industrial countries see, e.g., Duncan Lyall Burn, *The Economic History of Steel Making, 1867–1939: A Study in Competition* (Cambridge, 1940), p. 28; Peter Temin, *Iron and Steel in Nineteenth-Century America: An Economic Inquiry* (Cambridge, Mass., 1964), pp. 221–23, 274–77; Jeremy Atack and Jan K. Brueckner, "Steel Rails and American Railroads, 1867–1880," *Explorations in Economic History* 19 (1982): 339–59; Thomas Esper, "Industrial Serfdom and Metallurgical Technology in 19th-

Century Russia," *Technology and Culture* 23, no. 4 (October 1982): 593, n. 48; Patrick O'Brien, *The New Economic History of the Railways* (New York, 1977), pp. 65–66; and M. C. Urquhart and K. A. H. Buckley, eds., *Historical Statistics of Canada* (Cambridge and Toronto, 1965), p. 528.

33. Jogendra Nath Sahni, *Indian Railways: One Hundred Years, 1853 to 1953* (New Delhi, 1953), p. 72; George Walter Macgeorge, *Ways and Works in India: Being an Account of the Public Works in that Country from the Earliest Times up to the Present Day* (Westminster, 1894), pp. 270, 346, 405–406, 418; Maurice A. Harrison, *Indian Locomotives of Yesterday (India, Bangla Desh, and Pakistan)*, Part 1: *Broad Gauge* (Bracknell, England, 1972), p. 6.

34. *Statistical Abstract Relating to British India,* various years.

35. Morris David Morris and Clyde B. Dudley, "Selected Railway Statistics for the Indian Subcontinent (India, Pakistan, and Bangladesh), 1853–1946–47," *Artha Vijnana* 17, no. 3 (September 1975): 193.

36. N. S. R. Sastry, *A Statistical Study of India's Industrial Development* (Bombay, 1947), pp. 105, 111 (note: the 1900–13 figures are for the Bengal Iron and Steel Works); *Statistical Abstract Relating to British India from 1926–27 to 1935–36* and *Statistical Abstract Relating to British India from 1930–31 to 1939–40.*

37. For iron and steel statistics, see Urquhart and Buckley, pp. 484–86; U.S. Bureau of the Census, *Historical Statistics of the United States: Colonial Times to 1970* (Washington, 1975), pp. 599–600; and B. R. Mitchell, *European Historical Statistics, 1750–1970* (New York, 1975), pp. 581–88.

38. On this point see also M. D. Morris, "Toward a Reinterpretation of 19th Century Indian Economic History," *Journal of Economic History* 23, no. 4 (1963): 614–16.

39. Leland Hamilton Jenks, *The Migration of British Capital to 1875* (New York, 1927), p. 227; B. R. Tomlinson, *The Political Economy of the Raj, 1914–1947: The Economics of Decolonization in India* (London, 1979), p. 3; Burn, p. 80.

40. Sabyasachi Bhattacharya, "Cultural and Social Constraints on Technological Innovation and Economic Development: Some Case Studies," *Indian Economic and Social History Review* 3, no. 3 (September 1966): 252–53; Manoranjan Chaudhuri, *The Iron and Steel Industry of India: An Economic-Geographical Appraisal* (Calcutta, 1964), p. 30; Hitesranjan Sanyal, "The Indigenous Iron Industry of Birbhum," *Indian Economic and Social History Review* 5, no. 1 (March 1968): 101–8; John Keenan, *A Steel Man in India* (New York, 1943), pp. 16–18; M. D. Morris, "The Growth of Large-Scale Industry to 1947," in *The Cambridge Economic History of India*, vol. 2: *c. 1753–c. 1970*, ed. Dharma Kumar (Cambridge, 1983): 559–61.

41. Mahadev Govind Ranade, "Iron Industry—Pioneer Attempts," in *Essays on Indian Economics: A Collection of Essays and Speeches*, 2nd ed. (Madras, 1906), pp. 170–92; Reginald Henry Mahon, *A Report upon the Manufacture of Iron and Steel in India* (Simla, 1899), appendix 2; F. R. Mallet, "Iron" in *A Dictionary of the Economic Products of India*, ed. George Watt, 4 vols. (London and Calcutta, 1890), 4: 504; Morris, "Growth," pp. 584–85; Chaudhuri, pp. 29–30.

42. Chaudhuri, p. 30; Ranade, p. 183; Sanyal, p. 107.

43. Sunil Kumar Sen, *Studies in Economic Policy and the Development of India (1848–1926)* (Calcutta, 1966), pp. 41–42; Chaudhuri, pp. 31–33; Mallet, p. 504; Ranade, p. 187–88.

44. Mallet, p. 504.

45. India, Geological Survey, *Centenary of the Geological Survey of India, 1851–1951: A Short History of the First Hundred Years* (Calcutta, 1951), pp. 2–4; Deepak Kumar, "Patterns of Colonial Science in India," *Indian Journal of History of Science* 15, no. 1 (May 1980): 107; Bishnupada Guha, "The Coal Mining Industry" in *The Economic History of India,* ed. V. B. Singh (New Delhi, 1965), pp. 308–11; Vera P. Anstey, *The Economic Development of India,* 4th ed. (London and New York, 1952), p. 25.

46. William Arthur Johnson, *The Steel Industry of India* (Cambridge, Mass., 1966), p. 28; Guha, pp. 305–16; Sahni, p. 95; Anstey, pp. 25, 235–41, 609.

47. Ritter von Schwartz, *Report on the Financial Prospects of Iron Working in the Chanda District* (Calcutta, 1882), cited in Frank Reginald Harris, *Jamsetji Nusserwanji Tata: A Chronicle of his Life,* 2nd ed. (Bombay, 1958), pp. 150–51; Ranade, pp. 184–90; Mallet, p. 104; Chaudhuri, p. 31; Johnson, p. 9; Geological Survey, p. 27.

48. Sen, pp. 16, 46; *Statistical Abstract Relating to British India,* various years (railroad materials import statistics were discontinued after 1928).

49. A history of the stores policy can be found in India, Stores Purchase Committee, *Report of the Stores Purchase Committee,* 2 vols. (Simla, 1920–21), 1: 163–92; Sen, pp. 16, 46.

50. Sunil Kumar Sen, *The House of Tata, 1839–1939* (Calcutta, 1975), pp. 29–30.

51. Sen, *Economic Policy,* p. 19.

52. Morris, "Growth," pp. 585–87.

53. Sen, *Economic Policy,* pp. 75–76; Ranade, p. 189.

54. Daniel H. Buchanan, *The Development of Capitalistic Enterprise in India* (New York, 1934), pp. 281–84; Amiya Kumar Bagchi, *Private Investment in India, 1900–1939* (Cambridge, 1972), p. 300, n. 4; Chaudhuri, pp. 31–34; Johnson, p. 9; Keenan, p. 25.

55. Sen, *House of Tata,* p. 22.

56. Lovat Fraser, *Iron and Steel in India: A Chapter from the Life of Jamshedji N. Tata* (Bombay, 1919), pp. 10–13; Harris, p. 151; Buchanan, pp. 184–85.

57. S. N. Sen, "The Character of the Introduction of Western Science in India during the Eighteenth and the Nineteenth Centuries," *Indian Journal of History of Science* 1, no. 2 (1966): 112–22.

58. Kumar, pp. 107–8; Sen, *Economic Policy,* p. 117; Harris, p. 153; Fraser, pp. 13–15.

59. B. S. Saklatvala and K. Khosla, *Jamsetji Tata* (New Delhi, 1970), pp. 87–88; Sen, *House of Tata,* pp. 34–35; Harris, pp. 155–56.

60. Saklatvala and Khosla, p. 113.

61. J. C. Carr and Walter Taplin, *History of the British Steel Industry* (Cambridge, Mass., 1962), pp. 151–54.

62. Saklatvala and Khosla, pp. 116–18; Keenan, pp. 28–30; Harris, pp. 157–68; Sen, *House of Tata,* pp. 35–36.

63. *Records of the Geological Survey of India* 20, pt. 4 (1887): 167–70.

64. Jogesh Chandra Bagal, *Pramatha Nath Bose* (New Delhi, 1955), pp. ix–71, 103–8.

65. Harris, pp. 175–77; Johnson, p. 27; Buchanan, pp. 282–83.

66. P. N. Bose, "Notes on the Geology and Mineral Resources of Mayurbhanj," *Records of the Geological Survey of India* 31, pt. 3 (1904): 167–73.

67. Saklatvala and Khosla, p. 122; Harris, pp. 179–82; Bagal, pp. xiii–xiv, 77–78; Sen, *House of Tata*, p. 37; Fraser, pp. 30–45.

68. Samuel B. Saul, *Studies in British Overseas Trade, 1870–1914* (Liverpool, 1960), pp. 198–99; Sen, *House of Tata*, pp. 42–43; Anstey, p. 243.

69. Verrier Elwin, *The Story of Tata Steel* (Bombay, 1958), p. 35.

70. Beginning in the seventeenth century, the Parsis of Bombay had achieved wealth and entrepreneurial experience as middlemen between the British-dominated international trade and the Indian market. Ashok V. Desai, "The Origins of Parsi Enterprise," *Indian Economic and Social History Review* 5, no. 4 (December 1968): 307–18; Morris, "Growth," pp. 590–91; Johnson, pp. 30–31, 243–49; Fraser, pp. 51–54.

71. Fraser, pp. 54–64.

72. India, Indian Industrial Commission, *Report of the Indian Industrial Commission, 1916–18* (Calcutta, 1918), pp. xvii–xviii.

73. India, Indian Industrial Commission, "Confidential Inspection Notes," p. 7 in India Office Records, L/E/7/855 file 8417/15.

74. India, Department of Statistics, *Large Industrial Establishments in India* (Calcutta, 1920), p. 10; Indian Industrial Commission, p. 20; Fraser, pp. 58–86. On TISCO's coal mines see C. P. Simmons, "Vertical Integration and the Indian Steel Industry: The Colliery Establishment of the Tata Iron and Steel Company, 1907–1956," *Modern Asian Studies* 11, no. 1 (1977): 127–48.

75. Elwin, pp. 52–54; Keenan, pp. 70–71; Buchanan, p. 286.

76. "The Iron and Steel Position in India," *Asiatic Review* 17 (1921): 689–94; Carr and Taplin, p. 234; Elwin, p. 55; Buchanan, pp. 284, 293; Keenan, pp. 63–70.

77. Dispatch of November 26, 1915 to secretary of state for India, quoted in Madan Mohan Malaviya, "Note," in Indian Industrial Commission, pp. 314–15.

78. Ibid., pp. 148–49.

79. Stores Purchase Committee, 1: 20–22, 125.

80. R. S. Bisht and M. P. N. Namboodripad, "Iron and Steel Industry" in Singh, pp. 204–5; Keenan, p. 79.

81. P. J. Thomas, *India's Basic Industries* (Bombay, 1948), pp. 5–7; Sen, *Economic Policy*, pp. 119–21, and *House of Tata*, pp. 48–60; Morris, "Growth," p. 626; Buchanan, p. 292; Johnson, p. 12.

82. W. Arthur Lewis, *Tropical Development, 1880–1913* (London, 1970), pp. 27–28.

83. Indian Industrial Commission, pp. 80–82.

84. Keenan, p. 26.

85. Bagal, pp. 94–95.

9

Technical Education

The transfers of technology described so far have consisted of major development projects the Europeans undertook in order to bring their colonies into the world economy. If this book were to end here, a naive reader might believe that the colonizers were successful in transferring technology to the tropics; this is precisely the impression given by writers like George Macgeorge who fondly refer to the great construction projects as "monuments."

Technology, however, is not monuments but knowledge and activity. Its transfer, to be complete, involves the spread of activity and knowledge not only from one area to another, but also from one people to another. Hence our question becomes: How much, and for what reasons, did Asians and Africans learn about Western technology in the colonial period? And what was the role of schooling in that learning process?

Education, schooling, and learning are a minefield into which the historian steps with trepidation. Statistics can easily mislead. A larger number of students does not mean a more educated population. A more educated population does not mean one that is better able to solve its problems, or even understand them. Technical education sits uneasily between the school and the workplace. Educators are seldom craftsmen, and technicians rarely teach. Educational bureaucrats do not mix easily with businessmen, engineers, and workers. Learning technical subjects by face-to-face interaction with experienced technicians and hands-on contact with machines and physical processes is alien to traditional schooling.

The words "technical education" cover a spectrum of activities, and their meaning has changed over time to cover every sort of work-oriented learning. The lowest level of technical education was the training of children in simple rural skills of farming and primitive crafts, a process which required no literacy and took place without European assistance. During the colonial period, Europeans were involved in disseminating preindustrial crafts such as carpentry, bricklaying, ironworking, and sewing, and rural techniques such as plowing, crop rotation, and animal husbandry; the teaching of such skills often took place in mission stations and was associated with religious and elementary education.

Colonial governments and enterprises also taught the basic skills needed by industry, public works, railroads, and the like in trade schools and apprenticeship programs. Alongside these modern skills, other programs taught "artistic" crafts indigenous to each region: brasswork and pottery, carpet weaving and embroidery, leather work, and other crafts that colonial officials feared were being threatened by cheap factory-made imports. Such training was also associated with a primary education, usually in the vernacular. For the sake of clarity, let us call this vocational training.

What we will call technical education was training for such mid-level jobs in the modern sector as surveying, typography, draftsmanship, telegraphy, and machine repair; this training was usually associated with a primary or even a secondary education in the language of the colonial rulers. Graduates of such programs acquired two skills, one of which—their literacy in a European language—opened up more and usually better opportunities in administration and office work than their technical knowledge.

College-level engineering education was an even more sophisticated training for the few subordinate engineering jobs open to non-Europeans. The pinnacle of technical education, postgraduate engineering or scientific training leading to management, planning, or research positions, was practically nonexistent in the colonies before 1940.

Since our focus is the transfer of Western technologies, we will not dwell upon the indigenous rural skills and the arts and crafts. Our question is: What education in the Western technologies was available to colonial subjects, and what impact did it have on the development of the colonies?

The kinds of technical education offered in the colonies, and to whom, depended on a complex mix of factors, including the level of indigenous technology and the goals of the Europeans and their colonial subjects. The technological levels of colonial societies ran the gamut from advanced to primitive. Parts of India and Egypt were on the verge of becoming industrial in the mid-nineteenth century. Other parts of these two countries, as well as Java, Indochina, and North Africa, had mechanisms, energy sources, and craft techniques reminiscent of those found in early-modern Europe. Much of sub-Saharan Africa had an iron-age technology more primitive than that of medieval Europe. Papua-New Guinea, Borneo, and parts of Africa were still in the stone age.

The perceptions Europeans had of the culture of their colonial subjects certainly influenced the kind of education they offered. Thus in Papua, "a major unarticulated reason for developing technical education was that this was all that Papuans could cope with or aspire to. Academic education was considered to be beyond them."[1]

If attitudes toward education were related to the background of indigenous societies, they were also very strongly shaped by the goals and expected outcomes of colonial rule. Here of course the perceptions of colonizers and colonized differed sharply. From the rulers' point of view, administering a colony and increasing its production of export commodities required new technologies and new forms of labor, hence a certain amount of basic technical education. Yet the colonial rulers also wanted to preserve the social status quo, which they rightly feared was threatened by contact with the West; as the Advisory Committee on Education in the Colonies explained in 1935: "Economic forces and the onrush of new ideas are tending to loosen social bonds and weaken traditional restraints and to encourage an unregulated individualism which is destructive of the best elements of communal life."[2] This apprehension led them to favor the teaching of crafts and agrarian skills over academic education.

The colonized, however, demanded an academic education rather than training in crafts. The conflict between the plans of the colonizers and the ambitions of the colonized was almost universal in colonial history. Edwin Atkinson and Tom Dawson, authors of a report on technical education in India in 1911, wrote: "The general disinclination for hard physical labour on the part of the average educated Indian is the chief cause of failure in the technical education of the India of to-day."[3] In Egypt, technical education suffered because "by and large, only students who could not qualify for admission into academic institutions were enrolled."[4] In the Gold Coast, the British thought Africans had a cultural bias toward academic education and against manual occupations.[5] And a report on education in the British Empire noted that "there is even in Jamaica a touch of the feeling that work is degrading and unbecoming a scholar, and industrial work has been hampered accordingly."[6] Such statements could be multiplied a hundredfold.

The causes of this bias varied. Philip Foster has argued that Gold Coast Africans preferred an academic to a technical education for a practical reason: "The financial rewards and the employment opportunities for technically trained individuals were never commensurate with opportunities in the clerical field."[7] Denise Bouche found the same to be true of Senegal:

> It is true that manual crafts, especially those of wood and iron, are almost everywhere in Africa reserved for inferior castes, and are looked down upon even more than elsewhere. But the colonisers did nothing, in practice, to raise them in the esteem of their subjects. To students graduating together from the Ecole Pinet-Laprade, the public services offered a salary of four francs for draftsmen and 2.5 francs for fitters or carpenters. Who, quite apart from any prejudice, would not have preferred the first job?[8]

In India and, to a certain extent, in other Asian and North African colonies, the situation was more complex. There too, European missionaries and government officials set up programs to teach rural and preindustrial crafts. These were much less popular than the schools that offered an academic education in a European language. Again, the Europeans blamed the indigenous cultures for rejecting manual skills. This explanation was perhaps more valid in India than in Africa; Foster admits that cultural inhibitions may have played a part due to "a much older tradition of Brahmanic intellectualism."[9] The trouble there, as in many other parts of the world, was that Western technology cut across the neat distinctions between high-prestige intellectual jobs and low-prestige manual ones.

Like art and music, technology requires hands-on experience as well as book learning. Laying hands on a locomotive, a smelter, or a power loom is manual labor as well as a learning experience. A good part of technical learning is physical, takes place out-of-doors or in dirty and unpleasant places, and requires the student to spend some time as an apprentice, temporarily doing tasks associated with the lower classes. But it also requires book learning and the attitudes of thought and behavior of a middle-class person.

Ever since the Benedictine monks went about their daily chores with the motto *laborare est orare*—to work is to pray—European civilization has blurred the social distinctions between manual and intellectual work. It is in part the willingness of some Europeans (though by no means all) to combine manual and intellectual work which has stimulated the advance of Western technology. Yet even in the industrial West, self-made men who have risen from the working class are rarities, overshadowed by the many middle-class persons who have done just enough manual work to learn a technical profession. The education of an engineer or technician thus simulates social mobility, putting middle-class youths temporarily into the lower class, then out again.

Atkinson and Dawson expressed the Western ideal of on-the-job training in these words: "Every technically trained student must be prepared to start on the lowest rung of the ladder, show his superiority by hard work and technical knowledge, and having made himself indispensable and a commercial asset to his employer, he will then rise by the natural laws of supply and demand."[10] Such a situation is most likely to exist in countries in which there is social mobility and where prevailing values do not inhibit the educated from doing some manual labor.[11] Colonial India was not such a country. Members of the higher castes readily took to English legal and literary education because it was similar to the book learning that was highly prized in traditional Indian society. But they were averse to manual labor, as an official report on education explained: "Individual *bhadralog* (high caste people) do not, in fact, wish their sons to be *mistris* (carpenters). Each thinks that the sons of others, not his own son, may be diverted from the competition for employment in the clerical and professional market."[12] And one engineer observed:

> Our trouble in India is that the practical side of industry is not at present considered an honourable calling by any but a fraction of the section of the Indian community who should be attracted to our large industries, and until there is more inclination on the part of the Indian student to "take his coat off," the advantages of technical education are bound to be to a great extent nullified.[13]

Yet even if North African and Asian cultures were biased against manual work, we should not accept this as a sufficient explanation. For the people of those colonies were also practical-minded, and they found not only a more congenial atmosphere, but also higher pay and better opportunities in office work than in manual work.

Where technical training programs led to high-status and well-paid careers, there was no shortage of candidates. At the college and postgraduate levels of technical education (a level which did not exist in Africa), the tables were turned. The colonized eagerly sought such an education, both for personal advancement and as an expression of national aspirations. It was the Europeans who were reluctant to admit Asians and North Africans into their engineering schools and scientific institutes, and into the corresponding careers. Here their arguments that non-Europeans had a cultural bias against technology was not an explanation but an excuse, or even a weapon. Hence one has to agree with Robert Crane when he states that the British view was "at best a half truth" and a cliché that "tended to stifle experiments in technical training."[14]

Like all half-truths, the cliché served a purpose. In European eyes, a lethargic non-Western society was more attractive than one in the throes of cultural modernization and social mobility. Social rigidity suited the personal circumstances of Europeans in the colonies. For whatever social mobility may have existed back home, in the colonies Europeans were divided into rigid classes. The class in which a man arrived was the one in which he remained during his stay. No one rose from sergeant to officer, or from subordinate to covenanted civil servant. In Indian eyes there was no social mobility among the Europeans, only another caste system. Though European rule was based on Western technological superiority, the social mobility and the hands-on learning that had helped create that technological advantage were not transferred to their colonies. This was a major obstacle to the diffusion of Western technological culture.

Technical Education in Egypt

Surprisingly, Western technology came to nineteenth-century Egypt by diffusion first, and by relocation later. Mohammed Ali, ruler from 1805 to 1849, tried to make Egypt militarily strong and independent of the Ottoman Empire by emulating the Western powers, especially France. This in turn required a cadre of technicians and engineers. In the 1820s and 1830s, he opened a series of schools of mineralogy, munitions, applied chemistry, signaling, irrigation, agriculture, engineering, and translation. Most of these were ephemeral and had difficulty recruiting students, as Egyptian parents rightly feared that their sons, if educated, would only end up serving in the army. He also imported French experts and teachers and sent students to France to study technical and military subjects; from 28 in 1826, their number grew to 321 in 1849.

Yet this whole educational effort, based on Mohammed Ali's military ambitions, was vulnerable to a shift in the fortunes of war. When Egypt was defeated by France and Britain in 1841, the army was reduced to a tiny force, and both the need for technical experts and the funds to train them evaporated. Only the engineering college, Polytechnic, seems to have been spared the drastic retrenchment that followed. The revival of education had to await the reign of Ismail (1863–79), who lavished money on this as on everything else he undertook. The money, however, was borrowed from European bankers, and the debt crisis that followed led directly to the British invasion of 1882.[15]

Great Britain ostensibly conquered Egypt to restore order and fiscal integrity. One immediate consequence of the invasion was a drastic cut in the expenditures for education. Parsimony was reinforced by ideology, for Lord Cromer, the British resident from 1883 to 1907, believed too much education would create unemployed graduates who would turn into nationalist agitators, as in India. The number of students had to be strictly limited; only as many as could be employed in government offices, or as lawyers, doctors, engineers, teachers, and policemen, would receive an education. For the masses, Cromer's goal was "the three r's in the vernacular, nothing more." To control enrollments, scholarships were cut and tuition increased. Where 70 percent of students had received stipends in 1881, 27 percent did in 1892. As a result, only the well-to-do could afford to educate their sons beyond elementary school. The quality of education also changed, for the director of Public Instruction, Douglas Dunlop, stressed rote learning and strict discipline even more than before.[16]

Under Cromer, technical education suffered even more than the humanities. A survey taken in 1900 showed that 792 students had graduated from technical schools since 1889, of whom 615 worked for the government. The school of agriculture, writes Robert Tignor, was "a woeful institution" which "attracted only a handful of students;" as for the school of engineering, it was "the least attractive since the Public Works Department paid only small salaries and reserved most of the higher administrative and technical positions for British officials."[17] One of these officials, the hydraulic engineer Sir

Colin Scott-Moncrieff, expressed his opinion of Egyptian engineers in these words:

> Above all, they [the British engineers] have been hindered by the absence of trustworthy native engineers. . . . The Mohammedan has not yet learned to look on engineering as a learned profession worthy of a gentleman. The result is that, with a few exceptions, the Government engineers are very ignorant and lazy and not very honest, while their inferior social position makes them too timid to hold their own against unscrupulous pashas and madirs. The five English engineers have been obliged, however, to accept of them such as they are, and more than one has responded loyally to the new calls made on his brains and energies. . . . More than one rascally engineer has been brought to account and punished, and the effect has been good on the others.[18]

Before World War I, the inadequacies of the educational system were largely compensated for by importing large numbers of Europeans, who found Egypt more to their liking than the more distant tropical colonies. Their presence did not, however, prevent either the rise of Egyptian nationalism or the criticisms of outside observers. As a British investigating subcommittee pointed out in 1920: "No true social, economic, or political progress can be looked for without a complete revision of the educational system of Egypt."[19]

During the 1920s, as Egyptians gradually took over the internal administration of their country, the education system expanded again. By the early thirties, four intermediate agricultural schools and twenty-two "industrial" (i.e. crafts) schools enrolled 6,000 students. At the college level, the Polytechnic had 673 students and the higher school of agriculture, 489. As before, educated Egyptians considered technical fields inferior to the humanities and the law, and they aspired to government jobs. Their contribution was to replace some of the Europeans, rather than to diversify the Egyptian economy, then trapped in an excessive dependence on cotton exports to a world market in the midst of the Depression. The modern private sector of the economy remained weak and dominated by Europeans, as Amir Boktor explains:

> Technical schools are graduating a number of students annually, but the lack of factories and private enterprises makes it difficult for these graduates to earn a living. . . . There is also reason to believe that the technical schools (as well as secondary schools) do not take into account the needs of the country. . . . In the big cities, strange

to say, the majority of the people employed in repairing, oiling, and all kinds of work pertaining to motor-cars, even selling gasoline, are Europeans. . . . Likewise, electric, water, and gas companies supplying Cairo, Alexandria, and other cities with water, gas, and electricity employ Europeans. Tramway companies in Cairo and Alexandria are owned by foreigners; the motormen and the conductors, however, are Egyptians.[20]

The technological culture of the West came to Egypt in three waves. The first was a wave of cultural diffusion from France, started by Mohammed Ali to make his country militarily strong; it was slow and costly, and had only mediocre results. The second wave began in 1882. The British, in a hurry to modernize Egypt for their own benefit and that of the world market, abandoned the slow cultural diffusion of Western culture for a faster, more efficient import of European machines and experts. In the process, they interrupted the diffusion process for forty years. The third phase, a mixture of diffusion and relocation, began in the interwar period but only bore fruit after Independence.

[margin note: 3 waves of technological culture]

Vocational Education in West Africa

One celebrated complaint against European colonialists was that they educated only a pitifully small number of Africans beyond the secondary level, especially in technical fields. The first technical college in the Gold Coast opened in 1951. South of the Sahara, only one college was open to Africans before World War II, and that was in South Africa. In Portuguese Angola and Mozambique, the most backward of the colonies, 86 Africans attended secondary-level technical schools in the mid-1950s, and 2 had become engineers by 1961.[21] The French and British did better, but not by much: at Independence the British colonies in tropical Africa had only 150 university graduates in agronomy, and the French colonies only 4.[22]

The reason was the level of technological and economic development which the Europeans found when they conquered Africa and their fear of social upheaval if they introduced changes too fast. Even the West African colonies of France and Britain, which formed the most developed and commercialized region of sub-Saharan Africa and the one with the longest contact with Europe, were politically and economically much less developed than Egypt or India, and the question of education revolved around the training of clerks and craftsmen, rather than of engineers and agronomists.

The history of vocational education in French Africa dates back to the 1830s. When the first steamers reached Senegal, the colonial government found that European ship mechanics were prone to quit or die, and they decided to train Africans. Between 1838 and 1845, 11 Senegalese were sent to France to become apprentice mechanics, but most of them were mistreated, got into trouble, and were sent home. In the 1840s 4 more Senegalese studied in France, and another 22 between 1856 and 1866; they too were mistreated and did not benefit from their stay.

The idea naturally arose to train craftsmen in Senegal itself. In the 1850s, the Fathers of the Holy Spirit opened a workshop and taught 15 students various nonindustrial crafts. With the start of railroad construction in 1878, 8 Senegalese were trained in Saint-Louis, and 7 others were sent to France; like their predecessors, they ran into health and discipline problems. In 1886 the French navy began accepting a few Senegalese as apprentices but ended the program in 1897 when it had enough skilled workers. In the nineteenth century, the tiny demand for skilled workers in the modern sector did not warrant opening a school, and sending Africans to France failed for lack of proper supervision. The colony, like others in French Africa, fell back on imported personnel.[23]

In the first years of the twentieth century, as Dakar became a major port, the government founded two technical schools. The first one, the Ecole professionnelle Pinet-Laprade, started in 1903 without funds, building, or equipment. Its 10 students were barely literate and had to be given a belated elementary education and taught some simple handicrafts. Ten years later, it had gotten over its troubles, but its graduates, now literate in French, preferred office work to the jobs they were trained for.

In 1907 the colony's main employer, the French navy, founded the Ecole des pupilles-mecaniciens de la Marine to train ship mechanics. The students were required to spend three years at the school under military discipline, and two in the navy. Few Senegalese were willing to join under these conditions, and five years later, the school still only had 10 students.[24]

World War I gave an impetus to technical education in Senegal. By 1918 the Ecole Pinet-Laprade had 45 applicants for 43 places, and the navy mechanics' school received 29 applicants for 15 places.

A typographers' course, begun in 1914, got 65 applications for 20 places. By 1920 the three technical programs in the colony had a total of 150 students.[25] They attracted not only Senegalese, but also students from other French West African colonies. As a result, the other colonies did not initiate technical education programs until much later.

Gabon, a French outpost since the 1840s, had no government school until 1907 but relied on the Holy Ghost Fathers to train carpenters, woodworkers, masons, blacksmiths, farmers, and gardeners, while the Immaculate Conception Sisters of Castres taught girls to launder, iron, sew, cook, embroider, and garden. For other tasks the administration imported workmen from West Africa. After 1910, when Gabon, Congo, Ubangi-Shari, and Chad were joined to form French Equatorial Africa, the postal and telegraph service, and later the railroad, trained their employees on the job. Not until the 1930s was there a school to train postal clerks, radio operators, and public health workers.[26]

British West Africa followed a different path to a similar goal. Until World War I, education was left to missionaries who were more interested in spiritual than economic development. The skills and crafts they taught were those needed to build and sustain their mission stations.

Eastern Nigeria developed its educational system at about the same time as Senegal. Starting in 1846, small crafts programs trained nonacademic students in the skills needed by the Scottish Presbyterian Mission of Calabar. In the 1890s, missionaries founded several schools to train masons, boat makers, coopers, carpenters, and blacksmiths, and they taught girls domestic science and dressmaking. The railways, the Posts and Telegraphs, and the Public Works Department had apprentice programs for their lower-level employees. Much the same was true in Sierra Leone, where the first vocational and technical schools were opened by missionaries in 1911, offering the usual crafts and nonindustrial skills.[27]

In 1919 the Phelps-Stokes Fund of New York undertook to study education in Africa. The report of its investigation, published in 1922, aroused official interest in the question on the part of the Colonial Office.[28] As a result, in 1925 the Advisory Committee on Native Education in the British Tropical African Dependencies issued a memorandum on "Education Policy in British Tropical Africa."[29] It was a conservative document:

> Education should be adapted to the mentality, aptitudes, occupations and traditions of the various peoples, conserving as far as possible all sound and healthy elements in the fabric of their social life; adapting them where necessary to changed circumstances and progressive ideas, as an agent of natural growth and evolution.

It recommended that Africans be taught vocational subjects:

> It should be the aim of the educational system to instil into pupils the view that vocational (especially the industrial and manual) careers are no less honourable than the clerical, and of Governments to make them at least as attractive—and thus to counteract the tendency to look down on manual labour.

The goal of these policies was to discourage Africans from flocking to the cities and joining the hordes of office-seekers. These ideas were not new, for they can be traced back to the 1840s and have often been repeated since. Though popular with colonial officials, they aroused opposition from many Africans, who resented the closing of opportunities in better-paid clerical work.

The Gold Coast government was the most enthusiastic about these policies. In 1922 it set up trade schools for road foremen, carpenters, locomotive drivers, masons, postmasters, and others. Nigeria followed in 1932 with the Yaba Higher College, a secondary-level technical school. Like the schools of Dakar, these institutions did prepare Africans to do the middle-level work of the modern sector. But as this work was not well paid, they had no effect on the influx of Africans into cities or on the attractiveness of academic education and clerical jobs.[30]

Technical Education in India: Demand and Supply

From the point of view of technical education, India is the most interesting of colonies. Colonized earlier than other territories, India went through its colonial evolution sooner. Higher education was established, industries were created, and nationalism appeared. Despite its vast population of illiterates, India probably had more educated people than all the other colonies put together. In 1917–18 Bengal alone had as many people as the United Kingdom (about 45 million) and as many students preparing for university degrees (about 26,000). In education, India was thus decades ahead of other colonial areas.[31]

Furthermore, education was a major political issue, both for the British civil servants who had achieved their positions through education and for many Indians who saw education as a means of personal advancement and national liberation. One result of this tug-of-war between rival intelligentsias was an outpouring of statistics, reports, proclamations, books, and pamphlets of all sorts. To this day, education under British rule is a significant topic for historians of India.[32]

Finally, the case of India shows more clearly than any other the dual nature of technical education: as a response to a demand for technically trained people and as a means of developing the economy in the future. In a colonial setting, education responded to economic and political pressures from both the colonizers and the colonized. These pressures were not cumulative but often contradictory.

In the nineteenth century, the main employer of technically trained men was the Public Works Department. Until 1852 public works projects were undertaken by army officers who learned their trade on the job, among whom were the great hydraulic engineers Proby Cautley, Arthur Cotton, Richard Baird Smith, and Colin Scott-Moncrieff. After 1852 the newly established Public Works Department began to recruit civilian engineers. As a government agency, it modeled itself on other branches of the bureaucracy, which were divided into two strata: in the upper echelons were the "covenanted" civil servants hired in Britain under a covenant or contract; the rest, or "uncovenanted," included Indians, Eurasians, Europeans domiciled in India, and less educated men recruited in Britain. This class distinction was reinforced by a large gap in salaries and benefits: covenanted civil servants could retire after twenty-one years with a pension of £1,000 a year, while the uncovenanted could hope, at best, for a pension of £340 a year after thirty years. Nor was this the only distinction, for among the uncovenanted, the Europeans fared better than the Eurasians, and the Eurasians better than the Indians. One uncovenanted engineer showed both indignation and hypocrisy when he wrote:

> Such a gross anomaly . . . is neither more nor less than an imitation of caste privilege which is peculiar to India. It is not English. It is not just. . . . When applied to natives of India, whose social position and social needs are, as compared with the European, of

small importance, these rules cannot be deemed illiberal; but as applied to men who desire to pass their old age in their own country, and to continue in the position of gentlemen, they are so cruelly inapplicable as to render them practically inoperative.[33]

To maintain this social hierarchy, the Public Works Department spawned two educational systems, one in India and the other in England. The first engineering college was an outgrowth of the Ganges Canal. Named after the lieutenant governor of the North West Provinces who founded it in 1847, the Thomason Engineering College at Roorkee trained employees for the irrigation branch of the Public Works Department. It offered different curricula for different types of students: an engineering class for domiciled Europeans and a few Indians, an upper subordinates class to train British noncommissioned officers as construction foremen, and a lower subordinates class to train Indian surveyors. By the mid-1880s, the school had a hundred students, substantial buildings, and a reputation as an important center for the study of hydraulic engineering.

Soon after schools were set up in other parts of India to cater to the needs of the Public Works Department. In 1856 the Presidency College of Calcutta started a department of civil engineering, which became the Sibpur College of Engineering in 1880. It offered a more theoretical curriculum than Thomason. Its principal, S. F. Downing, observed:

> I have learnt from conversation with respectable educated natives that the fact of the department belonging to the Presidency College gives it a certain status in the eyes of native society; consequently a superior class is attracted to it than would be the case were a school attached to large Government workshops in which the students would have to work daily, such manual labour being, unfortunately, considered derogatory by upper class Bengalis. This appears to me to be an important point, because native Assistant Engineers, Public Works Department, have to associate officially with English gentlemen, and consequently the former ought, if possible, to be recruited from the upper middle class community.[34]

In 1859 some engineering classes offered at Poona, near Bombay, formally became the Poona College of Engineering. And in 1862 a surveyors' school in Madras, dating back to 1794, became the Madras Civil Engineering College. Like the Thomason College, they offered both civil engineering courses for a university-level degree and secondary-level training for lower subordinates, surveyors, and draftsmen. They had almost no European students, lower standards, and a much lower reputation than Thomason.

Immediately following the Rebellion of 1857–58, the Public Works Department expanded very fast to meet the demand for canals and government buildings. Its budget rose from £4 million in 1860–61 to £7.5 million in 1871, and its staff of engineers increased even faster, from 113 in 1840 to 545 in 1863, and to 896 in 1870. The increase consisted mainly of cilivians, whose numbers rose from 5 in 1850 to 533 in 1870.[35]

The department did not expect the four Indian engineering colleges to meet this sudden demand, nor did it want them to. By tradition, their graduates could not rise to the higher ranks of the department, reserved for the covenanted. After the Rebellion, Indians were regarded with suspicion, and the graduates of the four colleges were poorly motivated. As Col. George Chesney of the Royal Engineers complained in 1870:

> The Government guarantees eight appointments yearly to qualified students of the Roorkee College, the native members of which receive, in addition to a gratuitous education, a scholarship or stipend sufficient for all expenses. But a large proportion of the available scholarships have lapsed from not being sought for, and the taste for civil engineering is likely to be of slow growth among the people of India. The qualified students of the Calcutta Civil Engineering College . . . have, I believe, all obtained appointments on completing their course of study, but the class of Bengalee youths which frequents the College is not apt at engineering, and can take the place of European engineers but very gradually. The out-turn from Poona and Madras has hitherto been scarcely appreciable, but here, as elsewhere, the degree of facility afforded by Government to its native subjects has been in advance of the desire manifested by the latter to avail themselves of it.[36]

To fill the gap, the India Office recruited young engineers by open competition in Britain and sent them on to Roorkee for further training. They were called "Stanley engineers" after Secretary of State Lord Stanley, who devised this scheme. These men also proved disappointing, for they were poorly educated, in Colonel Chesney's opinion:

> The present mode of training an engineer, where a young man pays a fee to a civil or mechanical engineer for permission to work in an office or workshop, and pick up such crumbs of knowledge as fall in his way, is not education, and the cases must be very rare where

persons, after undergoing such a training, will be found able to pass an examination involving any knowledge of the principles of mathematics or theoretical mechanics. They usually take no knowledge of that subject into the office, and gain none there. . . .

Thus the result of the present system is, that we are not getting engineers, and that the qualifications of the persons who do enter the service in this way may be of the slenderest kind.[37]

Not only were the "Stanley engineers" poorly educated, they were also lower-class:

The class of men asked for are not such as is expedient to introduce in large numbers into India, where, in the intercourse between Europeans and Natives, more consideration for the feelings and prejudices of the latter is desirable than such of the former as have not had their manners softened by education are much in the habit of showing.[38]

In other words, a conflict had arisen between the needs of the Indian economy and that delicate balance of social classes that was the Raj. And at the heart of that conflict was the peculiar nature of modern technology which requires an engineer to be, at one and the same time, a gentleman and a worker.

Given these multiple dissatisfactions, the Public Works Department had one last option; in Chesney's words, "It seemed plain that the only course open to the Government was to revert to its original intention, and to take the preparation of the candidates under its own supervision."[39]

Chesney's lobbying paid off. In 1870 the Duke of Argyll, secretary of state for India, appointed him to found the Royal Indian Engineering College at Cooper's Hill near London. It opened in August 1872 with 42 students. Applicants had to be between seventeen and twenty years of age, "of sound constitution and of good moral character," and able to pass a test in mathematics; natural science; Latin; Greek; French; German; the works of Shakespeare, Milton, Johnson, Scott, and Byron; and English history from 1688 to 1756; in other words, they had to be young gentlemen. Furthermore, candidates had to be British, though the president of the college was authorized to admit 2 "natives of India" each year "if there is room." Toward the end of the century, Cooper's Hill also admitted 4 or 5 Siamese students each year and a small number of Egyptians.[40]

By all accounts, the students received a fine education. Of the

1,623 men who graduated from the college between 1872 and 1903, when it closed, 1,010 served in India, and of these, 764 served in Public Works. The rest went instead into the Indian Telegraph and Forestry departments, the Royal Engineers and Royal Artillery, the Admiralty, the Egyptian government service, and the Uganda Railway, among others.[41]

By the late 1880s the Public Works Department had achieved a stable system of recruitment. Each year it recruited approximately 15 graduates of Cooper's Hill and 6 junior officers from the Royal Engineers; these men formed the Imperial Service of Engineers. In addition, 9 to 10 graduates of the Indian engineering colleges were hired as engineers in the Provincial Engineering Services. Altogether, in 1886, the Public Works Department had 1,015 engineers, of whom 86 were Indians, 119 were Eurasians and domiciled Europeans, and 810 were Europeans recruited in Europe. Graduates of the Indian engineering colleges, other than the fortunate 9 a year, ended up as upper subordinates (i.e. overseers, surveyors, and supervisors) at a salary of 60 rupees a month, roughly one-ninth the average salary of the engineers; a few others found work in municipal government or in the Native States. Upward mobility between the strata was extremely rare.[42]

In India as elsewhere, civil engineering became professionalized in the late nineteenth century. The Indian experience differed from others, however, because professionalization was accompanied by the conquest of the upper echelons by the British middle class and the exclusion of Indians, Eurasians, and domiciled and working-class Europeans.

What was true of the Public Works Department was also true of other branches of government which employed technical and scientific personnel. The Geological Survey of India is a case in point. In 1858 the governing body of Calcutta University had opposed the introduction of geology into the curriculum. Thirty years later the GSI had seventeen covenanted members, one of whom, P. N. Bose, was a British-educated Indian who had been given a place in the GSI to get him out of England. In 1886 the GSI had also promised posts of deputy superintendent to two Hindus. Because geology was not taught in India, the Public Service Commission could declare itself "satisfied that the Government of India has done all that it would be justified in doing to secure the employment of Natives of India in this Department."[43]

The Telegraph Department went through a different evolution.

Unlike the Post Office, which grew out of an existing bureaucracy, the telegraph system was created by William O'Shaughnessy with the support of Governor-General Lord Dalhousie. O'Shaughnessy, who was in a hurry, could not recruit telegraphers and installers in Britain, so he had to train his own. Some of them were British soldiers, but most were Indians. His former assistant at the Calcutta Mint, Seebchunder Nandy, followed him into the telegraph service, where he supervised the laying of several important lines and became inspector of the line, second in command after O'Shaughnessy himself. This met with the approval of the East India Company's Court of Directors, who wrote the governor-general in 1856: "We are desirous that continued efforts should be made to qualify Natives to undertake the duties which in so many instances have to be performed by European agencies."[44]

For many years after the Rebellion of 1857, no more Indians were admitted into the higher ranks of the department, though many were recruited into the lower ranks. In 1887 the higher ranks, or Superior Establishment, consisted of 97 "gazetted" (i.e., covenanted) civil servants, of whom 2 were Eurasians and the rest Europeans recruited in Britain. The second tier, called the Signalling Branch, had 1,286 employees, of whom 889 were Eurasians, 147 were domiciled Europeans, and 250 were Indians. Thus the Telegraph Department, which began by recruiting whoever was available, soon emulated the quasi-caste system of the Indian bureaucracy. Not until 1896 did another Indian, Ganen Roy, become an officer in the Telegraph Department; he was one of the rare Indian graduates of Cooper's Hill, and in 1925 he became the first Indian director general of Posts and Telegraphs.[45]

The Imperial Forest Service also recruited its officers in Europe. At first, forestry students had to study on the Continent because of a lack of facilities in the United Kingdom. After 1885 they were trained at Cooper's Hill, with periodic visits to the Ecole Forestière at Nancy in France.[46] The service also admitted other candidates who had "obtained a degree with honours in some branch of natural science in a University of England, Wales or Ireland, or the B.Sc. degree in pure science in one of the Universities of Scotland." This made it difficult for Indians to qualify. As of 1912, of the 407 officers in the upper levels of the GSI and the Agricultural, Civil Veterinary, Forestry, and Railway departments, 6 were Indians.

The Indian government kept a careful record of the number of railway employees, divided by race. Table 9.1 summarizes these

Table 9.1   Railway Employees by Race, 1860–1910

Year	Europeans		Eurasians		Indians		Total
	Number	%	Number	%	Number	%	
1860	1,137	(6.1)	n.a.		17,652	(93.9)	18,789
1870	5,048	(7.3)	n.a.		64,185	(92.7)	69,233
1880	3,749	(2.4)	3,569	(2.3)	146,790	(95.3)	154,108
1890	4,494	(1.8)	5,366	(2.1)	244,658	(96.1)	254,518
1900	5,181	(1.5)	6,815	(2.0)	326,045	(96.5)	338,041
1910	7,207	(1.4)	8,862	(1.7)	502,284	(96.9)	518,353

statistics up to 1910.[47] The figures show several developments: an enormous increase in the number of railway employees; a slower increase in the number of Eurasians; and a fluctuating number but a declining proportion of Europeans. What the figures do not show is the positions the different racial groups occupied: Europeans held the managerial and higher technical jobs, Eurasians the midlevel skilled and supervisory positions, and Indians were on the bottom, in the unskilled jobs.

The policy of employing Europeans was costly. European locomotive drivers in India were paid three or four times the wages of drivers in England, or about ten times as much as their Indian counterparts. European supervisors and skilled workers earned roughly twice as much as back home and, in addition, received free passage, medical care, a family allowance, and sometimes housing as well. This caused concern in government circles because it weighed upon the treasury and created balance of payments problems for India. Already in 1878–79 a Select Committee of Parliament recommended reducing the number of European railway employees as a means of lowering the cost of state railway maintenance. Yet the railways were slow to Indianize. At the turn of the century, some began using Indian locomotive drivers for freight trains and shunting work. Express trains, however, were still commonly driven by Europeans until the 1930s.[48]

The persistence of the ethnic division of labor had two causes. One was the guarantee system, which put no pressure on companies to cease recruiting in Britain. The other was the belief that Indians would or could not learn to do technical work as well as Europeans. For many years it was debated whether Indians could be trained for skilled work. In 1874 the government announced:

Natives who can manage a steam-engine have, for many years, been found in the Presidency towns. There are even now a few natives capable of driving a locomotive engine, though many Engineers of experience express doubts whether the nerve and readiness of mechanical resource required to make a good driver are likely to be found largely amongst the natives of India.[49]

In a similar vein, the secretary of the Railway Department of the India Office, Juland Danvers, wrote in 1877 that Indians

are apt learners in mechanical art, and have been trained to very useful work in the locomotive shops. It will take time to qualify them for the more arduous duties of locomotive drivers, which require coolness, courage, and decision, but some have already shown themselves to be equal to such employment.[50]

To maintain their equipment, the railways built workshops at major rail intersections. At Parel near Bombay, the workshops of both the Great Indian Peninsula and the Bombay, Baroda, and Central India railways gave rise to a large industrial suburb. Similarly, large workshops at Lahore in Punjab, Jamalpur in Bihar, and Karagpur and Kanchrapara in Bengal attracted sizeable communities of workers. They employed an average of 1,300 workers apiece, while the largest, Jamalpur, employed 11,000.[51]

The workshops were centers for the diffusion of technical culture. In 1875 three railways—the East Indian, the Oudh and Rohilcund, and the Scinde, Punjab, and Delhi—operated apprenticeship programs to train European and Eurasian boys to become foremen, fitters, engine erectors, and locomotive drivers. Of the 61 students enrolled in EIR's apprenticeship program at Jamalpur in 1900, 40 were Europeans and Eurasians who were housed in a company hostel, while the 21 Indians in the program had to fend for themselves in town.[52]

Racial discrimination in railway apprenticeship programs remained constant almost to the end of the colonial period. During 1916 and 1917 the Indian Industrial Commission visited many railway workshops. Some of the testimony they gathered illustrates the situation. The locomotive superintendent of the GIP in Bombay explained the difference in the salaries of Indian and European trainees as follows:

Well, of course that is a question of market value. They may pass the same examination, but whether you get the same work out of them in the twelve months as the European is doubtful. . . . There are different market values for both classes, the one can live on R.35 a month, while the other cannot. Don't you think that this should be taken into consideration?[53]

Similarly, a British locomotive superintendent wrote to the chief mechanical engineer of the Bengal-Nagpur Railway:

> My experience of Bengalis has certainly been unsatisfactory. I have had two in the shops, and both have now left after a comparatively short period. They seem very unsettled, very unsatisfactory from the point of view of timekeeping, and of not very marked ability; and I am afraid that this will be my last experiment in accepting individuals of that nationality as apprentices.[54]

After visiting many workshops and interviewing dozens of experts, the commissioners concluded:

> We were forcibly struck, when visiting the large railway and private workshops throughout India, with the almost complete absence of Indians from the ranks of foremen and chargemen—the noncommissioned officers of the great army of engineering artisans. At present these posts are filled almost entirely by men imported from abroad. The railway companies are endeavouring to supply this deficiency by training European and Anglo-Indian youths, the sons of their own employees as a rule, and with fair prospects of success.[55]

What was true at the level of foremen and technicians was even more true for engineers. The railway companies did not hire graduates of the four engineering colleges, and very few of them found work on the state railways either. Of the 117 members of the Institution of Locomotive Engineers working in India in 1926, only 5 were Indian.[56]

Industries which were owned and operated by Europeans discriminated as much as the railways. The jute industry, owned and managed by Scots, employed almost no Indians "in positions of trust in mills in Calcutta, where chiefly Europeans were employed. These men were very clannish, and it would be difficult for an Indian to retain his position, if appointed."[57] Even though this industry could have cut costs by employing Indian technicians, its stockholders evidently preferred to sacrifice the extra profit for the feeling of security or racial solidarity they felt with European technicians.[58]

In Madras, the Chamber of Commerce wrote in 1904:

> The Chamber fears that the difficulty you mention of attracting suitable native candidates for them will prove insurmountable. The

reasons are that practically all manufacturing industries in India are at present run by Europeans, and they, when requiring men with expert knowledge for responsible posts (such as an ex-scholarship holder would naturally aspire to), would almost certainly prefer to employ a European, whose capacity and general reliability they could better form an opinion of.[59]

This attitude is borne out by the employment figures. Of the 107 managers, engineers, and chemists in Madras industries who earned more than 100 rupees a month in 1908, 66 were Europeans, 23 Eurasians, and 18 Indians.[60]

The Politics of Technical Education in India

A constant in the history of technical education in British India was the contrast between the government's oft-repeated policy of educating Indians in Western science and technology and its hesitation in carrying it out. The policy dated back to Thomas Macaulay's *Minute on Education* of 1835 and to Governor-General William Bentinck's order that "the object of the British Government should be the promotion of English literature and science."[61] The East India Company's *Despatch on Education,* drawn up by John Stuart Mill in 1854, emphasized mass vocational education:

> Our attention should now be directed to a consideration . . . namely how useful and practical knowledge suited to every station in life may be best conveyed to the great mass of the people who are utterly incapable of obtaining any education worthy of the name by their own unaided efforts; and we desire to see the active measures of Government more especially directed for the future to this object, for the attainment of which we are ready to sanction a considerable increase of expenditure.[62]

Yet when Britain seriously undertook to modernize the infrastructures of India, expenditures on education were limited, and Indians were kept in subordinate positions. The situation did not pass unnoticed. Even before Indian nationalists became aroused, the India Office showed concern, perhaps because covenanted engineers were so costly to support and pension off, and Cooper's Hill ran a deficit of £15,000 a year which was charged to the Indian treasury. In 1879, Secretary of State for India Viscount Cranbrook wrote Viceroy the Earl of Lytton:

Your Excellency's letter of the 11th November last has acquainted me with the steps taken by your Government to further the establishment of schools in India for the training of Natives for the position of foremen mechanics in the Public Works Department. . . . I see that your letter alludes more especially to the promise of success up to 1873 in regard to European and Eurasian lads. I do not gather how far the success has since extended also to the training of Native lads; but this is manifestly a matter of the highest consequence, and I shall be glad to have some recent information on the subject.[63]

A year later, Cranbrook brought up the matter once again:

In the last paragraph of the letter under reply your Government express a fear "that so long as Coopers Hill is maintained the extended employment of Natives of India in the superior grades of the Public Works Department will be practically impossible." This apprehension is based apparently on the assumption that those superior grades are desired by Her Majesty's Government to be "almost exclusively recruited from Coopers Hill," an assumption obviously incompatible with my recently expressed desire that "all reasonable facilities for entering on a career as Civil Engineers in the service of Government should be offered to Natives," and with numerous other passages to the like effect in Despatches of my predecessors in office as well as myself. A far more serious bar to the employment of Native Engineers is the fact that, as I lately observed, "the operation of Thomason College has been to add to the strength of the Europeans in the Department, rather than to increase the proportion of Native members." It is plain that this tendency should be at once arrested, and I must accordingly request that, in so far as may be consistent with pledges already given, no engineering appointments be henceforth guaranteed to any but Natives, at either the Thomason or any other Indian College; and also that no Europeans, other than Royal Engineer officers, be guaranteed such appointments without the previous sanction of the Secretary of State.[64]

This adamant view, so contrary to the traditions of British rule in India, was bound to meet resistance. Four years later a new secretary of state, Lord Kimberley, proved himself much more pliable on the issue; as he wrote Lord Ripon in July 1884:

Your Excellency's Public Works Letter, no. 14, dated 21st April last, respecting the reservation of the appointments made from the Indian Colleges to the Engineer Establishment of the Public Works Department, for Natives of pure Asiatic origin, has received my careful consideration in Council. . . . for a time, at any rate, it would be desirable, in order to secure the admission of Natives of pure Asiatic origin, to the higher branch of your Public Works Department, that one-half of the appointments made to the Department from the Colleges in India should be reserved for them. . . . but if, as your Excellency feels assured, this object can be gained by open and unrestricted competition, I am not disposed to insist on the maintenance of a distinction which may cause certain disadvantages, both to individuals and to the Service.[65]

Under Kimberley's new policy of tolerance for discrimination, the burden of change was deftly shifted to the private sector. A government resolution dated October 23, 1884 declared:

Every variety of study should be encouraged, which may serve to direct the attention of Native youths to industrial and commercial pursuits. . . . Efforts should be made to call forth private liberality in the endowment of scholarships not only in Arts colleges, but for the encouragement of Technical Education.[66]

The concern of secretaries of state for the admission of Indians into the upper echelons of the Public Works Department, as in other branches of the civil service, was a distant one. In India itself, the issue was taken up in 1886–87 by the Aitcheson Commission, which recommended that the civil service be broken up into an imperial and several provincial services, and that the provincial ones be opened to Indians. The British bureaucratic elite remained in possession of the higher positions in the Indian Civil Service.[67]

Technical education became a topic of public discussion in the 1880s. In Britain, the Royal Commission on Technical Education reported in 1884 that insufficient technical education was to blame for Britain's industrial decline vis-à-vis its Continental rivals. This report caught the eye of Sir Mountstuart Grant Duff, governor of Madras, who asked his director of public instruction, Mr. Grigg, to submit proposals to improve scientific and technical education in the presidency. Grigg, an educator of the old school, believed that "to institute public examinations in any suitable branches of knowledge is to create a demand for instruction in them." Examinations were therefore developed in such subjects as commerce, drawing, civil engineering, agriculture, and sanitary science, and thousands of students flocked to take them. The results were disappointing, as a later director of public instruction, A. G. Cardew, explained:

The attempt . . . to create examinations without first providing qualified teachers and adequately equipped training institutions was

to reserve the true order of progress. . . . scholarships too often
fell to youths who had not the least intention of following the industry as a livelihood and who merely drew the scholarship while it
lasted and then betook themselves to the role of clerk, peon, police-
man or whatever the customary occupation of their class might be.
It is thus not surprising that though a large number of persons have
passed the Technical examinations which Mr. Grigg established, the
general effect of the scheme on the industrial progress of the country
has been slight.[68]

In 1888 the Indian government asked local governments to survey
their industries and set up industrial schools. But local officials, for
the most part classically trained English educators with a literary
degree from Oxford and little experience in technical matters, were
not motivated to improve the situation. There the matter would have
stayed for many more years, had it not been for the rising tide of
Indian nationalism.

Indian nationalists took up the question of technical education
because they thought it was a precondition for economic development. Among the first to raise this issue was P. N. Bose in his 1886
pamphlet entitled *Technical and Scientific Education in Bengal*. In
it he decried the lack of training for industries such as dyeing, tanning, mining, soap and glass manufacture, sugar milling, and electrical engineering: "The Calcutta University is primarily responsible
for this highly unsatisfactory state of things. It takes cognizance of
theoretical knowledge only, ignoring most lamentably the principle
now universally recognized that practical tests should form the dis-
tinctive feature of Science Examinations."[69] He went on to advocate
a more experimental education, apprenticeships in Europe as well
as India, local purchases by government agencies, and the founding
of a Central Science and Technological Institute. He did not, however, simply assume graduates would find jobs awaiting them:

The work of Government will practically cease with training up the
men. The further work of starting factories, or of working mines
should be undertaken by us. With a large variety of raw materials in
abundance, and scientific men to properly utilize them, and with
cheap labour, there are good many industries which with judicious
management are bound to yield an adequate return. It will be the
duty of the practical technologists to point out the openings for
profitable investments, and capital even in such a poor country will
be forthcoming. One or two successful enterprises will lead to
others.[70]

Bose was by no means the only one to propose technical education as a means of industrializing India, for such ideas were in the air.[71] In 1887 the Indian National Congress, meeting at Madras, passed a resolution "that having regard for poverty of the people, it is desirable that the government be moved to elaborate a system of technical education." Similar resolutions were repeated at Congress meetings in 1892, 1898, 1900, and subsequently. Nationalist newspapers also took up the theme and demanded that the government spend more on technical education.[72] From then on, the insufficiency of technical education in India was a frequent theme in both nationalist publications and government reports.

The turn of the century saw education suddenly politicized in the struggle between British authority and an awakening nationalism, with technical education as one of the themes. The opening moves were made by George Curzon, viceroy from 1899 to 1905, a dynamic and authoritarian modernizer in the mold of Lord Dalhousie. The hornets' nest he stirred up, like the Rebellion that broke out after Dalhousie's reign, gave conservatives arguments aplenty to prove that reform and colonial rule were incompatible.

Though Curzon was given to sonorous and dogmatic pronouncements, on the question of technical education for India he was decidedly confused. On the one hand, he was upset that after so many "platitudes in viceregal and gubernatorial speeches," so little had been accomplished since 1880. On the other hand, he was convinced that Indian demands for technical education were "native clamourings for things about which they know nothing" and that "this rage for so-called technical education in India is merely one more aspect of the craze for posts, for the finding of billets, for young men of the educated classes, who, if they fail, as nine out of ten will fail, will only add to the discontented hordes."[73] Like many other imperialists, Curzon believed in helping India through education but distrusted educated Indians.

It was characteristic of his way of thinking that when he decided to do something about Indian education, he did it in secret and by command. In 1901 he summoned the provincial directors of education and the vice-chancellors of the universities to an educational conference at Simla; only one participant was not a government official, and no Indian was invited. Curzon presided over the entire two-week conference and personally drafted all 150 resolutions, which were passed without dissent. He attacked the system of university examinations and rote learning by which the Indian intelligentsia gained access to government posts and hoped to gain increasing influence in their country's destiny. He condemned it for being too literary and too weak in experimental science and technology: "If technical education is to open a real field to the youth of India, it is obvious that it must be conducted on much more business-like principles." He also opposed the demands for more engineering and technological institutes, when the greater need was mass education: "To start with Polytechnics, and so on, is like presenting a naked man with a top-hat when what he wants is a pair of trousers!"[74] The "Resolutions of the Simla Conference (1901) on Technical Education" found that existing industrial schools had little impact on education or on the economy, but it warned that schools should not engage in commercial enterprise under the guise of practical training.[75]

For all the resolutions that came out of the Simla conference, the results were disappointing. Their major result seems to have been the appointment of a Committee on Industrial Education, chaired by J. Clibborn. After much deliberation this committee recommended that industrial schools be replaced with a system of apprenticeship "organized upon the model of the Casanova boy artisan school [of Naples]. This institution aims at giving the boys belonging to the poorer classes of a notoriously vicious population such mental, moral, and manual training as will turn them into good citizens, honest men, and skilful artisans."[76]

The following year the central government asked the provincial governments for suggestions on how they would use grants for technical education. Most of the replies were negative. Assam and Madras wanted no grants; the United Provinces said technical schools were "unnecessary" and handicraft schools "would serve no useful purpose"; only Bombay and Bengal were interested. The responses of the provincial officials convinced Curzon that there was no demand for technical education; as usual, he considered Indian public opinion irrelevant.[77]

Finally, in January 1904, the government issued a resolution which rejected the Clibborn committee's preposterous idea of equating technical education with reform schools for the children of criminals. It declared that "the matter has not yet passed the stage at which many experiments must be tried and a proportion of failures must be expected"; and it recommended that the provinces establish technical schools in industrial centers and crafts schools in the lesser towns.[78] In effect, the government had no clearer ideas on technical education than before Curzon assembled the educational administrators at Simla.

Perhaps the most promising official initiative in the area of technical education occurred in Madras, the scene of Mr. Grigg's failure. In 1898 the presidency government appointed the principal of the School of Art, Alfred Chatterton, head of the Provincial Department of Industries. For a British official, Chatterton held some rather unusual views:

> Government have definitely accepted the principle that industrial development must precede technical education. . . . At the outset European experts should be freely employed, but it should be recognized that they will only be required for a few years, and as a rule therefore they should be got out on terminable agreements which should not be renewed. . . . it seems desirable that Government should establish small factories worked on a commercial basis to demonstrate that they can be run on a profit.

Putting these ideas into practice, Chatterton started small factories for handloom weaving, chrome tanning, and the manufacture of aluminum ware. At first other colonials regarded the experiment with a mixture of admiration and skepticism. John Hewett, then secretary to the Home Department, wrote in 1901:

> Mr. Chatterton has been successful in his efforts to develop this new industry—a result which the Government of India regard as extremely satisfactory in itself. . . . the Government of India wish it to be distinctly understood that commercial enterprises, such as this, must not be undertaken as a part of the scheme of Technical education in India.[79]

By 1907–8 Chatterton's little industries had aroused strong opposition from the British community and were denounced as "a serious menace to private enterprise and an unwarrantable intervention on the part of the state in matters beyond the sphere of government." Finally in 1910 Secretary of State for India Lord Morley ordered the Department of Industries abolished and forbade the Madras government to establish pioneer industries or enter into commerce.[80]

In the period 1900–14, technical education was stalemated by conflicting ideas. Indian nationalists demanded more of it, believing it would stimulate, or at least facilitate, economic development. British employers refused to hire Indian graduates of technical schools because they were poorly prepared and averse to manual labor. The government, seeing the difficult job situation facing Indian graduates, resisted appeals to increase technical education on the grounds that more graduates, far from creating more jobs, would only end up unemployed. The tricky nature of this question is still reflected in the literature. For example Robert Crane, writing in the 1960s, asserted: "Nor has it been possible to arrive at a satisfactory answer to a subsidiary question which arises whenever the role of technical education in economic growth is discussed: i.e., whether technical education breeds industry, or industry calls forth technical education."[81]

Despite appearances, it is not a chicken-and-egg situation, for many other factors are involved, and among them none are so important as government policies. One way the deadlock could have been broken, by setting up government-sponsored enterprises, was rejected on ideological grounds. Another, making employers hire and train Indians for technical positions, still lay in the future. Until World War I, India was still the land of laissez-faire economics and racial discrimination.

Foreign Study and Independent Schools

The flood of Indian students abroad, which is so visible today, began in the late nineteenth century. The pioneers came under private auspices or on their own; Cambridge admitted its first in 1865 and Oxford in 1871. In 1890 there were 207 Indian students in Britain, and in 1894, 308. After that, the numbers began to swell, reaching 700 in 1907 and 1,700 to 1,800 by 1912.[82]

While most students were financed by their families, the value of a foreign education became a matter of public recognition in the early years of this century. This recognition came from both the government and the nationalists. The Simla Conference of 1901 had recommended that the government award scholarships to 10 students every year to study "subjects connected with industrial science or research." Medicine, law, veterinary science, and forestry were excluded, as was engineering on the grounds that "there were enough Indians already clamouring for entrance to Cooper's Hill College." The program began in 1904. By 1912 it had sent 66 students to Britain, and 113 by 1917. Two-thirds of them studied textiles or mining, and the rest learned other industrial trades.[83]

Though the State Technical Scholarship program was well publicized, it affected but a fraction of the Indian students who went abroad, for the demand for a foreign education was much larger than what the government was willing to supply. In response to this need, a Bengali nationalist organization, the Association for the Advancement of Scientific and Industrial Education of Indians, started a more ambitious program. In 1904, its first year, it gave out 16 scholarships, and more every year thereafter. By 1916–17, some 300 Indians had been sent abroad, of whom 140 had returned.[84]

Reports of their experiences vary. Those who went to Japan, as a few did, suffered from language problems. In Britain, the Morison Committee on State Technical Scholarships reported that students "appear . . . to be quite the equals of their British fellow-students in capacity . . . rather above the average at book work and in the class room . . . less good at experimental work in the laboratory . . . somewhat deficient in initiative."[85] The trouble came when these students tried to supplement their book learning with practical experience in industry. The Morison Committee reported:

> Are British manufacturers willing to give Indian students the opportunity of studying the actual conduct of their business? . . . in some [industries], there is an insuperable objection to admitting Indian or any foreign students inside the factory, in others they are admitted freely, in others again there is no regular system of apprenticeship, but some employers are willing, if properly approached, to admit a few Indians upon the broad patriotic ground that they are fellow-subjects.[86]

Alfred Chatterton, voicing his personal views, wrote more critically:

> English manufacturers look upon Indian technical students as possible future competitors and naturally they will extend to them none of the facilities or privileges without which experience cannot be gained. Foreign manufacturers, especially in Germany, welcome Indian students and afford them greater facilities but only because they regard them as possible future customers.[87]

Thus most Indian technical students in Britain received a good theoretical education but not the practical experience to go with it. This was at the root of a misunderstanding when they returned home, degree in hand, as the Chamber of Commerce of Upper India pointed out:

> These students would presumably expect to be installed in positions of trust and importance, and as it is not considered that it would be possible for them in the time at their disposal to gain more than a limited and circumscribed acquaintance with the practical details of the particular industries they had selected for their studies, it would

be extremely doubtful that the heads of important concerns would regard them as qualified to replace European experts, possessing years of practical experience, in the most responsible appointments. On the other hand, the very fact of their being favored by such special selection at the hands of Government . . . would be calculated to render them unfitted in their own estimation for the more subordinate positions in mills and factories.[88]

In 1907 Sir John Hewett, lieutenant governor of the United Provinces, complained: "The question of technical and industrial education has been before the Government and the public for over twenty years. There is probably no subject on which more has been written, or said, while less has been accomplished."[89] The long delay had a dual cause: the natural lethargy and procrastination of official life in India, and the belief of Indian officials, from the viceroys on down, that technical education was only meant to meet existing demands, and anything more would only flood the labor market with unemployable graduates. Nationalists had a much more positive view of technical education, and they were finally driven to action by another of Curzon's decisions: the Partition of Bengal in 1905. Undertaken for reasons of administrative convenience, Partition was interpreted by the Hindu intellectuals of Calcutta as a divide-and-rule tactic to thwart their growing influence. Their indignation, inflamed by the news of the Japanese victory over Russia, soon spread to the large student population of Bengal and from there to the countryside and to other parts of India. One of the rallying cries of the protest was swadeshi, "our own country."

The swadeshi movement was economic and educational as well as political. Its principal method was the boycott of British goods, especially cotton cloth, which nationalists considered the main culprit in the ruin of India's handloom industry and the impoverishment of India generally. The idea of a boycott had a long history dating back to the 1880s when the Indian government abolished duties on Lancashire cottons and later imposed a countervailing duty on Indian cotton goods in the name of free trade. But in 1905, the boycott caught the popular imagination and provided the first link between nationalist intellectuals and the Indian masses.[90]

Soon sales of English cotton goods dropped by 75 percent. The Indian cotton industry took advantage of this opportunity to raise prices and expand production; between 1904 and 1910 India added thirty-nine new mills, 36,304 looms, and almost a million spindles.[91] Bengali nationalists, not content with providing a great boon to the industries of western India, tried to turn the boycott into a means of industrializing Bengal itself. S. D. Mehta explained their failure:

> Soap manufactories, match factories and cotton mills comprised the ranks of industrial flotations. . . . Most of these institutions were started by men whose main asset was patriotism, and whose chief snag was paucity of financial resources. Many had no industrial training or financial experience worth the name. . . . Once the flush of optimism was over, it was found that many of the new companies had been started by men whose enthusiasm had outrun their organizing and financial abilities.[92]

What was lacking, the leaders of the movement believed, was education. Hence they founded the National Council of Education in 1905, which set out to organize a Bengal National College for students who did not want to study under the government's rules. This group was too radical for some, and in June 1906 a more moderate reform group split off and formed the Society for the Promotion of Technical Education. A month later this new society founded the Bengal Technical Institute, which attracted students "who could not afford to receive such training or education in any other way, and those who passed from it could earn their livelihood independently of the posts and professions controlled by the Government."[93]

The first principal of the Bengal Technical Institute and its rector until 1920 was P. N. Bose. He steered it on a moderate course, emphasizing practical training in tanning, soapmaking, dyeing, ceramics, electroplating, lithography, carpentry, and elementary mechanical and electrical engineering. The institute did not attempt to compete with the engineering colleges but provided Bengal, for the first time, with the equivalent of Bombay's Victoria Jubilee Technical Institute. By 1909 it had 124 students. Of all the educational ventures that sprang up in the revolutionary days after the 1905 Partition, the Bengal Technical Institute alone survived, becoming in 1929 the Jodavpur College of Engineering and Technology.[94]

Meanwhile, changes were taking place at the other end of the educational spectrum. The industrialist J. N. Tata had long been interested in education. As early as 1889 he sent the educational administrator Burjonji Padshah to study European universities. Convinced that "science was the hand maid of industry," he dreamed of establishing a scientific research institute combining the best of German seminars, French lecture classes, and Anglo-American laboratory methods. In 1898 he announced his plan and presented it to the newly arrived viceroy, Lord Curzon, offering to contribute 3 million rupees (£200,000) for buildings, equipment, and an endowment fund. The Indian National Congress and the nationalist press hailed the idea, but Curzon hesitated for several years. In 1903 he wrote Secretary of State Hamilton: "There are in India neither the students for the work nor the places for them when they have completed their studies and I shall not myself be at all surprised, though I should bitterly regret, if in another ten or fifteen years time the costly experiment will turn out to have been a dismal failure."[95]

Not until 1909, long after Tata had died and Curzon gone home, did the government of India agree to the scheme. The Maharajah of Mysore offered land in Bangalore, 500,000 rupees, and a pledge of 50,000 rupees a year, and the Indian government granted 250,000 rupees and 90,000 a year. The new Indian Institute of Technology, the first postgraduate scientific research institution in India, opened its doors in 1911.[96]

Technical Education after World War I

Among the arguments against investing in technical education, one of the most often used was the bias of middle- and upper-class Indians toward book learning, academic studies, and clerical careers. This argument was strengthened by contrasting the number of students enrolled in the different schools (see Tables 9.2, 9.3, and 9.4).

Table 9.2 Students in Public Secondary Schools

	Secondary[a]		Technical[b]		Other Special[c]		
Year	Number	%	Number	%	Number	%	Total
1894–95	525,303	95.7	5,480	1.0	17,881	3.3	548,672
1899–1900	592,829	94.9	4,878	.8	27,048	4.3	624,755
1904–5	679,769	93.4	6,482	.9	41,347	5.7	727,598
1909–10	863,993	85.8	9,597	.9	133,536	13.3	1,007,106
1914–15	1,102,864	83.3	12,145	.9	208,265	15.7	1,323,274
1919–20	1,281,810	90.7	14,202	1.0	117,390	8.3	1,413,402
1924–25	1,541,705	85.8	21,456	1.2	233,712	13.0	1,796,873
1929–30	2,246,208	87.2	30,336	1.2	300,808	11.7	2,577,352
1934–35	2,362,004	90.2	29,433	1.1	227,843	8.7	2,619,280
1939–40	2,659,201	85.4	39,472	1.3	415,345	13.3	3,114,018

Notes: (a) Both English and vernacular schools; (b) engineering, surveying, and technical-industrial schools; (c) normal schools, reform schools, and schools of art, commerce, medicine, etc.

Table 9.3 College and University Students

Year	Arts	Law	Medi-cine	Engi-neering	Agricul-ture[b]	Other[c]	Total
1894–95	14,422	2,534	844	571	47	66	18,484
1899–1900	16,287	2,375	1,151	813	47	71	20,744
1904–5	19,752	3,228	1,655	998	153	317	26,103
1909–10	23,184	2,879	1,569	1,203	260	434	29,529
1914–15	41,956	4,479	1,755	1,268	397	836	50,688
1919–20	52,482	5,991	3,446	1,355	1,252	1,390	65,916
1924–25	64,996	7,960	3,816	1,864	1,005	2,635	82,276
1929–30	79,085	7,426	3,846	2,131	1,511	3,167	97,166
1934–35	91,785	7,256	5,028	2,074	1,229	4,436	111,808
1939–40	119,536	6,749	5,640	2,821[a]	2,280	7,878	144,904

Notes: (a) Includes technology; (b) includes forestry and veterinary science from 1914–15; (c) education and (from 1914–15 on) commerce.

Table 9.4 College and University Students (Percentages)

Year	Arts	Law	Medi-cine	Engi-neering	Agricul-ture[b]	Other[c]
1894–95	78.0	13.7	4.6	3.1	.3	.4
1899–1900	78.5	11.4	5.5	3.9	.2	.3
1904–5	75.7	12.4	6.3	3.8	.6	1.2
1909–10	78.5	9.7	5.3	4.1	.9	1.5
1914–15	82.8	8.8	3.5	2.5	.8	1.6
1919–20	79.6	9.1	5.2	2.1	1.9	2.1
1924–25	79.0	9.7	4.6	2.3	1.2	3.2
1929–30	81.4	7.6	4.0	2.2	1.6	3.3
1934–35	82.1	6.5	4.5	1.9	1.1	4.0
1939–40	82.5	4.7	3.9	1.9[a]	1.6	5.4

Notes: (a) Includes technology; (b) includes forestry and veterinary science from 1914–15; (c) education and (from 1914–15 on) commerce.

The statistics show two trends. The student population increased tremendously over the forty-five-year period, nearly sixfold in the case of secondary students, almost eightfold in that of college students. But there was hardly any change in the proportion of students majoring in the technical fields. In other words, one can find in these statistics arguments for both sides: that the vast majority of Indian students preferred the academic fields, and that the number of technical students was rising fast. We can only conclude that the solution to this puzzle lies not in trying to fathom the aspirations of students, but in looking at the demand for their skills and the policies of the government and the major employers.

In education as in so much else, World War I marked a turning point for India. The relationship between India and Britain proved more vulnerable to war than anyone had suspected. Britain withdrew many of its engineers and technicians, and even equipment, which it could not replace. The Indian economy, lacking a sufficient technical and industrial base, was not able to take up the slack and contribute much to the war effort other than manpower, raw materials, and used railroad equipment.

Concern over India's weakness led to the appointment of the Indian Industrial Commission in 1916. The commission deplored the inadequacy of technical education and training in India and the resulting need to import trained personnel. It criticized the government for sponsoring "literary and philosophic studies to the neglect of those of a more practical character"; for limiting its efforts to meeting the needs of the Public Works Department; and for having put so much effort into holding conferences and issuing reports, and so little into practical results.[97] These criticisms were nothing new, having been repeated in numerous official reports since the 1880s. What was new was that the commission tied technical education to industrial development and gave the government the responsibility for both.

This new attitude filtered into the actions of the government during the twenties and thirties. Slowly, hesitantly, the Indian government groped toward a policy of industrial development, putting the interests of the Indian economy ahead of those of free trade or British exports. Import duties were raised to protect Indian industries; thus the cotton tariff, which was practically nil in the 1890s, climbed to 50 percent in 1934, giving the Indian mills two-thirds of their domestic market.[98]

Similarly, in government employment, the reforms of the civil service demanded by nationalists since the 1880s were gradually implemented in the 1920s. The recommendations of the Royal Commission on Public Service of 1912, to open more posts in the Civil Service to Indians, were accepted after the war, albeit with deliberation.[99] Thus in 1921 the new Indian Service of Engineers, which recruited its members in India as well as Britain, replaced both the Imperial Engineering Service, which had recruited only in Britain, and the Provincial Engineering Services, which had recruited in India. This gave Indians a greater access to the higher posts in Public Works.[100]

The same was true of the Posts and Telegraphs, according to former director general Sir Geoffrey Clarke, who wrote in 1927:

> In the Post Office indianization began many years ago, and there has never been any distinction of race in the matter of promotion to the higher appointments. In fact, most of these are at present held by Indians, and very efficient and trustworthy officers they have proved. I think it is purely due to this elimination of race distinction both in the matter of appointment and pay that the Department has been able to work so smoothly in troubled times.[101]

Education was transferred to provincial governments responsible to elected legislatures. Five new engineering colleges were established. The Benares Hindu University opened a department of engineering during the war. After the war the Bihar College of Engineering (Patna), the Maclagan College of Engineering (Lahore) and the N. E. Dinshaw Civil Engineering College (Karachi) were founded, as was a department of engineering at Rangoon University in Burma, then a part of India. The number of students rose proportionately, from 865 in 1901–2 to 1,443 in 1922 and to 2,253 in 1937. Other branches of engineering made their appearance; of the 323 bachelor's degrees in engineering given in 1939–40, half were in civil engineering, and the rest in electrical, mechanical, mining, or metallurgical engineering, all new fields.[102]

Beside these colleges, several specialized technical institutes were also founded. At Cawnpore the government opened the Harcourt Butler Technological Institute in 1920 to train men for the tanning, vegetable oil, soap, paint, and varnish industries. Next to their steel mill, the Tata Iron and Steel Company established the Jamshedpur Technical Institute in 1921 to train students in metallurgy, electrical and mechanical engineering, and other skills at the secondary and college levels. At Dhanbad in Bihar the government founded the Indian School of Mines in 1926. In Bombay the Victoria Jubilee Technical Institute, specializing in textiles since the 1880s, branched out into chemistry, plumbing, and sanitary engineering. The 1920s saw technical schools springing up all over. The number of secondary-level technical and industrial schools rose from 242 in 1911–12 to 536 in 1936–37, and the number of students also increased, from 12,064 in 1911–12 to 30,548 in 1936–37.[103]

Agricultural, veterinary, and forestry education also expanded. New agricultural colleges with experimental farms were established at Coimbatore, Poona, Naini, and Lyallpur. The number of students in agricultural colleges rose from 219 in 1901–2 to 1,000 in the twenties and to 1,500 in the thirties. This meant three graduates each year for every million in the farming population. Of these, however, only 2 or 3 percent returned to farming—the rest preferred working in government offices. The same was true in other fields. As of 1937 the five veterinary colleges had 605 students. India, the land of cows, had one veterinarian for every 125,000 head of cattle. The Forest Research Institute and College at Dehra Dun was closed in 1937 for lack of students.[104]

After World War I, young Indians went abroad to study in far greater numbers than ever before, an estimated 500 or more each year to Great Britain alone. A symptom of this trend was the publication of official guides for students seeking a foreign education, such as R. K. Sorabji's *Facilities for Indian Students in America and Japan* (Calcutta, 1920) and Edward Sandes's *Report on Civil Engineering Education in Great Britain, in 1924. With Remarks on the Training of Indian Students in Engineering* (Roorkee, 1925). Another was the appointment in 1921 of a Committee on Indian Students in England, chaired by the Earl of Lytton, parliamentary under secretary of state for India.[105]

The British had never pursued a policy of de jure racial discrimination in India. Instead, they had simply recruited at home for the upper levels of the Indian bureaucracy, on the grounds that the training for such posts was better in Britain. The obvious consequence, as the Lytton Committee noted, was that Indian "feel that by obtaining their education in the United Kingdom they have a better chance of securing suitable employment in India, especially in the Indian Public Services."[106]

Like its predecessor the Morison Committee, the Lytton Committee found that the biggest problem facing Indian students in Britain was practical training:

> In Engineering, the metallurgical industries and the manufacture of machinery, there should be little difficulty in normal times in securing the practical training required, but in other industries . . . the difficulties are much greater, and employers contend that the necessity of guarding their trade secrets, and the fear of trade competition would prevent them from admitting Indian students to their works even if they had plenty of work to do.[107]

The Lytton Committee concluded that it was often wasteful and unnecessary to train Indians in Britain, that foreign study should be reduced for cases where training was unavailable in India, and that therefore, more and better technical education should be made available in India itself.

On the railways, Indianization began right after the war as the government acquired the various railway companies and influenced their personnel policies. The results were twofold: to replace de facto racial discrimination with overt racial quotas and to replace European personnel with Eurasians, a group that had long been associated with the railways and telegraphs. The overall change is reflected in the aggregate figures in Table 9.5.[108]

Table 9.5 Railway Employees by Race, 1920–1941

Year	Indians		Eurasians		Europeans		Total
	Number	%	Number	%	Number	%	
1920–21	708,639	(97.4)	11,404	(1.6)	7,141	(1.0)	727,184
1930–31	736,536	(97.6)	13,567	(1.8)	4,647	(0.6)	754,750
1940–41	712,116	(97.9)	13,238	(1.8)	2,142	(0.3)	727,486

Right after the war the Acworth Committee, appointed to investigate the railways, commented on the employment patterns it found:

> At the date of the last report there were employed on the railways of India about 710,000 persons; of these, roughly 700,000 were Indians and only 7,000 Europeans, a proportion of just 1 per cent. But the 7,000 were like a thin film of oil on the top of a glass of water, resting upon but hardly mixing with the 700,000 below. None of the highest posts are occupied by Indians; very few even of the higher. . . . That they have not been advanced to higher posts, that even in the subordinate posts of the official staff there are not more of them, has been a standing subject of complaint before us. . . . Until recently opportunities for the technical training of Indians were lacking. And in the absence of opportunities, naturally few Indians were able to reach the standard required for the superior posts.[109]

As late as 1936 the Wedgwood Committee, appointed to seek ways to save money, wrote:

> Supervision in workshops was defective. This defect can be removed by providing better facilities for training Indian apprentices, instead of relying on the European supervisory staff, which is usually very costly. . . . If the Indian railways have still to depend upon a large

number of European supervisors for running their workshops, it is largely due to the indifference they have shown in the past, in training Indian apprentices.[110]

During and after World War I, the railways replaced many of their midlevel imported European personnel with resident Europeans and Eurasians. Between 1910 and 1930–31 the number of Europeans employed on the railways dropped from 7,207 to 4,647, while the number of Eurasians rose from 8,862 to 13,567. At the same time, more and more of the Europeans were residents of India, the sons of railroad men. Frank D'Souza, appointed in the late thirties by the Railway Board to investigate minority representation, described this middle layer of employees:

> Domiciled Europeans and Anglo-Indians . . . for several generations, have occupied the better paid subordinate appointments in certain Departments. In the Traffic Department, the higher grades of Guards, Assistant Station Masters and Station Masters and subordinate supervisory posts, such as those of Traffic Inspectors, were practically the monopoly of this community, members of which were recruited generally either as Ticket Collectors or probationary Guards. Similarly, in the Locomotive (or Mechanical Engineering) Department, Domiciled Europeans and Anglo-Indians largely recruited as Firemen naturally constituted the majority of the better-paid Drivers. From the ranks of the latter the senior subordinate supervisory posts of Loco Foremen and Loco Inspectors were filled.[111]

In the 1920s, in response to political agitation, the Indian government began recruiting Indian students as apprentice railroad engineers. These students were given four years of technical education and practical training at the Jamalpur workshop. Among this first generation of Indian railroad engineers was D. V. Reddy, who remembered:

> Jamalpur was a Railway colony housing British officers and Anglo-Indian supervisors with a sprinkling of Indian officers and supervisors. It took the former sometime to get used to the Indianisation idea and they were, therefore, lukewarm to the special class apprentices in the earlier years. Four years in Jamalpur, a workshop town, was a trial of patience.

After Jamalpur, the trainees were placed in railroad workshops in Britain for two more years. There Reddy was apprenticed to a chief mechanical engineer who was, to his amazement, "an exceptionally nice man who had in his early years served on the Indian Railways. . . . I realized the truth for the first time in what was often said 'that there was a world of difference between the Englishman at home and the Englishman in the Colonies.' "[112]

By the 1930s the problem had shifted from racial barriers against Indians to barriers between them. The Railway Board sought to deflect criticisms of hiring practices by setting quotas for each ethnic group. Thus in the upper ranks, 25 percent of vacancies were reserved for Europeans recruited in England by the Office of the High Commissioner for India; 2.5 percent for domiciled Europeans and Eurasians; 25 percent for Muslims; 6 percent for other minority communities such as Sikhs, Parsis, and Indian Christians; 21.5 percent for Hindus; and 20 percent for promotions from the ranks.[113]

Despite these quotas, designed ostensibly to increase the opportunities for Indians, the Railway Board was able to fill only nine of the twelve vacancies allotted to Europeans in the years 1935–39, "the revised scales of pay having proved unattractive."[114] It was the retrenchment policies of the Depression era, as much as nationalist agitation, which changed the ethnic composition of the railway staff. India was no longer as attractive a place for Britons to work as it had been in the heyday of imperialism before World War I.

For sixty years, the question of technical education in India was bedeviled by a false dilemma: would it stimulate economic growth, or just create more unemployed graduates? The answer came during World War II. As European personnel were withdrawn for the war effort and the demand for goods and services rose, the Indian economy once again was strained to the utmost. The last official report on technical education in British India, issued in 1943, shows the change:

> The experience of the war, however, has already led to a number of salutary changes; it has compelled a large expansion of industry and created a greatly increased demand for technicians of all grades, while at the same time the urgent need for skilled and semi-skilled workers had led to almost every technical institution in the country becoming a centre for Technical Training Schemes. Many young men, who would otherwise not have embarked on a technical career have been recruited under these schemes and the prejudice against industrial employment has been steadily breaking down.[115]

And as for the future,

> New industries are springing up everywhere and old ones are moving forward under rapidly changing conditions. In order that this process may continue with ease and efficiency it seems desirable to

evolve a system of technical education which can cope with the ever changing requirements of industry.[116]

And in fact that is what happened: from 1937 to 1947 the number of engineering colleges jumped from nine to sixteen, while their student population more than doubled, from 2,253 to 5,162.[117] Thanks to a new combination of rising demand, government support, and Indianization, technical education and economic growth had broken the deadlock of colonial pessimism.

Conclusion

As colonialism aged, widely differing opinions on the purposes of technical education caused conflict and confusion. From one perspective, technical education served to meet current demands for craftsmen, engineers, and skilled workers. This was the short-run view often espoused by colonial officials. Constrained by tight budgets and pessimistic about the impact of government policies on the economy, they were reluctant to invest in technical education for fear of overshooting the mark, thereby creating that bugbear of colonialists, unemployed graduates. If the educational system did not meet the needs of the economy, the difference could easily be made up by importing technicians from Europe. This, and colonial views on the superiority of Europeans, led to the two-tier system of covenanted and subordinate staff, imperial and provincial services. In Africa, with its small demand for college-educated technicians, all the higher positions could be filled with Europeans, and only the lower-level jobs were left to Africans.

Most of the colonized also took the short-run view. With the best positions reserved for Europeans, the demand for technically trained personnel was much smaller than the demand for the academically educated. Whatever cultural biases existed among Asians and Africans were reinforced by the Western monopoly on the higher levels of modern technology.

But education does not just meet present demands, it also prepares youths for the future. Visions of the future differed sharply between colonizers and colonized. To the colonizers—at least before World War I in India and Egypt, and before World War II in Africa—the future was an indefinite extension of the present. Changes would come but gradually; indeed, one of their goals was to prevent too-rapid changes. An expanding but not diversifying colonial economy would need craftsmen, skilled workers, and subordinate technicians, while the managerial and creative aspects of technology could safely be left to Europeans.

To nationalists in the colonies, in contrast, the future was filled with radical changes. Economic development and diversification would accompany political independence; in India the push for economic independence (*swadeshi*) even preceded the struggle for political independence (*swaraj*). To nationalists, technical education should prepare the future by training men to replace the Europeans and create a new economy.[118]

The educational system, of course, was under government control. In India and Egypt, it offered training in the various trades up to civil engineering but only began to offer the more modern mechanical, electrical, and chemical engineering specialties at the very end of the colonial period. In Africa the educational system did not even get as far as civil engineering before 1940. The contrast between Africa and India reveals the limitations of colonial rule in the field of education. In Africa, Europeans argued convincingly that they had created both the demand for educated Africans and the schools to educate them. In India, despite employment discrimination, the schools fell behind the demand by the turn of the century, and Indians went abroad or founded their own schools to obtain the education they required. Colonial rulers educated their subjects up to a point. Beyond that point, they withheld the culture of technology.

To natives of the colonies driven by a personal or political ambition to go beyond what their country's educational system offered, the answer lay elsewhere: to work in a noncolonial setting or create enterprises of their own. Let us now look at some attempts by colonial subjects to compete with the Europeans on their own terms.

Notes

1. Tony Austin, *Technical Training and Development in Papua, 1894–1941* (Canberra, 1977), p. 5.

2. Great Britain, Advisory Committee on Education in the Colonies, Colonial No. 103, *Memorandum on the Education of African Communities* (London, 1935), quoted in Philip Foster, *Education and Social Change in Ghana* (Chicago, 1965), pp. 160–61.

3. Lt.-Col. Edwin Henry de Vere Atkinson and Tom S. Dawson, *Report on the Enquiry to Bring Technical Institutions into Closer Touch and More Practical Relations with the Employers of Labour in India* (Calcutta, 1912), p. 10.

4. Joseph S. Szyliowicz, *Education and Modernization in the Middle East* (Ithaca, N.Y., 1973), p. 195.

5. Philip Foster, "The Vocational School Fallacy in Development Planning" in *Education and Economic Development,* ed. C. A. Anderson and Mary Jean Bowman (Chicago, 1965), p. 148.

6. United States Bureau of Education, *Education in Parts of the British Empire* (Washington, 1919), pp. 35–36.

7. Foster, "Vocational School Fallacy," p. 145. See also Foster, *Education and Social Change,* p. 151, and Remi Clignet and Philip Foster, "Convergence and Divergence in Educational Development in Ghana and the Ivory Coast," in *Ghana and the Ivory Coast: Perspectives on Modernization,* ed. Phillip Foster and Aristide R. Zolberg (Chicago, 1971), p. 280.

8. Denise Bouche, *L'enseignement dans les territoires français de l'Afrique occidentale de 1817 à 1920: Mission civilisatrice ou formation d'une élite?,* 2 vols. (Lille and Paris, 1975), 2:543–44.

9. Foster, "Vocational School Fallacy," p. 148.

10. Atkinson and Dawson, p. 1.

11. Revolutionary movements in older societies like Russia or China have often had to encourage (or force) their young intellectuals to do some manual labor.

12. J. A. Richie, *Progress of Education in India, 1917–1922. Eighth Quinquennial Review,* 2 vols. (Calcutta, 1923, and London, 1924), 1: 90.

13. India, Indian Industrial Commission, *Confidential Evidence,* pp. 6–7, in India Office Records, L/PARL/2/404D.

14. Robert I. Crane, "Technical Education and Economic Development in India before World War I" in Anderson and Bowman, pp. 180–81.

15. On technical education in Egypt before 1882, see James Heyworth-Dunne, *An Introduction to the History of Education in Modern Egypt* (London, 1939), pp. 141–77, 225–39; Mohammed K. Harby, *Technical Education in the Arab States* (Paris, 1965), p. 12; Robert L. Tignor, *Modernization and British Colonial Rule in Egypt, 1882–1914* (Princeton, N.J., 1966), pp. 38–42; and Szyliowicz, pp. 32–35, 102–104.

16. Tignor, pp. 320–25; Szyliowicz, pp. 122–27.

17. Tignor, pp. 323, 331, 347.

18. Colin C. Scott-Moncrieff, "Irrigation in Egypt," *Nineteenth Century* 17 (February 1885): 344–45.

19. Szyliowicz, p. 132.

20. Amir Boktor, *School and Society in the Valley of the Nile* (Cairo, 1936), p. 233.

21. Helen Kitchens, ed., *The Educated African: A Country-by-Country Survey of Educational Development in Africa* (New York, 1962), pp. viii–ix, 243–44.

22. Montague Yudelman, "Imperialism and the Transfer of Agricultural Techniques" in *Colonialism in Africa, 1870–1960,* ed. Peter Duignan and L. H. Gann, vol. 4: *The Economics of Colonialism* (Cambridge, 1975), p. 356.

23. Bouche, 1: 218–48.

24. Ibid., 2: 525–48.

25. Ibid., 2: 845–46, 879.

26. David Gardinier, "Vocational and Technical Education in French

Equatorial Africa (1842–1960)," in *Proceedings of the Eighth Annual Meeting of the French Colonial Historical Society, 1982* (Lanham, Md., 1985), pp. 113–23; Pierre Labé, "L'éducation agricole et rurale en Guinée française," *L'Afrique française* 46 (July 1936): 404–8.

27. D. L. Sumner, *Education in Sierra Leone* (Freetown, 1963), pp. 163–64; Victor I. Nwagbaraocha, "Vocational and Technical Education in Eastern Nigeria during the Colonial Period" in *Proceedings of the Eighth Annual Meeting of the French Colonial Historical Society*, pp. 124–36.

28. The first Phelps-Stokes report was: Jesse Jones, *Education in Africa: A Study of West, South, and Equatorial Africa by the African Education Commission* (New York, 1922). Three years later the Phelps-Stokes Fund issued a report on East Africa: Jesse Jones, *Education in East Africa* (New York, 1925).

29. *Parliamentary Papers* 1925 (Cmd. 2374) vol. 1, p. 192.

30. Foster, "Vocational School Fallacy"; Nwagbaraocha, pp. 124–36; Clignet and Foster, p. 280.

31. Aparna Basu, *The Growth of Education and Political Development in India, 1898–1920* (Delhi, 1974), p. 107.

32. See Monica Alice Greaves, *Education in British India, 1698–1947: A Bibliography and Guide to the Sources of Information in London* (London, 1967).

33. U., "Engineers in India," *Fraser's Magazine* 97 (November 1878): 562–63.

34. "Note by Mr S. F. Downing, Principal of the C. E. Dept. of the Presidency College," in "Technical Education in Bengal: Selected Papers from the Records of the Bengal Secretariat, P. W. Dept., compiled by Mr F. J. E. Spring, Under-secretary, P. W. D. (Calcutta, 1886)," in India Office Records, V/27/865/5, p. 82.

35. A. G. Cardew, "Note on Technical Education in Madras," in Madras, Government of Madras, *Papers Relating to the Industrial Conference held at Ootacamund in September 1908* (Madras, 1908), p. 15; Edward Warren Caulfield Sandes, *The Military Engineer in India*, 2 vols. (Chatham, 1933–1935), 2: 353–59; Syed Nurullah and J. P. Naik, *A History of Education in India (during the British Period)*, 2nd ed. (Bombay, 1951), pp. 160–61, 378–79; India, Home Department, *Papers Relating to Technical Education in India, 1886–1904* (Calcutta, 1906), in India Office Records, V/27/865/1, pp. 6–8; "Technical Education in Bengal," pp. 64–83.

36. G. Chesney, "The Civil Engineering College for India," (dated India Office, October 7, 1870) in India Office Records, L/PWD/8/9, pp. 3–4.

37. Ibid., p. 6.

38. "Circular no. 35 P.W., dated Simla, the 29th June 1870. Resolution by the Government of India, Public Works Department. Training in India of Natives for the Position of Foremen Mechanics," in India Office Records, V/27/865/5.

39. Chesney, p. 7.

40. India Office Records, L/PWD/8/243: "Natives of India as Candidates for R.I.E.C."; L/PWD/8/247 file 200: "Report on Siamese students (1890–1906) at R.I.E.C. Cooper's Hill"; and L/PWD/8/10: "Alphabetical List of Students who have entered the College and particulars as to their Dis-

posal Dating from opening of College"; John Gordon Patrick Cameron, *A Short History of the Royal Indian Engineering College, Cooper's Hill* (Richmond, 1960), pp. 5–8; Anthony Farrington, *The Records of the East India College Haileybury & Other Institutions* (London, 1976), p. 135–36; Whitworth Porter, *History of the Corps of Royal Engineers* (London, 1889), pp. 172–77.

41. India Office Records, L/PWD/8/9: "Memorandum from the Board of Visitors of the Royal Indian Engineering College to the Right Honourable the Marquis of Hartington, M.P., Secretary of State for India" (Cooper's Hill, May 1882) and "Royal Indian Engineering College, Cooper's Hill" (Report by Thomas W. Keith, Accountant General, India Office, 20 July 1885); Cameron, p. 12; Farrington, pp. 137–38; Sandes, 2:349.

42. India, Public Service Commission, *British Attitude toward the Employment of Indians in Civil Service. Report of the Public Service Commission (1886–1887) headed by Sir Charles U. Aitcheson* (reprint, Delhi, 1977); Nurullah and Naik, p. 378; Basu, pp. 89, 103.

43. Public Service Commission, pp. 104–5; Deepak Kumar, "Patterns of Colonial Science in India," *Indian Journal of History of Science* 15, no. 1 (May 1980): 108.

44. Krishnalal Jethalal Shridharani, *Story of the Indian Telegraphs: A Century of Progress* (New Delhi, 1956), pp. 12–16.

45. Ibid., p. 148; Public Service Commission, p. 136; Sir Geoffrey R. Clarke, "Post and Telegraph Work in India," *Asiatic Review* 23 (1927): 79–108. Murty S. Kalyanasundaram, *The Autobiography of an Unknown Telegraphist* (Kodaikanal, 1973), describes relations between Europeans and Indians in the Telegraph Department.

46. Sandes, 2: 349; India Office Records, V/6/312, pp. 124–25.

47. Figures from Morris David Morris and Clyde B. Dudley, "Selected Railway Statistics for the Indian Subcontinent (India, Pakistan, and Bangladesh), 1853–1946-47," *Artha Vijnana* 17, no. 3 (September 1975): 202–4.

48. Fritz Lehmann, "Railway Workshops, Technology, and Personnel in Colonial India," *Journal of Historical Research* (Ranchi, India) 20, no. 1 (August 1977): 49–61; J. N. Westwood, *Railways of India* (Newton Abbot, England, and North Pomfret, Vt., 1974), pp. 31, 81–82; Ramswarup Deotadin Tiwari, *Railways in Modern India* (New York, 1941), pp. 62–63.

49. "Resolution by the Government of India, Public Works Department: Training of Natives for employment as Engine-drivers, Mechanics, and Plate-layers on State Railways (Dated Fort William 4th June 1874)," in "Technical Education in Bengal," pp. 85–86.

50. Juland Danvers, *Indian Railways: Their Past History, Present Condition, and Future Prospects* (London, 1877), p. 39, quoted in Lehmann, p. 5.

51. Sunil Kumar Sen, *Studies in Economic Policy and the Development of India (1848–1926)* (Calcutta, 1966), pp. 7–8; India, Department of Statistics, *Large Industrial Establishments* (Calcutta, 1920), pp. ii, 3, 11; Jogendra Nath Sahni, *Indian Railways: One Hundred Years, 1853 to 1953* (New Delhi, 1953), pp. 102–3.

52. Lehmann, p. 56; Clement Hindley, "Indian Railway Developments," *Asiatic Review* 25 (1929): 645; Crane, pp. 167–201.

53. India, Indian Industrial Commission, *Confidential Evidence,* pp. 187–88, in India Office Records, L/PARL/2/404D.

54. Ibid., pp. 60–61.

55. India, Indian Industrial Commission, *Report of the Indian Industrial Commission, 1916–18,* in *Parliamentary Papers* 1919 (Cmd. 51), vol. 17, pp. 118–20.

56. Lehmann, pp. 52–53.

57. Atkinson and Dawson, p. 24.

58. Morris D. Morris, "The Growth of Large-Scale Industry to 1947," in *The Cambridge Economic History of India,* vol. 2: *c. 1757–c. 1970,* ed. Dharma Kumar (Cambridge, 1983), p. 572.

59. R. Nathan, *Progress of Education in India 1897/98–1901/02* (Calcutta, 1904), p. 192.

60. Cardew, p. 20.

61. Crane, p. 168.

62. *Papers Relating to Technical Education,* p. 2.

63. "Despatch from Viscount Cranbrook to the Earl of Lytton, dated February 6, 1879," in India Office Records, V/6/307, p. 370.

64. "Despatch no. 1 of January 8, 1880," in India Office Records, V/6/308, p. 311.

65. "Despatch no. 38 of July 24, 1884," in India Office Records, V/6/312, p. 571.

66. *Papers Relating to Technical Education,* p. 18; Krishna Dayal Bhargava, ed., *Selections from Educational Records of the Government of India,* vol. 4: *Technical Education in India, 1886–1907* (Delhi, 1968), pp. 1–2.

67. Bradford Spangenberg, "Aitcheson Commission: An Introduction" in Public Service Commission, pp. xiii–xvii.

68. Cardew, pp. 14–15.

69. Pramatha Nath Bose, *Technical and Scientific Education in Bengal* (Calcutta, 1886), p. 2.

70. Ibid., pp. 9–10.

71. There were others, e.g. Dinshah Ardeshir Taleyarkhan, *How to Introduce National Technical Education in India* (Bombay, 1886), and Alexander Tomory, *Technical Education in Europe. A Possibility for Bengal. A Lecture* (Calcutta, 1892).

72. Basu, pp. 86–87.

73. Ibid., pp. 82, 93. See also her "Technical Education in India, 1900–1920," *Indian Economic and Social History Review* 4, no. 4 (December 1967): 361–74.

74. David Dilks, *Curzon in India,* 2 vols. (London, 1969–70), 1: 244; Michael Edwardes, *High Noon of Empire: India under Curzon* (London, 1965), pp. 144–45.

75. *Papers Relating to Technical Education,* pp. 251–52.

76. Ibid., pp. 257–58.

77. Basu, *Growth of Education,* pp. 84–85.

78. Bhargava, 4: 5; John P. Hewett, "Inaugural Address to the Industrial Conference, Nainital (UP), August 19–31, 1907," in *Bhargava,* 4: 272; *Papers Relating to Technical Education,* pp. 257–59.

79. *Papers Relating to Technical Education,* pp. 249–50.

80. Basu, *Growth of Education,* pp. 89–90; Indian Industrial Commission, *Report,* p. 78.

81. Crane, pp. 167–68.

82. F. H. Brown, "Indian Students in Great Britain," *Edinburgh Review* 217 (January 1913): 138–56.

83. India, State Technical Scholarships Committee, Sir Theodore Morison, Chairman, *Report of a Committee Appointed by the Secretary of State for India to Inquire into the System of State Technical Scholarships Established by the Government of India in 1904; with Appendices,* in *Parliamentary Papers* 1913 (Cd. 6867), vol. 47, p. 7.

84. Ibid., p. 10; Basu, *Growth of Education,* p. 86; R. C. Majumdar, *History of the Freedom Movement in India,* 3 vols. (Calcutta, 1963), 2: 32.

85. State Technical Scholarship Committee, pp. 12–13.

86. Ibid., pp. 21–22.

87. Alfred Chatterton, *Industrial Evolution in India* (Madras, 1912), p. 152.

88. Nathan, p. 192.

89. Hewett, p. 269.

90. Bipan Chandra, *Rise and Growth of Economic Nationalism in India: Economic Policies of Indian National Leadership, 1880–1905* (New Delhi, 1966), pp. 132–40; Majumdar, 2: 25–40; S. D. Mehta, *The Cotton Mills of India, 1854 to 1954* (Bombay, 1954), pp. 86–90.

91. Mehta, p. 91.

92. Ibid., p. 93. See also Majumdar, 2: 35.

93. Haridas Mukherjee and Uma Mukherjee, *The Origins of the National Education Movement* (Calcutta, 1957), pp. 47–50; Majumdar, p. 80.

94. Jogesh Chandra Bagal, *Pramatha Nath Bose* (New Delhi, 1955), pp. 103–8; Basu, *Growth of Education,* p. 87; Amiya Bagchi, *Private Investment in India, 1900–1939* (Cambridge, 1972), p. 154.

95. Quoted in Basu, *Growth of Education,* pp. 80–81.

96. Lovat Fraser, *India under Curzon and After* (London, 1911), p. 324; Sunil Kumar Sen, *The House of Tata, 1839–1939* (Calcutta, 1975), pp. 26–28; Dinsha Edulji Wacha, *The Life and Work of J. N. Tata,* 2nd ed. (Madras, 1915), pp. 51–54; B. S. Saklatvala and K. Khosla, *Jamsetji Tata* (New Delhi, 1970), pp. 46–47, 75, 93–99.

97. Indian Industrial Commission, *Report,* pp. 71, 104–7, 121–22.

98. Angus Maddison, *Class Structure and Economic Growth: India and Pakistan Since the Moghuls* (New York, 1972), pp. 55–57.

99. Michael Kidron, *Foreign Investments in India* (London, 1965), p. 18.

100. Richie, 1: 158–59.

101. Geoffrey R. Clarke, "Post and Telegraph Work in India," *Asiatic Review* 23 (1927): 97–98.

102. Bankey Bihai Misra, *The Indian Middle Classes: Their Growth in Modern Times* (Oxford, 1961), p. 336.

103. Basu, *Growth of Education,* p. 97; G. E. Fawcus, *Note on Education at Jamshedpur in Bihar and Orissa* (Calcutta, 1930), pp. 1–3; John Sargent, *Progress of Education in India, 1932–1937. Eleventh Quinquennial Review,* 2 vols. (Delhi, 1940), 2: 154; A. T. Weston, "Technical and Vocational Educa-

tion," *The Annals of the American Academy of Political and Social Science* 145 (1929): 151–60.

104. Arthur Mayhew, *The Education of India: A Study of British Educational Policy in India, 1835–1920, and its Bearing on National Life and Problems in India Today* (London, 1926), p. 168; Nurullah and Naik, pp. 690–94, 793–94; Ramanbhai G. Bhatt, *The Role of Vocational and Professional Education in the Economic Development of India from 1918 to 1951* (Baroda, 1964), pp. 130–32.

105. The *Reports and Appendices* and *Evidence* are in the India Office Records, V/26/864/13.

106. *Report of the Committee on Indian Students in England, 1921–22*, pp. 8–9.

107. Ibid., p. 19.

108. Figures from Morris and Dudley, p. 204.

109. India, Railway Committee, 1920–21, *Report of the Committee appointed by the Secretary of State for India to enquire into the administration and working of the Indian Railways* (London, 1921), pp. 58–59.

110. Tiwari, p. 117.

111. Frank D'Souza, *Review of the Working of the Rules and Orders Relating to the Representation of Minority Communities in the Services of the State-Managed Railways* (New Delhi, 1940), p. 24.

112. Duvur Venkatrama Reddy, *Inside Story of the Indian Railways: Startling Revelations of a Retired Executive* (Madras, 1975), p. 27.

113. D'Souza, p. 1.

114. Ibid., pp. 6–7.

115. India, Board of Education, "Report of the Technical Education Committee of the Central Advisory Board of Education in India, 1943, together with the decisions of the Board thereupon," pp. 2–3 in India Office Records, V/26/865/5.

116. Ibid., p. 25.

117. Bhatt, p. 117; Nurullah and Naik, pp. 795–96.

118. After independence, new governments went too far in the other direction and created more engineers than their economies could absorb. See Clement Henry Moore, *Images of Development: Egyptian Engineers in Search of Industry* (Cambridge, Mass., 1980).

10

Experts and Enterprises

Technical education is but one way in which the culture of technology spreads. Another way is through enterprises and experience. European enterprises and government agencies restricted non-Europeans to the lower jobs until they were forced to do otherwise by political pressures at the very end of the colonial era. Enterprises owned by Asians and Africans, in contrast, had every incentive to use their own people, for reasons of ethnic solidarity as well as economy.

It is sometimes asserted that underdeveloped countries lack entrepreneurs, or that cultural factors—the caste system, otherworldliness, parasitic landlordism, and the like—inhibit the entrepreneurial spirit.[1] This is perhaps a problem of definition. As travelers noted, India, West Africa, the Arab world, and Southeast Asia teemed with eager traders and purveyors of myriad goods and services. Many of them had the entrepreneurial qualities of business acumen and willingness to take risks, and some even had access to venture capital. Whole communities like the Marwaris, Chettiars, and Armenians were known for their entrepreneurial culture.[2]

But entrepreneurship alone does not lead to economic development. The kinds of enterprises that could have led the colonies toward economic development required other elements that were in short supply in the colonial world. To create modern industries and businesses, entrepreneurs also needed information about foreign machines, technical processes, and business practices, information which was not forthcoming from the educational system. In other words, they had to be importers or creators of technological culture.

Only in rare instances did Asians and Africans enter occupations where they competed with Europeans on their own ground. A few did so in the technical professions. Others operated businesses in fields in which small size was not an impediment but an advantage. Rarest of all were the non-European entrepreneurs able to acquire the machines and expertise to compete on an industrial scale.

This chapter includes a number of examples of entrepreneurship and expertise, ranging from small-scale farmers to project engineers and sizeable corporations. While very diverse, they demonstrate the efforts that Africans and Asians made to go beyond both their traditional economies and the subordinate jobs available in colonial governments and European firms. They also illustrate the obstacles colonialism placed in the way of native enterprises.

The Smallholders

Though modern technology tends to gigantism, there are activities in which small size, simple techniques, and hard work can outcompete large organizations. We have already encountered one such case, that of the tin miners of Malaya, who held their own against large, well-capitalized Western firms until World War I. Their technology was not Western but Chinese, and their success rested as much on their more efficient exploitation of labor as on technical superiority. Yet the enterprising spirit of the Chinese miners and the geology of tin ores (in contrast to copper) coincided to favor them, for a time, against their Western rivals.

The realm of mining and minerals yields few such cases. More typical is the fate of the Indian ironworkers and Katangan copper miners, whose occupations vanished when Western enterprises got underway. It is rather in the realm of agriculture that one finds real competition between European-owned plantations and Asian or African small holders.

Our first case is that of the rubber growers of Malaya and the Dutch East Indies. The research conducted by the botanical gardens of Singapore and Peradeniya, the rubber planters' association of East Sumatra, and the Rubber Research Institute of Malaya (which we saw in Chapter 7) was designed to help European-owned estates. But seeds, seedlings, and methods of tapping and cultivation could not be reserved for an elite, and they soon found their way into the possession of small-scale farmers.

Until 1907 Asians hesitated to enter the rubber business, put off by the long delays and by bad memories of the coffee boom and bust of the 1890s. In Malaya the authorities also did what they could to discourage smallholders. Choice lands near roads and rail lines were reserved for companies. Ostensibly to protect the traditional Malay society, the Malay Reservation Enactment forbade the sale of Malay village lands to non-Malays. Not until World War I did the Agriculture Department employ Malays as extension agents.

Despite all this, Asians caught the rubber fever too. Beginning in 1907 they bought land and planted heveas. Compared to those of the European companies, their plots were tiny. In 1909–10, while 70 percent of the land grants to sterling companies were over 200 hectares, 63 percent of grants to Chinese were under 16 hectares, as were half the grants to Indians. Two-thirds of the grants to Malays were 4 hectares or less. Nevertheless, the share of hevea lands granted to Asians, which was virtually nil in 1906, grew faster than the European estates until 1915, after which it leveled off at 45 percent.[3]

Smallholders' methods differed radically from those of the estates. They did not prepare the ground by burning or clean-weeding as Europeans did, but planted the heveas amid the underbrush alongside their other crops, two or three times more densely packed than on the plantations. They invested only one-quarter as much per hectare as the companies did. Theirs being family farms, they could afford to tap their trees much sooner and more often than the cost-conscious estates. To estate managers and scientific experts, Asian smallholdings looked like little jungles, and the heveas on them seemed neglected and sickly. Yet their yields were higher than those of the estates, and their costs were lower. The smallholders' processing methods reproduced those of the estates on a smaller scale. While the resulting rubber was often dirty and of uncertain quality, the lower price they received for it was compensated for by their much lower expenditures on equipment. In good years, the rubber provided a welcome supplement to a family's income; when prices fell, as they did in the early 1920s, smallholders could turn to other crops or forest products and leave the heveas to recover until better times.[4]

The Depression almost ruined the natural rubber business and drove many Western-owned estates to bankruptcy. In 1934, in response to the drop in prices, the rubber-producing countries of Asia joined together in an International Rubber Regulation Scheme. Exports were strictly limited, and new plantings were forbidden except in Indochina. This brought relief to the planters and the big rubber companies, but at the expense of smallholders. During the late twenties, smallholders had planted half a million hectares with selected seeds supplied by the government. In allocating export quotas to European and Asian-owned plantations, the Malayan and Dutch East Indian agriculture departments underassessed the yields of the smallholdings because they looked less productive than they really were. Though smallholders produced half the rubber in those two colonies, they were only allocated four-tenths of the exports, and they were forbidden to start new groves.[5] Sir Andrew McFadyean, chairman of the British North Borneo Company and a member of the International Rubber Regulation Committee, summed up this policy: "One of the primary objects of the Rubber Control Scheme was to protect European capital in Malaya, Borneo, and the Netherlands East Indies from competition arising from the production of rubber by the natives at a fraction of the cost involved on European-owned estates."[6] In the end, Malayan and Indonesian smallholders held their own against the rubber estates only because they were willing to wait out the hard years of the Depression at a bare subsistence level.

The cocoa farmers of the Gold Coast were even more successful than their Asian counterparts at standing up to European competition. One reason is that cocoa is easier to raise and requires less skilled labor than rubber; another is that the Gold Coast government, unlike those of Malaya and the Dutch East Indies, seems to have favored smallholders.[7]

The Dutch tried to plant cocoa in the Gold Coast in 1815, and so did the Basel Mission in 1843 and 1857, but they failed. The first successful introduction of the cocoa tree is credited to Tete (or Tetteh) Quarshie, a Ga blacksmith who obtained seedlings from the island of Fernando Po in 1879. Soon he had a cocoa nursery at Mampong and was distributing pods and seedlings to other farmers. The first consignment of Gold Coast cocoa was shipped to Europe in 1885.

Sir William Brandford Griffith, governor of the Gold Coast from 1886 to 1895, took a special interest in agriculture, perhaps because of his West Indian background. It was he who established a botanical garden at Aburi in 1890, where cocoa, vanilla, coffee, nutmeg, and other crops were successfully raised. From Aburi, cocoa seedlings were distributed to farmers.

According to the economic anthropologist Polly Hill, the Akwapim people possessed some unusual characteristics. They were migrant farmers, willing to buy and sell land and operate farms in several localities at once; and they were eager entrepreneurs who saw property as a means to make profits, and profits as a means to buy

more property. Cocoa, a crop that does not need the constant attention that rubber does, suited them well.[8]

By 1900, cocoa was spreading throughout the southern Gold
Coast. From then until 1930 was a golden age. The Gold Coast became the world's foremost cocoa producer, and cocoa was its foremost crop. To take advantage of this situation, two European companies, African Plantations Ltd. and Scottish Co-operative Wholesale
Society Ltd., opened cocoa plantations with, respectively, 16,000 and
25,000 trees. They failed, as Seth Anyane explains, because "the local people had already discovered that it was far more profitable to
produce cocoa themselves on a peasant basis rather than work for
wages on the European plantation."[9] As a result, Europeans withdrew from production, though they remained in marketing. By the
1930s, the Gold Coast had some 150,000 African-owned farms aver-
aging 2.5 hectares apiece and producing an average of 1.5 tons of
cocoa a year.

The Gold Coast government, like other British colonial governments in West Africa, favored small-scale peasant agriculture and
discouraged settlers and plantations. It also provided such services as
extension agents, entomological and plant-pathological research, and
marketing. A few Africans were trained in agronomy; by the 1920s
two of them had risen to the rank of assistant superintendent of agriculture, a position never before occupied by Africans.[10]

The success of cocoa exposed the Gold Coast farmers to the
same troubles that befell the rubber smallholders of Asia: a collapse
in the price of their product during the Depression. A very commercialized agriculture, with its concomitant exposure to the vicissitudes
of the world market, made the people of the Gold Coast the most
highly politicized of all Africans by the end of the thirties, and the
first to demand and achieve independence after World War II.

Even when they successfully competed with European enterprises, small-scale miners and agriculturalists were marginally involved in the transfer of Western technology. More often, they adapted
a traditional technology like tree farming or irrigation to a new situation, or developed their own devices and techniques with local resources. This, of course, is the very "appropriate" or "intermediate"
technology which has been so highly extolled in the past few years. If
there ever was a time when intermediate technologies could have
flourished, it was in the colonial era when the gap between crafts and
industry was not as wide as it has since become. Yet we find in the
record of colonial history very few examples, other than those we

have just described, of successful intermediate technologies. Despite sporadic official backing, British attempts to make iron with charcoal in India were defeated by industrial coke-iron. Gandhi notwithstanding, the spinning wheel was a political, not a commercial device. Only a few artistic crafts such as pottery, brasswork, leather work, and handloom weaving survived the industrial onslaught, but precariously. In protected economies, such as those that existed in parts of Asia and Africa before and after the colonial era, other alternative technologies might have developed, given enough time and investments. Colonial economies, however, were open to world trade or at least to the products of their metropolis's industries. This limited the opportunities for intermediate technologies, differing from both the traditional and the industrial, to emerge. The most successful enterprises that arose in the colonies did so not by creating alternative technologies, but by acquiring the technologies of the West.

The Experts

Two sorts of Asians and Africans acquired Western technical culture: the experts through their education and experience, and the entrepreneurs by purchasing the expertise they needed.

Before 1940 there were almost no African engineers. The only one I have encountered was Herbert S. H. Macaulay, a member of the elite of Lagos, Nigeria. Patrick Cole, the historian of Lagos, describes his career:

> Macaulay was sent to England to study survey, civil engineering and piano tuning by the Lagos government as an appreciative gesture for the work of his grandfather, Bishop Ajayi Crowther. . . . When Macaulay returned to Lagos in 1893 he was employed as surveyor of Crown grants at the inadequate salary of £120 per annum, despite Governor Carter's recommendation that his minimum pay should be £250. His salary rose gradually to £200 a year, at which point he resigned from the civil service because of allegations by his white superior that he had abused his official position for private ends, and engaged in private practice.[11]

After that, he entered political life and did not practice engineering again.

In contrast to Africa, India had hundreds of working engineers even in the late nineteenth century. Almost all of them were subordinates. The very few who rose beyond that level did so by leaving government service and rising to prominence either in private enterprise or in the service of the Native States. Three cases will illustrate this process.

Several times already we have run across the name Pramatha Nath Bose. By all reports an outstanding geologist, he resigned from the Geological Survey of India in 1903 because of the preferential treatment accorded a European rival, his junior, Sir Thomas Holland. He subsequently worked for the Maharajah of Mayurbhanj and became famous for leading the Tatas to the iron ore deposits of Gurumaishini. In 1905 he joined the swadeshi movement; though anticolonialist, he continued to spread the culture of Western technology as the director of the Bengal Technical Institute. Finally, at the end of his career, in which he had contributed so much to the industrialization of India, he turned against "industrialism, with the consequent evils of militarism, mammonism, merciless exploitation of the weaker peoples, etc." and called for a return to nature and rural simplicity.[12]

Similar compromises and career changes are also present in the biographies of the two most prominent Indian engineers of the colonial period, R. N. Mookerjee and M. Visvesvaraya. Born in 1854, Rajendra Nath Mookerjee learned English at missionary schools and obtained his engineering degree from Presidency College in Calcutta, which prepared young men to become overseers and subordinates in the Public Works Department. As his biographer explains, "the cumulative pressure of the cribbed and confined education received by the youths of Bengal resulted in an almost complete obliteration of the adventurous spirit."[13]

Unlike his classmates, Mookerjee somehow retained that adventurous spirit; instead of entering government service, he struck out on his own as a private contractor. In Calcutta he was fairly well received and obtained a maintenance contract on the Palta waterworks. Upcountry, however, British officials were more entrenched in their prejudices. Though he submitted the lowest bid to build the Agra waterworks, he was refused:

> Hughes [Waterworks Chief Engineer] recommended acceptance, but the Chairman of the Municipality opposed on two grounds: one, that the firm was Indian and accordingly could not be relied upon for satisfactory execution of the work, and secondly, that it was a Bengali firm and Bengal in the eyes of the U.P. [United Provinces] civilian was, even in those days, an undesirable factor in politics.[14]

In order to get the contract to build the waterworks at Allahabad, Mookerjee joined a British engineer, Acquin Martin. Mookerjee supervised the work, while Martin imported the equipment from Britain. Upon the completion of the project in 1892, the two men decided to form a partnership:

> Several of Rajendra Nath's friends advised him to insist on the firm being called "Martin Mookerjee & Co.", but after careful deliberation and with a shrewd sense of the realities of commercial India in those days he did not make the suggestion, for he felt that association with an Indian name publicly would prejudice the chances of the firm in so far as Government patronage was concerned. . . . The decision was thus made that the firm be called just "Martin & Co."[15]

Subsequently, Martin and Company went on to build many other municipal waterworks, streetcar lines, and buildings, and to manage numerous industries around Calcutta. After Martin's death in 1906, Mookerjee became the sole senior partner. Official recognition followed his business success. During World War I he was appointed to the Indian Industrial Commission and became its most active investigator. Though often at odds with its other members because of his nationalistic views, he was nonetheless appointed to several other important bodies such as Acworth's Railway Committee in 1920 and Lord Inchcape's Retrenchment Committee of 1924. He summed up his views in these words: "Our political friends are busy translating their aspirations into terms of constitutions and are thinking about majorities, electorates and votes. I . . . picture an India of busy workshops, smoky factories, sanitary dwellings for the workpeople and eager money-getters."[16]

Like Mookerjee, Mokshagundam Visvesvaraya was a civil engineer. He graduated from the Poona College of Science in 1883. From then until 1909 he worked as an assistant engineer in the Bombay Public Works Department, supervising small irrigation projects, water-supply systems, and sewerage works. His autobiography is coy about the reasons for his resignation in 1909:

> For several years I had superseded a large number of seniors, at one time about 18 in number, on account of the special offices to which I was appointed. Government reverted some two or three of these officers to their former positions and I also learnt that there was some discontent on account of supersession. Remembering that there was political feeling in the country at the time, I thought there

was little chance of Government appointing me Chief Engineer except when my regular turn came according to my original rank. I thereupon decided to retire from the service of the Bombay Government.[17]

He soon found much more rewarding work in the Native States, first in Hyderabad where he spent a few months planning a flood-protection dam and a sewerage system, and then in Mysore where he became chief engineer. In that position, he assisted the Maharajah in developing the state's industries. His main work, though, was building a dam at Mettur on the Cauvery River. This dam created the Krishnaraja Sagar, India's largest reservoir. It was among the first multiple-use water projects in the world, providing irrigation for almost 50,000 hectares and electricity for the Kolar gold mines and the sugar mills of Mysore.[18]

In 1912 Visvesvaraya was appointed Dewan (prime minister) of Mysore. A convinced Westernizer, he tried to bring his backward state into the twentieth century. Like Bose, he devoted much of his time to technical education. He later wrote:

> I had been impressed in my previous travels abroad with the importance which the Western nations attached to education. I was convinced that the unsatisfactory economic condition in Mysore was due chiefly to neglect of education. My travels in Japan towards the closing years of the nineteenth century had created a deep impression on me in this respect. The Japanese leaders had found out the secret that education was the basis of all progress. The object which the Japanese Education Department has steadily kept in view was the training of the native mind to European ways of thinking and working.[19]

Under his Dewanship, a number of technical schools were founded, including schools of agriculture and mechanical engineering in Bangalore and a technical institute in the city of Mysore. In 1921 the Bombay government appointed him chairman of the Technical and Industrial Education Committee, which split along racial lines: the 7 Indian members recommended an institute of technology, but the 10 Europeans refused. In the end, the Bombay government was interested in apprenticeship programs, not higher education.

What conclusions can we draw from these cases? It was not impossible for an Indian to achieve great prominence in a modern technical profession, but it was extraordinarily rare, and it could only happen outside the normal career path, whether by associating with a British engineer as Mookerjee did, or by entering the service of a Native State, as did Bose and Visvesvaraya. In India the British did little to encourage native achievement but tolerated a certain amount of it. Nowhere else in the colonial world did this happen.

The Cotton Mills

The Indian cotton industry has attracted more attention than any other, and for good reason. It was the first Western-style industry on the subcontinent; it was developed mostly by Indian entrepreneurs; it competed successfully, without tariff protection, with the powerful cotton industry of Lancashire; and it managed to adapt Western methods to Indian labor conditions. Not surprisingly, entrepreneurship, protection, and labor conditions have dominated most discussions of this industry.[20] How the technologies of spinning and weaving came to India is our concern here.

A few Europeans tried to set up cotton mills in India in the early nineteenth century but failed. The first successful venture was that of the merchant and financier Cowasjee Nanabhoy Davar, whose Bombay Spinning and Weaving Company opened in 1854. Davar obtained ideas and advice from Messrs. Platt Brothers and Company of Oldham, manufacturers of textile machinery. Along with 5,000 throstles, Davar brought over an engineer, William Whitehead, and four other Lancashiremen to supervise the carding, spinning, and weaving departments.[21]

This set a pattern. While the American cotton industry was created by skilled immigrants, the Indian industry was built by Indian entrepreneurs who imported English machinery and hired English expatriate technicians.[22] Capital for Davar's mill and those of his many imitators came from the wealthy traders of Bombay. The first entrepreneurs were mostly Parsis, who had long been active in shipping and the import-export business. But it was not long before they were followed by other communities. Of the 95 mills in Bombay in 1914, 34 were controlled by Parsis, 27 by Hindus, 15 by Europeans, 10 by Muslims, and 5 by Jews. Joint ownership and intercommunity cooperation were also common.[23] The cotton mill industry grew rapidly, first in Bombay, then in Ahmedabad, and later in Madras, Cawnpore, Nagpur, and other centers. By 1914 India had 271 mills and exported yarn to much of the Far East.

For the first thirty years or so, the mills relied on carding, spinning, and weaving masters and other technicians from Lancashire. These men did not adapt easily to life in India. Writes S. D. Mehta, historian of the industry:

> The difficulties of language were unusually great, not only in rela-
> tion to the workers but frequently also in relation to the employers
> and other members of the latter's office. . . . He [the English
> worker] was not always readily accepted, being a man of relatively
> poor education and means, and devoid as he was of sophisticated
> manners by other classes of highbrow Englishmen who had set them-
> selves up into a caste of super-Brahmins.[24]

Yet they came, lured by good salaries of 250 to 450 rupees a month, plus allowances for housing, fuel, and servants. Their contracts stipu-lated that they should train Indians; for example, "the said John Smith shall impart all information, practical or theoretical, in all branches of his duties and to the best of his ability, to all native ap-prentices and jobbers, and to the workpeople under his charge without additional fee."[25] These Lancashiremen, however, were not schooled but apprenticed, and considered their skills to be trade secrets which they were most reluctant to pass on. Even Indian mill owners did not press them to because of what Mehta calls

> the halo that surrounded the mechanical genius of the European. It
> was almost universally accepted that the Englishman had a natural
> aptitude for the mechanical arts, and was fitted, as such, to occupy
> all the higher posts. . . . The rest of the Indians, according to ideas
> current then, did not possess any such aptitude, and therefore were
> discouraged from taking to these occupations.[26]

Nor was there, in the first thirty years of the industry, much pressure from educated Indians to compete for jobs with the skilled workers from Lancashire, despite their attractive pay. The reason, according to Mehta, was caste prejudice: "The conditions of work, involving as they did contact with classes of men who had been socially stratified differently for ages, would have been as unwelcome as it would have been abrupt."[27] And Rutnagur adds: "The life in a cotton mill or a workshop was looked upon as inferior and humiliating for the sons of the better class families and it was only when a youngster lagged behind in a school or college that he was relegated to a mill or fac-tory." Some young Indians did apprentice themselves to skilled work-ers; they obtained these apprenticeships through family or friends, and often paid 2,000 or 3,000 rupees for the privilege of learning one of the textile trades.[28] Nonetheless, it was not until the late 1880s, when the first Indians who had learned textile technology in school came on the scene, that the Lancashiremen began to give way to Indian cadres.

British machinery exporters and their agents in India supplied the mills on easy terms; some mills were even started by machinery importers like Greaves, Cotton and Company; Bradbury, Brady and Company; and the Wadias. Mill owners, like other capitalists in India, often contracted out the supervision of their mills to managing agents, some of whom were also machine importers. In such situations, equipment purchases benefited the agent rather than the mill. Arno Pease, a British textile expert who visited India, noted that some mills "are museums of all kinds of machinery from different machinists, as every time a mill was taken over the new mill agent was anxious to get some of the machinery into the mill which was made by a firm of textile machinists that he happened to represent at the time."[29] Though a commercial rival of Lancashire, the Indian cotton industry was, from the start, its technological satellite.

Importing ready-made machinery is the most cost-effective way of starting an industry in a late-industrializing country. But entrepreneurs still have a range of choices which can make it profitable to be technologically daring. The spirit of innovation set one mill owner, J. N. Tata, apart from the others and provided him with the fortune with which his firm later pioneered in the steel and hydroelectric industries.

Tata was a frequent traveler, driven by curiosity about foreign industrial methods. He first encountered the Lancashire cotton industry during a trip to England in 1864. Five years later, with money earned supplying the British expedition to Abyssinia, he bought a derelict oil mill and installed spinning machines in it. Dissatisfied with this venture, he sold it at a profit and went back to England to study textile machinery. Upon his return in 1874 he started the Central India Spinning, Weaving and Manufacturing Company. After a long search, he settled on the city of Nagpur in the Central Provinces as the site of a new mill, which opened in 1877 under the name of Empress Mill.

To operate it, Tata recruited a former railway employee, Bezonji Dadabhai Mehta, as general manager and a Lancashireman, James Brooksby, as technical manager. The Empress Mill was large by Indian standards, with 15,552 throstles, 14,400 mule spindles, and 450 looms. Unfortunately, the first machines were cheap and of poor quality. Two years later, having learned from the experience,

Tata replaced his machines with more expensive ones of the latest model.[30]

In 1883 Brooksby, on leave in England, sent two ring-spinning frames to Nagpur. The ring frame, an American invention, was faster than the throstle and required only unskilled labor. It was not well-received in Lancashire; according to Mehta, "the English millowners, who were then revelling in a spirit of smug self-assurance, prevented ring spinning from being given a fair trial in England."[31] Actually, as Tata realized, the ring spindle suited India with its short-staple cotton, unskilled labor, and large demand for coarse yarns better than it did the needs of Lancashire, which specialized in high-count yarns and fine cloth. After much experimentation, Tata decided to replace his throstles and mules with ring frames. When his usual supplier, Platt Brothers, refused to make them, Tata turned to a rival machine maker, Brooks and Doxey, and suggested improvements. The Empress Mill was thus the first in India to adopt the new ring frames. Thanks to these and other innovations, it was able to double its output and show an average profit of 20 percent during its first eighteen years. Other mills, both Indian and European-owned, soon followed suit, and within ten years India had a million ring spindles.[32]

In contrast to the Scottish jute mill owners of Calcutta, the Indian mill owners of Bombay tried to assert their independence from British expertise. In this, they were partially successful. With funds donated by the wealthy mill owners, Sir Dinshaw Petit and Sir Jamsetjee Jeejeebhoy, and the municipality of Bombay, Nowrosjee Wadia, a mill manager with an engineering background, started the Victoria Jubilee Technical Institute in 1889. At first it taught textile technology and mechanical engineering; later it added courses in electrical engineering and industrial chemistry. It was the first technical school in India with a clear industrial vocation, designed to train both managers and skilled workers. It had a predominantly English faculty. Though it admitted seventy-five students each year, its impact was slow in coming. For a long time, mill owners believed that apprenticeship on the job was better than schooling. Only after World War I did mill owners find "practical men" too slow in adapting to new techniques, and they began demanding diplomas of their technicians.[33]

Gradually, apprentices and graduates of the Victoria Jubilee Technical Institute replaced the foreigners. By 1895, 58 percent of the managers, engineers, and carding, spinning, and weaving masters in the Bombay mills were Indian. The trend continued after that, as the proportion of Europeans declined to 28.4 percent in 1925 and to 16.4 percent in 1940.[34]

Indianization of the managerial and technical staff was not accompanied, however, by sufficient innovations in equipment or processes. The Indians who took over the industry had been trained by Lancashiremen, sometimes in Lancashire itself, at a time when England was slipping badly on the world cotton market. India inherited not only British technology, but British technological obsolescence as well.[35]

Until World War I the Indian cotton industry had only one serious competitor, Lancashire. Their coexistence in an open market was due to the fact that they produced somewhat different products: Lancashire specialized in fine yarns and cloth, India in coarse yarns for both the power and the handloom industries. Already before the war, the Japanese were able to eat into Indian sales of coarse goods to China. After 1918, Japanese goods flooded the Indian market, to the consternation of Indian manufacturers who clamored for protection. Their appeals were heard, and by the Cotton Textile (Protection) Act of 1930 they effectively obtained control of their domestic market. In the process they abandoned their one-time foreign customers to the Japanese.

The reasons for the victory of the Japanese over the Indian cotton industry include labor conditions, finance, and management. But the heart of the matter was a growing gap in the productivity of labor. Before 1914, Japanese productivity was on a par with that of the better Bombay mills; it took one worker to supervise each loom. In the interwar period, Japanese productivity soared while Indian productivity stagnated. Between 1926 and 1935 the output of yarn per Japanese worker rose 63 percent, and cloth output 122 percent. By the 1930s a Japanese worker supervised, on the average, 6 looms or 56 spindles, while an Indian worker could handle only 2 looms or 32 spindles.

The rise in Japanese productivity was, in turn, due to technical advances by a more competitive management and a highly developed machine industry. In particular, the automatic Toyoda looms and new ring frames replaced older models, while new techniques of fiber blending, bleaching, dyeing, and printing made Japanese cloth as cheap as the Indian and as fine as the English. In a survey taken in 1930, Arno Pease found only two mills, the Buckingham and Carnatic in Madras, that had automatic looms. The rest of the industry, especially its oldest branch in Bombay, remained wedded to Lancashire methods and machines and lost out to Japan.[36]

Behind the growing technological gap was India's lack of a textile-machine industry. As with railway equipment, India provided a large enough market to warrant one, but not in open competition with British machines.[37] In peacetime, the cotton mills were better off with imported than with locally made machines. But the war, which stimulated all Japanese industries, left the Indian mills unable to replace their aging equipment, which prevented them from taking advantage of the sudden surge in demand.[38] As Nathan Rosenberg has shown, the capital-goods industry plays a disproportionately large part in developing technical skills and creativity, and this is precisely what labor-abundant, underdeveloped economies lack most.[39]

The Indianization of the Indian cotton mills was thus only partly successful. The entrepreneurs and the capital were predominantly Indian, to be sure (in contrast to the Calcutta jute industry, which remained in foreign hands until the 1920s). Technical expertise was gradually transferred from Englishmen to Indians by way of apprenticeships and schooling. But the machines were all imported, and the technological creativity needed to keep the industry competitive did not develop in time to rescue it from the decline of its role model, Lancashire.

Indian Shipping and Shipbuilding

Shipping is an unusual industry in that most of its factors of production are mobile. For a country to have a maritime trade, it needs only harbors and inland transportation; all the rest can be in foreign hands. In particular, it does not need a shipbuilding industry. Under colonial conditions, shipping and shipbuilding industries can disappear even while maritime trade grows. India is a case in point.

For centuries, Indians had built oceangoing ships. Until the mid-nineteenth century, Indian shipowners sent their vessels throughout the Indian Ocean and as far as China and the Indonesian archipelago.[40] Their ships were made of teak, a material that resists the worms and rot of tropical waters better than any northern wood. So good was the teak, and so skilled the Indian carpenters who worked it, that the East India Company and other Europeans based in India had many of their ships built locally. In the early days of steam, Indian shipyards even built a few steamers, though the engines were imported from Britain.[41]

What destroyed Indian shipbuilding was British iron. Without modern iron and mechanical industries, Indian shipbuilders could not compete with the Europeans after midcentury. Indian shipyards continued to build small wooden craft but no more oceangoing ships; some closed, while others became dry docks for foreign-built steamers. Not until after Independence did India launch its first metal ship.[42]

Indian shipping was almost completely eclipsed after 1850 by ships owned by Europeans. Even in the coastal trade of India, the competition from the P&O and British India lines and from British tramp steamers was more than Indian shipowners could withstand. The problem was not simply one of economic efficiency. The world of shipping was always highly political, and if French, Italian, or Japanese lines survived, it is because they had considerable help from their respective governments. Potential Indian shipowners found their government consistently favoring their strongest competitors, the British lines.[43]

A few locally owned shipping companies were able to coexist with the major British lines by associating Indian and Anglo-Indian business interests. Thus the Bombay Steam Navigation Company, founded in 1845 by a mixed board of British and Parsi members, was at various times managed by J. A. Shepherd (1860–1898), Hajee Ishmael and Hajee Ahmed (1898–1906), and Killick Nixon and Company (1906–1939). It had twenty-three steamers and ran a regular service along the Malabar coast, stopping at small ports that the British India Line ignored.

Another company which found a niche was the Bombay and Persia Steam Navigation Company, founded in 1877 by a group of Bombay Muslims to carry their coreligionists to and from Arabia, a trade that the British India disdained. It survived under the name "Mogul Line," after the Muslim emperors of India.

Not so lucky were those that dared confront the major lines in their own territory. In 1894, J. N. Tata, finding the P&O's freight rates for cotton yarn and piece goods to the Far East too high, decided to fight back. First he switched his business to the Italian Rubattino and the Austrian Lloyd lines, until they joined the Bombay–Far East Conference and raised their freight rates 46 percent. He then made an agreement with Nippon Yusen Kaisha, chartered two freighters, and offered to carry cotton to Japan at 12 rupees for every 3.7 cubic meters, compared to the P&O's rate of 19 rupees. The P&O retaliated by lowering its rate to 1.5 rupees, then offered to carry cotton free of charge for all customers who avoided Tata's ships. Tata wrote the secretary of state for India:

> With scores of Liners, English and foreign, plying in these waters which our petted and much glorified Anglo-Indian Company can afford, and perhaps found it good policy to tolerate, it is only jealous of a small enterprise like ours, and while it can lovingly take foreigners and possible future enemies of England to its bosom, it discards the poor Indian for whose special benefit it professes to have come to India and from whose pocket it draws the greater part of its subsidy.

This eloquent protest notwithstanding, the government did not interfere in the dispute, and Tata was forced to close his shipping enterprise. Thereupon the P&O raised its rates to 16 rupees.[44]

A similar fate befell the Swadeshi Shipping Company of Tuticorin, which tried in 1906 to compete with the British India line on the Ceylon-India route. The British India responded by cutting freights below cost and offering to take passengers free of charge. This time, however, Indian traders boycotted the British India. The authorities then stepped in. Swadeshi ships were rammed and port authorities held up their clearances. Finally the police arrested V. O. Chidabaram Pillai, its director and a disciple of the nationalist Bal Tilak, for attending a political meeting, whereupon the company collapsed.[45] As a result of these and similar incidents, Indian shipping companies had no more than 10 percent of India's coastal trade and 2 percent of its foreign trade before 1918.

As the political relations between India and Britain changed after World War I, so did shipping. In 1919 Walchand Hirachand, Narottam Morarjee, and other Bombay merchants bought the hospital ship *Loyalty* and founded the Scindia Steam Navigation Company. Walchand sailed his ship to London. For the return trip he had to book passengers and freight himself, because British travel agents and shippers boycotted the new line. The British India started the usual freight war. Its director, Lord Inchcape, then chairman of the powerful Retrenchment Committee of the Indian government, offered to buy the Scindia line, but Walchand refused his offer. In 1923 Scindia was admitted to the Indian Coastal Conference on condition that it carry only coastal freights and agree not in increase its fleet by more than seven ships in ten years. George Blake, the official historian of the British India Line, commented:

> The whole story of the Indian effort in the field of shipping, and of the little freight and rate wars that inevitably ensued, is long and really rather tedious. . . . Indian shipping did not get on its feet until the formation in 1929 . . . of the now well-established Scindia Company. James Lyle Mackay [Lord Inchcape] had his brushes with this concern, but the differences were smoothed out by . . . the Delhi Agreement of March, 1923: an instrument that could be regarded as distinctly liberal towards the Scindia interests.[46]

What Blake referred to as "distinctly liberal," Walchand Hirachand called "a slavery bond."[47]

Indian nationalists hailed the very survival of the Scindia line as a victory, and so it was. If the British India and P&O did not crush the Scindia as they had the Tata and Swadeshi lines, it was because the political climate had changed. Inside the Legislative Assembly and out, nationalists were becoming dangerously vociferous, demanding the reservation of coastal trade for Indian-owned ships, the training of Indians for the merchant marine, and government support for Indian shipping and shipbuilding. In 1921–22 the new Legislative Assembly adopted a resolution to encourage an Indian merchant marine but had no power to enforce it. In 1923 the Indian Mercantile Marine Committee went to work, in the usual ponderous fashion of government committees, to investigate the problem, distribute questionnaires, and take testimony from 168 witnesses. It discovered Indians and Britons sharply divided on the question of shipping. In March 1924 the committee recommended a policy of coastal reservation, reserving the coastal trade for Indian-registered, -owned, and -managed ships, and that Indians be trained to become ships' officers.[48]

Training was a sore point, however. From the beginning of British rule in India, Indian sailors, or lascars, served on British ships. In the East India Company's Bombay Marine, about half the crews had been Indian. In 1911 the British merchant marine employed over 45,000 lascars and 205,000 British subjects. As one maritime historian explained: "Lascars were constitutionally better in the stokehold than Europeans when a vessel was in the tropics. . . . No one suggested that Europeans and lascars should work together."[49]

Ships' officers, however, did not come up from the ranks of

sailors, but from educated middle-class families. Of all the fields
which Indians went to Britain to study, maritime training was the
most difficult to break into, and the merchant marine training schools
the most hostile to Indians. In 1922 the Committee on Indian Students in England, chaired by Lord Lytton, reported:

> (1) that the Committee of Management of the Training Ship "Conway" regard Indians as ineligible for admission to that ship on the ground that they are not "British born," . . . they would be unwilling to take them because of "the difficulties of religion, caste and feeding; also the mixture of races and colours in the confined space on the ship.
> (2) that the authorities of the Thames Nautical Training College (H.M.S. "Worcester") would be prepared to admit a limited number of Indians. . . .
> (3) that the Managers of the Nautical College, Pangbourne, are not prepared to entertain the suggestion that Indian cadets should be admitted to their College.[50]

The British shipping companies, even those that employed lascars
in great numbers, refused to admit Indians into their officers' apprenticeship programs. J. W. A. Bell of Mackinnon Mackenzie and
Company, agents for the British India and the P&O in India, blamed
the class structure of India for this state of affairs; as he told the
Merchantile Marine Committee in 1923: "There does not appear to
have been, at any time, any desire evidenced by the youth of the
country, of the better class, to adopt a seafaring career."[51]

After much procrastination, the government agreed to a compromise. While rejecting coastal reservation as "flag discrimination,"
it decided in 1925 to turn the troop ship *Dufferin* into a training ship
for Indian youths. In its first ten years the *Dufferin* turned out 132
navigation officers. Shipping lines were under no obligation to hire
them, however; the Scindia line took on 32 of them, while British
lines took only 10.[52]

Nationalists would not accept the defeat of the principle of
coastal reservation, which became one of Mahatma Gandhi's "eleven
points." The British lines continued to engage in rate wars against
small coastal shippers. The Depression, in shipping as in other businesses, did not soften the conflict but redirected it. Japanese shipping companies began making inroads into Indian coastal shipping.
In 1935 the new Government of India Act specifically forbade any
discrimination against British-registered shipping in India. At the
same time, the first Indian Commerce Member in the Indian govern-

ment, Sir Joseph Bhose, allocated parts of the coastal shipping to Indian-owned companies, among them the Scindia line.[53]

Apart from the epic of Gandhi, Indian politics in the 1930s do not make exalting reading. The petty squabbling and byzantine procrastinations of the period may have benefited the well-established shipping lines but left India with a tiny merchant marine: sixty-three ships and 132,000 gross registered tons divided among ten small lines. In World War II, when the British Empire desperately needed ships, India lost half its small fleet and had no shipyard to replace them. It was an ironic legacy from that association of two of the world's great trading nations, Britain and India.

Indianizing the Steel Industry

In contrast to the railways, the jute industry, and even the cotton industry, the Indianization of the steel industry stands as a model of a successful technology transfer. In less than three decades an industry based on foreign expertise replaced its foreign technicians with Indians, and profitably.

During his last trip to America, J. N. Tata had written to his sons his views on the talents of different nationalities in the steel business. Americans, he believed, made the best managers and blast-furnace personnel, while Germans excelled in the open-hearth furnaces, Englishmen in the rolling mills, and Welshmen in the coke ovens.[54] Tata's heirs followed his advice. From the start, the Tatas showed a clear bias toward Americans. In addition to the original team of Weld, Perin, Kennedy, and Sahlin, the Tatas hired a series of American general managers for the plant: Robert G. Wells, A. E. Woolsey, Barton Shover, T. W. Tutwiler, C. A. Alexander, and John L. Keenan. Many of these men, in fact, had worked together in the steel mills of Gary, Indiana.

The Tatas had originally intended to have over 300 foreigners in a total workforce of 4,000. But foreigners were expensive. Between October 1911 and June 1912, TISCO employed 119 foreign workers at an average salary of 4,966 rupees, compared to an average of 158 rupees for its 2,759 local employees. In addition, it had to offer the foreigners free passage to and from India, free housing, and even a racetrack for their entertainment.

When the war broke out, the firm faced new problems. The Germans who operated the open-hearth furnaces were suspected of plotting to sabotage the steel works, and armed guards had to be posted next to each one until replacements could be brought in from America. As the Germans were paid less than other foreigners, their replacement raised the average wage level. Americans, unlike Britons, had a high turnover; of the original contingent of 28, only 8 remained over eighteen months. This too raised their costs. Between July 1915 and June 1916 the firm only employed 82 covenanted workers, but at an average of 6,862 rupees, an increase of 38 percent; in contrast, its local workforce of 4,156 workers were paid 199 rupees a year on the average, an increase of 26 percent. TISCO had begun replacing foreigners with Indians wherever it could.[55]

Yet in the early 1920s, TISCO's most valuable skilled workers were still foreigners. In the blast-furnace department, the superintendent, assistant superintendent, and 6 general foremen were foreign; the open-hearth department was run by 40 foreigners and 3 Indians. As the construction of the "greater extensions" proceeded, the plant required more, not fewer, foreign workers. Their number rose from 121 in 1914–15 to a peak of 229 in 1924. In 1921–22 their average salary was 13,527 rupees a year, 56 times that of the average Indian (240 rupees).[56] The high cost of foreign workers would have been a sufficient reason to replace them with Indians. Yet the company had other incentives. In 1920 the foreign workers went on strike for a week, and TISCO management saw its long-run goal of Indianization turned into an urgent necessity.[57]

Indianization involved all levels of employees, technical and supervisory personnel as well as workers. None was easy to obtain. It is sometimes asserted that underdeveloped countries have an abundance of unskilled labor. For the steel industry, this was a fallacy. Workers strong and fast enough to do steel mill work were hard to find, and they had high rates of turnover and absenteeism because of their residual attachment to seasonal agricultural work and their observance of religious holidays. The result was that TISCO's labor force was unproductive, and labor costs per unit of output were higher than in the West.

The Indianization of the technical and supervisory staff presented a more difficult problem, for such jobs involved skills that could only be acquired by deliberate schooling, not on the shop floor. The company went about this in two ways. The first method was to send key employees abroad and to hire foreign-trained Indians. This process began during World War I, when several of TISCO's Indian managers went to Japan to observe a Japanese steel mill. In 1915, S. Chose, a Bengali trained by the General Electric Company, became chief engineer of the plant's electrical department. Similarly, A. C. Bose, a graduate of Carnegie Institute of Technology, became chief chemist in the chemical laboratory. A number of other TISCO employees were sent to the United States for training at General Electric and Westinghouse. Jehangir Ghandy began as an apprentice in the blast-furnace department in 1921 and was later sent to Carnegie Tech to study metallurgical engineering; in 1937 he became the first Indian general manager of TISCO.[58]

But such a program was clearly too expensive for any but the top personnel. What was needed, the company realized, was a technical institute in Jamshedpur itself. The idea must have been discussed for some time, since the Indian Industrial Commission had commented favorably on it in 1918.[59] In 1919 Dorabji Tata announced plans for what was to become the Jamshedpur Institute of Technology. Two years later it admitted twenty-three students with a Bachelor of Science degree. At first it offered a three-year post graduate program combining theory with practical training at the steel plant. John Keenan commented on the education of the students: "The young Technical Institute boys were now taught that a steel man should go to England, watch steel-plant operation, and see what not to do; he should go to America and learn how to avoid spendthrift practice; then he should go to Germany and see the job done right."[60] In 1927 it added an Apprenticeship School, a five-year program that admitted fifteen- to eighteen-year-old boys, preferably the sons of TISCO employees, to become fitters, welders, machinists, pattern makers, and the like. Then in the early 1930s it added a Technical Night School with courses for nurses, masons, telegraph operators, and other skilled workers.[61] John Keenan, one of the last Americans at TISCO, wrote: "I doubt very much if Tata's could ever have achieved the Indianization of the technical personnel of the plant without the Institute."[62]

With the graduates of its school and on-the-job training of other workers, TISCO was able to reduce its foreign work force from 229 in 1924 to 122 in 1930, and to 64 in 1934. Indian skilled workers, supervisors, and engineers cost the company, on average, half as much as the foreigners they replaced. In addition, experience and better equipment made the remaining work force more productive. Indianization had several consequences. Labor productivity rose sharply, from 5.52 tons of finished steel per employee in 1923–24 to 36.32 tons in 1935–39. This brought TISCO's direct labor costs per ton of output in line with labor costs elsewhere, kept the steel plant competitive, and provided employment.[63] TISCO's actions constituted an industrial declaration of independence a decade before political independence, the reverse of what happened in other colonial territories.

Conclusion

The experts and enterprises we have considered in this chapter stood between two worlds. They were ethnically African or Asian, but they engaged in activities that used Western technology or competed with Western enterprises. In addition to the usual obstacles to economic activity in underdeveloped countries—cultural barriers, fluctuations in the world market, lack of capital, changes in Western technology—they also faced the attitudes of their colonial governments.

Some succeeded mainly on their economic merits. The rubber and cocoa smallholders and the tin miners succeeded for a time because their labor-intensive techniques were more cost-effective than the large, overcapitalized enterprises of their Western rivals. In the cotton industry, labor and capital could be substituted more smoothly than in other fields, and the commercial skills and capital available in Bombay were sufficient to defeat their European rivals. In contrast, the swadeshi enterprises of Bengal failed mainly for commercial and technical reasons.

Yet in all the activities we have seen, politics was a factor. Even the Indian cotton industry flourished on its own merits only in the manufacture of the coarser yarns, and it did not take over the rest of the Indian market until it got tariff protection in the 1930s. Politics also very obviously determined the success of the steel industry and the failure of shipping and shipbuilding.

Similarly, the lives of the eminent technicians were determined by politics. Mookerjee's success was not due only to his merits as an engineer and contractor, but also to his partnership with Martin, which gave him opportunities denied to other Indian entrepreneurs. Bose and Visvesvaraya found their niche in the employ of Native States, Indian entrepreneurs, and nationalist organizations.

Given these experiences, it is not at all surprising that Africans and Asians emerged from the colonial period convinced that the road to development led through politics, not through private entrepreneurship and laissez-faire.

Notes

1. There is a voluminous literature on the cultural obstacles to entrepreneurship and economic development in India. See, for example, Bankey Bihari Misra, *The Indian Middle Classes: Their Growth in Modern Times* (Oxford, 1961), p. 251, and India, Indian Industrial Commission, *Report of the Indian Industrial Commission, 1916–18* in *Parliamentary Papers,* 1919 (Cmd. 51), vol. 17, p. 66. On the communities, see Rajat K. Ray, *Industrialisation in India: Growth and Conflict in the Private Corporate Sector, 1914–47* (Delhi, 1979), p. 25, and Thomas A. Timberg, *The Marwaris, from Traders to Industrialists* (New Delhi, 1978), pp. 56–79. Parasitic landlordism is emphasized by Barrington Moore, *Social Origins of Dictatorship and Democracy: Land and Peasant in the Making of the Modern World* (Boston, 1966), pp. 354–70. Entrepreneurship is discussed in Amiya Kumar Bagchi, "European and Indian Entrepreneurship in India, 1900–1930," in *Elites in South Asia,* ed. E. Leech and S. N. Mukerjee (Cambridge, 1970), pp. 223–57, and *Private Investment in India, 1900–1939* (Cambridge, 1972), pp. 165–70. See also Francis G. Hutchins, *The Illusion of Permanence: British Imperialism in India* (Princeton, 1967), p. 129; Helen B. Lamb, "The 'State' and Economic Development in India," in *Economic Growth: Brazil, India, Japan,* ed. Simon S. Kuznetz et al. (Durham, N.C., 1955), pp. 464–95; and Morris David Morris, "Private Investment on the Indian Subcontinent, 1900–1939: Some Methodological Considerations," *Modern Asian Studies* 8, pt. 4 (October 1974).

2. See Philip D. Curtin, *Cross-Cultural Trade in World History* (Cambridge, 1984), chap. 9; and Timberg.

3. Colin Barlow, *The Natural Rubber Industry: Its Development, Technology, and Economy in Malaysia* (Kuala Lumpur, 1978), p. 26.

4. Ibid., pp. 25–41, 71; J. H. Drabble, *Rubber in Malaya, 1876–1922: The Genesis of an Industry* (Kuala Lumpur, 1973), pp. 70–77, 100, 205–9; T. A. Tengwall, "History of Rubber Cultivation and Research in the Netherlands Indies," in *Science and Scientists in the Netherlands Indies,* ed. Pieter Honig and Frans Verdoorn (New York, 1945), pp. 350–51.

5. Chong-Yah Lim, *Economic Development of Modern Malaya* (Kuala Lumpur, 1967), pp. 76–82; Barlow, chap. 3; Tengwall, pp. 349–50.

6. Barlow, p. 72.

7. On cocoa in the Gold Coast, see Seth La Anyane, *Ghana Agriculture: Its Economic Development from Early Times to the Middle of the Twentieth Century* (London, 1963), and Polly Hill, *The Migrant Cocoa-Farmers of Southern Ghana: A Study in Rural Capitalism* (Cambridge, 1963).

8. Hill, pp. 178–82.

9. Anyane, p. 40.

10. Ibid., pp. 15–20.

11. Patrick Cole, *Modern and Traditional Elites in the Politics of Lagos* (London, 1975), p. 111.

12. Pramatha Nath Bose, *National Education and Modern Progress* (Calcutta, 1921).

13. Kailash Chandra Mahindra, *Sir Rajendra Nath Mookerjee* (Calcutta, 1933), p. 110.

14. Ibid., p. 150.

15. Ibid., p. 168.

16. Ibid., pp. 236–37.

17. Mokshagundam Visvesvaraya, *Memoirs of My Working Life* (Bangalore, 1951), p. 29.

18. Ibid., pp. 42–49; Henry Cowles Hart, *India's New Rivers* (Calcutta, 1956), pp. 28–41.

19. Visvesvaraya, p. 65.

20. S. M. Rutnagur, *Bombay Industries: The Cotton Mills* (Bombay, 1927); S. D. Mehta, *The Cotton Mills of India, 1854 to 1954* (Bombay, 1954); Morris David Morris, *The Emergence of an Industrial Labor Force in India: A Study of the Bombay Cotton Mills, 1854–1947* (Berkeley, 1965).

21. Mehta, p. 100.

22. Rutnagur, p. 9.

23. Morris David Morris, "The Growth of Large-Scale Industry to 1947" in *The Cambridge Economic History of India,* vol. 2: *c. 1757–c. 1970,* ed. Dharma Kumar (Cambridge, 1983), pp. 580–81.

24. Mehta, p. 101.

25. Rutnagur, p. 291.

26. Mehta, p. 103.

27. Ibid.

28. Rutnagur, p. 573.

29. Arno S. Pease, *The Cotton Industry of India: Being the Report of the Journey to India* (Manchester, 1930), p. 60. See also Sung Jae Koh, *Stages of Industrial Development in Asia: A Comparative History of the Cotton Industry in Japan, India, China, and Korea* (Philadelphia, 1966), p. 102, and H. Fukazawa, "Cotton Mill Industry" in *The Economic History of India,* ed. V. B. Singh (New Delhi, 1965), pp. 223–27.

30. Frank Reginald Harris, *Jamsetji Nusserwanji Tata: A Chronicle of His Life,* 2nd ed. (Bombay, 1958), p. 29; Mehta, pp. 57–59.

31. Mehta, p. 43.

32. Sunil Kumar Sen, *The House of Tata, 1839–1939* (Calcutta, 1975), pp. 21–22; B. S. Saklatvala and K. Khosla, *Jamsetji Tata* (New Delhi, 1970), pp. 16–30; Dinsha Edulji Wacha, *The Life and Work of J. N. Tata,* 2nd ed. (Madras, 1915), pp. 28–35; Harris, pp. 14–35; Mehta, pp. 43–44.

33. Ramkrishna M. Chonkar, *Twenty Years of Technical Education in Bombay, Being a Record of Twenty Years' Progress of the Victoria Jubilee Technical Institute, Bombay, 1887–1907* (Bombay, 1908); P. N. Joshi, "Training Technical Personnel for the Textile Industry of India: The Role of the Victoria Jubilee Technical Institute, Bombay," in *The Indian Textile Journal, Special Souvenir Number* (Bombay, 1954), pp. 562–65; Morris, "Growth," p. 583; Indian Industrial Commission, *Report,* pp. 27–28, 105–6.

34. Rutnagur, pp. 288–94; Mehta, p. 58; Daniel Houston Buchanan, *The Development of Capitalistic Enterprise in India* (New York, 1934), p. 211.

35. Ibid., pp. 205–6; Bagchi, "European and Indian Entrepreneurship," p. 256.

36. Pease, pp. 118, 158–59; Gilbert Ernest Hubbard, *Eastern Industrialization and Its Effect on the West,* 2nd ed. (London, 1938), pp. 294–301; Koh, pp. 61–64, 157–61; Mehta, pp. 164–65.

37. Rutnagur, pp. 630–31.

38. Ray, pp. 63–64, 193–95.

39. Nathan Rosenberg, *Perspectives on Technology* (Cambridge, 1976), pp. 149–57.

40. T. S. Sanjeeva Rao, *A Short History of Modern Indian Shipping* (Bombay, 1965), pp. 46–54; Dipesh Chakrabarty, "The Colonial Context of the Bengal Renaissance: A Note on Early Railway-Thinking in Bengal," *Indian Economic and Social History Review,* 11, no. 1 (March 1974): 97; Radhakumud Mookerji, *Indian Shipping: A History of the Sea-Borne Trade and Maritime Activity of the Indians from the Earliest Times* (London, 1912), pp. 244–52.

41. Henry T. Bernstein, *Steamboats on the Ganges: An Exploration in the History of India's Modernization through Science and Technology* (Bombay, 1960), pp. 28–31; H. A. Gibson-Hill, "The Steamers Employed in Asian Waters, 1819–39," *The Journal of the Royal Asiatic Society, Malayan Branch* 27, pt. 1 (May 1954): 127–61; Daniel Thorner, *Investment in Empire: British Railway and Steam Shipping Enterprise in India, 1825–1849* (Philadelphia, 1950), pp. 127–33; Victor F. L. Millard, "Ships of India, 1834–1934," *Mariner's Mirror* 30 (1944): 144–53.

42. Rottonjee Ardeshir Wadia, *The Bombay Dockyard and the Wadia Master Builders* (Bombay, 1955), pp. 329–60; Max E. Fletcher, "The Suez Canal and World Shipping, 1869–1914," *Journal of Economic History* 18 (1958): 569–70; Sanjeeva Rao, pp. 134–35.

43. According to T. N. Kapoor, 85 percent of the coastal trade of India was in foreign hands in 1924–25; see his "Shipping, Air and Road Transport" in Singh, p. 351. In 1939, it was still 60 percent foreign, according to Sanjeeva Rao, pp. 128–30.

44. Sanjeeva Rao, pp. 67–70; Asoka Mehta, *Indian Shipping: A Case Study of the Workings of Imperialism* (Bombay, 1940), pp. 34–43; Harris, pp. 92 ff.

45. Ray, pp. 94–95; Sanjeeva Rao, pp. 71–73.

46. George Blake, *B. I. Centenary, 1856–1956* (London, 1956), p. 170.

47. N. G. Jog, *Saga of Scindia: Struggle for the Revival of Indian Shipping and Shipbuilding: Golden Jubilee Volume* (Bombay, 1969), pp. 17–43.

48. Asoka Mehta, pp. 49–64; Jog, pp. 44–73.

49. John Frederick Gibson, *Brocklebanks, 1750–1950,* 2 vols. (Liverpool, 1953), 2: 8–9; See also Millard, p. 144.

50. *Report of the Committee on Indian Students in England,* p. 51.

51. Sanjeeva Rao, p. 98.

52. Ibid., pp. 92–106; Jog, pp. 58–60; Ray, p. 105.

53. Sanjeeva Rao, pp. 107–30; Kapoor, p. 352.

54. John L. Keenan, *A Steel Man in India* (New York, 1943), p. 38.

55. Ibid., pp. 14, 41–46; William Arthur Johnson, *The Steel Industry of India* (Cambridge, Mass., 1966), p. 43.

56. Buchanan, pp. 211, 321.

57. Sen, *House of Tata,* pp. 66–67.

58. Ibid., p. 70; Keenan, pp. 108–9, 133–35.

59. Indian Industrial Commission, *Report,* p. 101.

60. Keenan, pp. 173–75.

61. Ibid., pp. 135–40; Sen, *House of Tata*, p. 67; Verrier Elwin, *The Story of Tata Steel* (Bombay, 1958), pp. 68–69.

62. Keenan, pp. 140–41.

63. Ray, p. 91; Johnson, p. 42; Morris, "Growth," pp. 628–29.

11

Technology Transfer and Colonial Politics

One of the boasts of the imperialists was that they removed the barriers to European trade and investment, opening up the non-Western world to such blessings of Western technological civilization as railways, telegraphs, and sanitation. In many parts of the European empires, the transfers were massive and had penetrated deeply even before 1914. Botanical research transformed the economies of Southeast Asia. Barrages and canals quadrupled the agricultural output of Egypt and large tracts of India. Mines brought the outside world into central Africa. Railways crisscrossed the Indian subcontinent, permitting millions of people to travel. Harbors, shipping lines, and the telegraph linked the tropical lands to the rest of the world. And at the nodal points of transport and communications, towns grew into great cities. Enough was achieved to justify the imperialists' claim that in just a few decades their work had overshadowed the monuments of past empires.

Imperialism is sometimes presented as a political system set up and maintained to serve the interests of the European business class. No doubt, many businesses profited from imperialism. But it was less because imperialism opened up business opportunities than because businesses were able to profit from the peculiar politics of empire in ways they could not have in independent non-Western countries. Trade did not so much follow the flag as come wrapped in it.

The various technologies we have considered here were all of a commercial or economic nature, and on a large scale. And almost all of them came wrapped in flags. British investors would not have built railways in India without a guarantee. Shipping lines got mail contracts. Cable companies got exclusive landing rights or subsidies.

Mining companies got concessions and favorable labor laws. Enterprises owned by colonial subjects flourished with government help, like TISCO, or struggled without it, like the Indian shipping companies. Rare were those, like cotton mills or tin mines, that ever lived in a free-market atmosphere. In contrast, many enterprises, like harbors, irrigation works, telecommunications networks, and botanical research stations, were largely public. In other words, the new economic activities that arose in the colonies, and the transfers of technology that made them possible, were essentially political in nature.

Economic and technological changes took a different path in the colonies than they did in the West or in the independent non-Western nations. This was not only because of the basic poverty of the tropics, or their economic exploitation by Europeans, or their traditional cultures, although these factors did play a part in the process. The peculiar shape of change occurred because of the motives of their political elites.

The Europeans who ruled the colonies were in an ambiguous position. On the one hand they represented a conquering civilization which obtained its power from ingenious innovations, and they certainly shared the Western love of new devices and the urge to proselytize their technomania among the "backward races" of the world. On the other hand, they were conservatives at heart who hoped to shield the societies they ruled from the dangerously disruptive social forces that came with exposure to the world market and to Western ideas.

Like other ruling elites, colonial officials pursued many goals: their prestige, authority, and security; the external security of their territories and of the empires they belonged to; the properity of their European homelands; the welfare of their subjects; and their personal wealth and comfort—though not necessarily in that order. They were also responsible to several different constituencies at once: the European government they served, their colleagues and other European colonials, and their indigenous subjects. Their multiple motives and constituencies help explain their attitudes toward particular technologies.

Europeans in the colonies were especially enthusiastic about the new means of transportation and communication, and they frequently joined interests in Europe in lobbying for these innovations. Railways, ships, and telegraphs served their needs as administrators and increased the prosperity of their country and the treasury of their colony. Their contribution to the welfare of their indigenous subjects (trade, pilgrimages, or famine relief) were a welcome side effect. A final reason is seldom found in official documents but appears vividly in literary accounts of colonial life, namely the comfort, convenience, and personal prestige which modern means of transportation and communications provided in tropical countries to those who could afford them.

Similarly, colonial officials supported the construction of European-style housing and official buildings in the colonial cities, and sometimes they built entirely new cities like Dakar and New Delhi that reflected their sense of power and prestige. The new colonial cities and neighborhoods incorporated Western technologies as well as Western styles: water supply and sewerage systems, electricity, streetcars, hospitals, railways, telephones, paved roads, and other amenities.

Those technology transfers which increased the production of tropical products and their export to the West received official blessing and administrative support from colonial officials, for such technologies benefited private European interests in the colony and back home, and improved the balance of payments and the tax base of the colony. To many colonial officials, however, rapid economic changes, especially those brought on by the greedier and more exploitative enterprises, were unsettling to the indigenous populations, hence politically risky. As self-styled guardians of the native peoples, colonial officials had mixed opinions about such enterprises, and one finds a spectrum of reactions, from wholehearted enthusiasm to extreme reluctance.

Some technologies served mainly the interests of the native populations: public health services, famine railways and irrigation systems, and municipal services in the native quarters, to name a few. These received official blessing, but only within the narrow limitations of tightfisted colonial budgets.

Lastly, those technology transfers which might have led to the growth of import-substitution industries were generally viewed with suspicion. Not only would they have competed with European manufactures, but they threatened to bring forth native industrialists, engineers, technicians, and factory workers who would have challenged the authority of the colonial regimes. Many colonial officials were from the English gentry and the French petite bourgeoisie, which had lost ground to industrialists and workers. They had gone to the colonies to become enlightened aristocrats—sahib, effendi, chef—ruling simple peasants. They did not yearn to preside over another industrial revolution.

Not only were colonial officials inclined toward some technological systems and away from others, they also favored the geographic relocation of technology over its cultural diffusion. Every time a new process or piece of equipment was introduced into a colony, it came with European experts to set it up and to operate it, and sometimes to pass their jobs on to their sons. How else can one explain that Europeans were still driving locomotives in India ninety years after the first railway was built? Of course it would have been prohibitive to use Europeans in semiskilled or unskilled jobs, and therefore the colonial governments encouraged technical training up to a certain level: to secondary schools in Africa, and to colleges in Egypt and India. But education to handle the newest technologies, or to install and manage complex systems, was reserved for Europeans as long as possible.

If the colonizers were in an ambiguous position, so were the colonized. Western technology had led to their defeat and captivity and threatened their culture and way of life. No one illustrates their ambivalent attitudes toward Western technology quite as well as Mohandas K. Gandhi, who wore handwoven garments made of homespun yarn but also used a watch, traveled by train, and kept in touch with his followers by telephone.

It is not surprising that Asians and Africans had ambivalent feelings about Western technology, or that many found it difficult to adjust to. What is more remarkable is that Europeans found it so easy to believe that cultural obstacles prevented Asians and Africans from learning to operate Western machinery. The European bias became a self-fulfilling prophecy when colonial subjects were denied the opportunity to study the highest levels of technology or, if they did, to compete in the technical professions with Europeans. To Europeans in the colonies, native cultural obstacles were simply too useful.

Despite their cultural ambivalence, many Asians and Africans saw Western technology as their key to power and prosperity, and they sought more machinery and knowledge than the Europeans offered them locally. Yet even under the best of circumstances, importing technology was hazardous. Small businesses like the Chinese tin mines of Malaya could not afford to keep up with the rising cost of Western machines. Politically inspired entrepreneurs, like the swadeshi businessmen, were crushed by the competition. Entrepreneurs with both business skills and capital were rare, and those who understood Western machines rarer still. If the Tatas have appeared several times in this book, it is because there were no others of their caliber in India before 1920, or elsewhere in the colonial world before 1940. Yet they too were at the mercy of government policies that tolerated their cotton mills, suppressed their shipping, and encouraged their steel plant.

Even when they sought Western technology, nationalists and entrepreneurs differed from their European counterparts in their attitude toward the transfer process. Unlike Europeans, they favored cultural diffusion over geographic relocation. Indian cotton mill owners were slow to train Indian technicians, but eventually they did so, in contrast to the Scottish jute mill owners. Several of the entrepreneurs went beyond meeting their own needs and founded technical schools and institutes of science and technology. Although their attempts to import Western technology were overshadowed by the massive relocations of machines and experts undertaken by Europeans, their efforts contributed to more distant goals of technological maturity and economic development.

But nationalists and indigenous entrepreneurs only began to influence policy after World War I in India and Egypt, and after World War II in other colonies. During the long delay, the policies of the European businessmen and colonial officials had unanticipated side effects. Traditional handicraft industries damaged by imports were often not replaced by modern industries. As the multiplier effects of a growing demand were leaked to the West, the tropics grew more dependent on Western machines and experts, and vulnerable to the whipsaw fluctuations of the world market and to the development of synthetic substitutes. Meanwhile, the tropical populations had begun their phenomenal expansion, creating ever more millions of poor and ignorant people without the skills to break out of their poverty.

We began this book by contrasting Karl Marx's prediction that India would soon industrialize with T. S. Ashton's remark, a century later, about "the lot of those who increase their numbers without passing through an industrial revolution," and we asked why the transfer of technology that was already evident in the mid-nineteenth century did not lead to the kind of economic development that the West had experienced

The difference between growth and development is a matter of innovation, entrepreneurship, and diversification; in other words, hu-

man capital. Simply stated, the reason the tropics experienced growth
but little development under colonial rule is that investments went
into physical not human capital, and that the transfer of technology
was more geographic than cultural.

Although reluctant to predict the future, historians sometimes
like to speculate about alternative pasts. What would have happened
to Asian and African societies if Europeans had not conquered
them? Perhaps they could have preserved their traditional ways for
a few generations more, like China, Afghanistan, or Arabia; or they
might have become modern and industrial, like Japan. But it is
unlikely that they would have become both modern and underdeveloped, as they did under colonial rule.

Bibliographical Essay

The sources I have consulted in the writing of this book have been listed in the notes. Among them, a few were sources not only of information, but also of ideas and inspiration. These are works I especially recommend to the reader wanting to learn more about technology transfer in the colonial world.

Chapter 1 Imperialism, Technology, and Tropical Economies

The economic development of the tropics has stimulated a large literature among economists, but most of it lacks historical perspective. Only a few writers have sought to connect tropical underdevelopment with the dynamics of Western technological change over the long run. The classic work of Paul Bairoch, *Révolution industrielle et sous-développement* (Paris, 1963), is one such attempt; though overly schematic and often controversial, it focuses attention on the gulf that separates the first industrial revolution, which spread easily, with later phases of industrialization, which have been cultural and economic minefields for latecomers. Nathan Keyfitz's seminal article, "National Population and the Technological Watershed," *Journal of Social Issues* 23 (January 1967): 62–78, calls attention to the relations between Western technology, the demand for tropical products, and the growth of tropical populations, and draws pessimistic conclusions. Three economic histories have presented a much more sanguine view of the fate of tropical countries: *Tropical Development, 1880–1913*, edited by W. Arthur Lewis (London, 1970), and the two books of A. J. H. Latham, *The International Economy and the Underdeveloped World, 1865–1914* (London and Totowa, N.J., 1978), and *The Depression and the Developing World, 1914–1939* (London and Totowa, N.J., 1981), emphasize the vitality of the tropics and conclude that neither colonialism nor Western technology held them back.

Chapter 2   Ships and Shipping

There is a huge literature on ships and shipping, especially British ship-
ping. Among the best general works are Carl E. McDowell and Helen M.
Gibbs, *Ocean Transportation* (New York, 1954), and Ronald Hobhouse
Thornton, *British Shipping* (London, 1939). The connections between
British maritime activity and the British Empire, especially India, are the
subject of two fine, if somewhat dated, books: Halford Lancaster Hoskins,
British Routes to India (London, 1928), and Daniel Thorner, *Investment
in Empire: British Railway and Steam Shipping Enterprise in India, 1825–
1849* (Philadelphia, 1950). Max E. Fletcher's "The Suez Canal and
World Shipping, 1869–1914," *Journal of Economic History* 18 (1958):
556–73, provides the best brief coverage of that important topic.

Chapter 3   The Railways of India

The Indian railways have received a good deal of attention from historians. The best general histories are M. A. Rao's *Indian Railways* (New
Delhi, 1975) and J. N. Westwood's *Railways of India* (Newton Abbot,
England, and North Pomfret, Vt., 1974). The early finances of the railways are covered in Thorner's *Investment in Empire,* and their construction and technical details in George Walter Macgeorge's *Ways and Works
in India: Being an Account of the Public Works in that Country from the
Earliest Times up to the Present Day* (Westminster, 1894). Two articles
by Fritz Lehmann have focused on the supply of locomotives: "Great
Britain and the Supply of Railway Locomotives to India: A Case Study
of Economic Imperialism," *Indian Economic and Social History Review*
2, no. 4 (October 1965): 279–306, and "Empire and Industry: Locomotive Building Industries in Canada and India, 1850–1939," *Proceedings
of the Indian History Congress* (40th Sessions, 1979, Waltair, India):
985–96.

Chapter 4   The Imperial Telecommunications Networks

Since World War II, scholars have paid surprisingly little attention to the
global telecommunications networks, and a general history of the subject
has yet to be written. A summary of the history of telegraph cables can
be found in Bernard S. Finn, *Submarine Telegraphy: The Grand Victorian Technology* (London, 1973). Three dated but still useful books are
Maxime de Margerie, *Le réseau anglais de câbles sous-marins* (Paris,
1909?); Charles Lesage, *La rivalité anglo-germanique, Les câbles sous-
marins allemands* (Paris, 1915); and Frank James Brown, *The Cable and
Wireless Communications of the World: A Survey of Present Day Means*

of International Communication by Cable and Wireless, containing Chapters of Cable and Wireless Finance (London, 1927). Two important aspects of cable communications are treated in Jorma Ahvenainen, *The Far Eastern Telegraphs: The History of Telegraphic Communications between the Far East, Europe and America before the First World War* (Helsinki, 1981) and P. M. Kennedy, "Imperial Cable Communications and Strategy, 1870–1914," *English Historical Review* 86 (1971): 728–52.

Chapter 5   Cities, Sanitation, and Segregation

The literature on colonial urbanization tends to focus on particular cities. See, for example, George B. Endacott, *A History of Hong Kong,* rev. ed. (London and New York, 1973); Geoffrey Moorhouse, *Calcutta* (London, 1971); and Assane Seck, *Dakar, métropole ouest-africaine* (Dakar, 1970). The best book on urbanization and public health in the Western world is George Rosen, *A History of Public Health* (New York, 1958). On health and urbanization in Africa, see Philip Curtin, "Medical Knowledge and Urban Planning in Tropical Africa," *American Historical Review* 90, no. 3 (June 1985): 594–613.

Chapter 6   Hydraulic Imperialism in India and Egypt

The historical literature on irrigation in Egypt and India is small but fortunately includes some excellent works. On India see especially D. G. Harris, *Irrigation in India* (London, 1923), and, for the northwestern part of the subcontinent, Aloys Arthur Michel, *The Indus Rivers: A Study of the Effects of Partition* (New Haven and London, 1967). On the Nile, see Harold E. Hurst, *The Nile: A General Account of the River and the Utilization of Its Waters,* rev. ed. (London, 1957), and John Waterbury, *Hydropolitics of the Nile Valley* (Syracuse, N.Y., 1979). And on the impact of irrigation on Egyptian society, the essential source is Robert L. Tignor, *Modernization and British Colonial Rule in Egypt, 1882–1914* (Princeton, N.J., 1966).

Chapter 7   Economic Botany and Tropical Plantations

There are very few works on the history and economics of botany and tropical agriculture. On the period of plant transfers, see the interesting but very contentious book by Lucile H. Brockway, *Science and Colonial Expansion: The Role of the British Royal Botanic Gardens* (New York, 1979). A clear summary of twentieth-century developments can be found in Robert Evenson, "International Diffusion of Agrarian Technology," *Journal of Economic History* 34, no. 1 (March 1974): 51–73. Certain tropical crops have received their biographies, notably Noel Deerr, *The History of Sugar,* 2 vols. (London, 1949–50) and J. H. Drabble, *Rubber in Malaya, 1876–1922: The Genesis of an Industry* (Kuala Lumpur, 1973). Science and agriculture in the Dutch East Indies are the subject, respectively, of Pieter Honig and Frans Verdoorn, eds., *Science and Scientists in the Netherlands Indies* (New York, 1945), and Clifford Geertz, *Agricultural Involution: The Process of Ecological Change in Indonesia* (Berkeley, Los Angeles, and London, 1963).

Chapter 8   Mining and Metallurgy

No work deals with mining and metallurgy in general. Instead, there are different literatures on the different metals and metalliferous regions. On Malaya, see Wong Lin Ken, *The Malayan Tin Industry to 1914* (Tucson, Ariz., 1965), and H. C. Chai, *The Development of British Malaya, 1896–1909* (Kuala Lumpur, 1964). On Central Africa, see Simon E. Katzenellenbogen, *Railways and the Copper Mines of Katanga* (Oxford, 1973), and Jean-Luc Vellut, "Mining in the Belgian Congo" in *History of Central Africa,* ed. David Birmingham and Phyllis M. Martin (London and New York, 1983), pp. 126–62. On the Indian industry, the best survey is William Arthur Johnson, *The Steel Industry of India* (Cambridge, Mass., 1966).

Chapter 9   Technical Education

While education in the colonial empires has received a fair amount of attention, technical education awaits its historian. On India see Aparna Basu, *The Growth of Education and Political Development in India, 1898–1920* (Delhi, 1974), and "Technical Education in India, 1900–1920," in *Indian Economic and Social History Review* 4, no. 4 (December 1967): 361–74; also Robert I. Crane, "Technical Education and Economic Development in India before World War I" in *Education and Economic Development,* ed. C. Arnold Anderson and Mary Jean Bowman (Chicago, 1965), pp. 167–201. For Africa, see Denise Bouche, *L'enseignement dans les territoires français de l'Afrique occidentale de 1817 à 1920: Mission civilsatrice ou formation d'une élite?* (Paris, 1975), and two works by Philip Foster: "The Vocational School Fallacy in Developmental Planning," in Anderson and Bowman, and *Education and Social Change in Ghana* (Chicago, 1965). On Egypt see J. Heyworth-Dunne, *An Introduction to the History of Modern Education in Egypt* (London, 1939); Joseph S. Szyliowicz, *Education and Modernization in the Middle East*

(Ithaca, N.Y., 1973); and Robert L. Tignor, *Modernization and British Colonial Rule in Egypt, 1882–1914* (Princeton, N.J., 1966).

Chapter 10   Experts and Enterprises

On rubber and tin, see the works listed under Chapters 7 and 8 (above). For cocoa, see Seth La Anyane, *Ghana Agriculture: Its Economic Development from Early Times to the Middle of the Twentieth Century* (London, 1963), and Polly Hill, *The Migrant Cocoa-Farmers of Southern Ghana: A Study in Rural Capitalism* (Cambridge, 1963). The subject of entrepreneurship and technology imports to India has a large literature. Among the better works are Amiya Kumar Bagchi, *Private Investment in India, 1900–1939* (Cambridge, 1972); Daniel H. Buchanan, *The Development of Capitalistic Enterprise in India* (New York, 1934); and Rajat K. Ray, *Industrialization in India: Growth and Conflict in the Private Corporate Sector, 1914–47* (Delhi, 1979). On particular industries, see S. D. Mehta, *The Cotton Mills of India, 1854 to 1954* (Bombay, 1954), and T. S. Sanjeeva Rao, *A Short History of Modern Indian Shipping* (Bombay, 1965). The role of the Tata family is described in Sunil Kumar Sen, *The House of Tata, 1839–1939* (Calcutta, 1975). And on the Swadeshi movement, see Haridas Mukherjee and Uma Mukherjee, *The Origins of the National Education Movement* (Calcutta, 1957). On individual entrepreneurs, see Frank R. Harris, *Jamsetji Nusserwanji Tata: A Chronicle of His Life*, 2nd ed. (Bombay, 1958); K. C. Mahindra, *Sir Rajendra Nath Mookerjee* (Calcutta, 1933); and M. Visvesvaraya, *Memoirs of My Working Life* (Bangalore, 1951).

Index